Environmental Injury to Plants

Environmental Injury to Plants

EDITED BY

Frank Katterman

Department of Plant Sciences
College of Agriculture
University of Arizona
Tucson, Arizona

Academic Press, Inc.
Harcourt Brace Jovanovich, Publishers
San Diego   New York   Boston
London   Sydney   Tokyo   Toronto

This book is printed on acid-free paper. ∞

Academic Press, Inc.
San Diego, California 92101

United Kingdom Edition published by
Academic Press Limited
24–28 Oval Road, London NW1 7DX

Library of Congress Cataloging-in-Publication Data

Environmental injury to plants / edited by Frank Katterman.
        p.   cm.
     Papers from a one-day symposium held in June 1988 in Toronto,
Ont., sponsored by the American Chemical Society.
     Includes bibliographical references.
     ISBN   0-12-401350-3 (alk. paper)
     1. Crops--Effect of stress on. 2. Plants, Effect of stress on.
3. Crops--Physiology.   I. Katterman, Frank.   II. American Chemical
Society.
SB112.5.E58   1990
632'.1--dc20   89-28741
                                                               CIP

Printed in the United States of America
90 91 92 93   9 8 7 6 5 4 3 2 1

Contents

CHAPTER 4
Anatomy: A Key Factor Regulating Plant Tissue Response to Water Stress
Kaoru Matsuda and Ahmed Rayan

CHAPTER 5
From Metabolism to Organism: An Integrative View of Water Stress Emphasizing Abscisic Acid
Katrina Cornish and John W. Radin

CHAPTER 6
Biochemistry of Heat Shock Responses in Plants
Mark R. Brodl

CHAPTER 7
Reduced Cell Expansion and Changes in Cell Walls of Plant Cells Adapted to NaCl
Ray A. Bressan, Donald E. Nelson, Naim M. Iraki, P. Christopher LaRosa, Narendra K. Singh, Paul M. Hasegawa, and Nicholas C. Carpita

CHAPTER 8
Gene Expression during Adaptation to Salt Stress
John C. Cushman, E. Jay DeRocher, and Hans J. Bohnert

CHAPTER 9
Use of Two-Dimensional Gel Electrophoresis to Characterize Changes in Gene Expression Associated with Salt Stress of Barley
William J. Hurkman

CHAPTER 10
Mechanisms of Trace Metal Tolerance in Plants
Paul J. Jackson, Pat J. Unkefer, Emmanuel Delhaize, and Nigel J. Robinson

CHAPTER 11
Spectrophotometric Detection of Plant Leaf Stress
Harold W. Gausman and Jerry E. Quisenberry

Contributors

Numbers in parentheses indicate the pages on which the authors' contributions begin.

Hans J. Bohnert (173), Department of Biochemistry, and Department of Molecular and Cellular Biology, University of Arizona, Tucson, Arizona 85721

Ray A. Bressan (137), Center for Plant Environmental Stress Physiology, Department of Horticulture, Purdue University, West Lafayette, Indiana 47907

Mark R. Brodl (113), Department of Biology, Knox College, Galesburg, Illinois 61401

Nicholas C. Carpita (137), Department of Botany and Plant Pathology, Purdue University, West Lafayette, Indiana 47907

Katrina Cornish (89), U. S. Department of Agriculture, Agricultural Research Service, Western Regional Research Center, Albany, California 94710

John C. Cushman (173), Department of Biochemistry, University of Arizona, Tucson, Arizona 85721

E. Jay DeRocher (173), Department of Molecular and Cellular Biology, University of Arizona, Tucson, Arizona 85721

Emmanuel Delhaize (231), CSIRO, Division of Plant Research, Canberra, Australia

Harold W. Gausman (257), U. S. Department of Agriculture, Agricultural Research Service, Cropping Systems Research Laboratory, Lubbock, Texas 79401

Charles Guy (35), Ornamental Horticulture Department, Institute of Food and Agricultural Science, University of Florida, Gainesville, Florida 32611

Paul M. Hasegawa (137), Center for Plant Environmental Stress Physiology, Department of Horticulture, Purdue University, West Lafayette, Indiana 47907

William J. Hurkman (205), U. S. Department of Agriculture, Agriculture Research Service, Western Regional Research Center, Albany, California 94710

Naim M. Iraki[1] (137), Department of Botany and Plant Pathology, Purdue University, West Lafayette, Indiana 47907

Paul J. Jackson (231), Genetics Group, Life Sciences Division, Los Alamos National Laboratory, Los Alamos, New Mexico 87544

[1]Present address: Department of Botany, Hebrew University, Jerusalem, Israel.

P. Christopher LaRosa (137), Center for Plant Environmental Stress Physiology, Department of Horticulture, Purdue University, West Lafayette, Indiana 47907

Daniel V. Lynch (17), Department of Biology, Williams College, Williamstown, Massachusetts 01267

Kaoru Matsuda (63), Department of Molecular and Cellular Biology, and Department of Plant Sciences, University of Arizona, Tucson, Arizona 85721

Donald E. Nelson (137), Center for Plant Environmental Stress Physiology, Department of Horticulture, Purdue University, West Lafayette, Indiana 47907

Jerry E. Quisenberry (257), U.S. Department of Agriculture, Agricultural Research Service, Cropping Systems Research Laboratory, Lubbock, Texas 79401

John W. Radin (89), U. S. Department of Agriculture, Agricultural Research Service, Western Cotton Research Laboratory, Phoenix, Arizona 85040

Ahmed Rayan (63), Department of Molecular and Cellular Biology, University of Arizona, Tucson, Arizona 85721

Nigel J. Robinson (231), Department of Biological Sciences, University of Durham Science Laboratories, Durham DH1 3LE, England

Narendra K. Singh (137), Department of Botany and Microbiology, Auburn University, Auburn, Alabama 36849

Peter L. Steponkus (1), Department of Agronomy, Cornell University, Ithaca, New York 14853

Pat J. Unkefer (231), Isotope and Structural Chemistry Group, Isotope and Nuclear Chemistry Division, Los Alamos National Laboratory, Los Alamos, New Mexico 87545

Preface

The chapters in this book discuss various facets of plant environmental stress that are of critical importance to those who are concerned with the production of horticultural or agronomical plants on a large scale. Although not all aspects of plant environmental stress have been included in this publication, (for example, anerobiosis, pathogens, and pesticides) the subjects covered are those of major concern to various commodity groups in given regional sections of the United States. For example, in the northeastern, northern, Rocky Mountain, and midwest areas, freezing temperatures are a major problem. As one moves toward the northwest, west, south central, and southeast regions, chilling damage seems to be an additional dominant factor. Finally, in the arid locations of the southwest, drought, heat, and salt stress take their toll on economically important plants. Many talented scientists are involved in the study of environmental plant stress. This book is an outgrowth of a one-day symposium in Toronto that was sponsored by the American Chemical Society in June of 1988 and represents a sampling of these research workers.

The selection of topics should be of interest both to the scientific layman and to the professional in a specialized subject area of plant stress. The former, who may be a novice to the discipline as a whole, will be able to obtain an overall idea (from the review nature of each chapter) of the recent trends in each of the specific stress categories. On the other hand, the emphasis or bias for a particular trend will bring the professional up to date in his or her particular area of interest.

The arrangement of subjects and chapters is as follows: The first three chapters are concerned with both freezing and chill injury. P. L. Steponkus uses a plant protoplast system to characterize the process of freezing injury and cell lysis during an induced freeze/thaw cycle. D. V. Lynch examines the role of membrane lipids in chilling injury and adds a new dimension to one of the major tenents of the original membrane hypothesis. C. Guy focuses on the ability of the plant to increase its tolerance to freezing stress through cold acclimation. He presents evidence and discusses the possibility that a small number of genes control the physiological processes which lead to a higher freezing tolerance of the plant cell.

Chapters 4 and 5 describe two different aspects of drought stress. K. Matsuda and A. Rayan show that the reaction of plant tissue to water stress depends upon both the physiological properties and the anatomical features that regulate the transmission of the water deficit effect to the cells in question. K. Cornish and J. W. Radin review the role of abscisic acid (ABA) in stomatal responses to drought stress. On the molecular level, evidence is presented that ABA causes the appearance of new transcripts that nearly match the set induced by drought stress.

Chapter 6 by M. R. Brodl represents a detailed review of heat shock responses in plants during the past decade. Some of his work points to a potential mechanism of induction for a heat shock response.

Chapters 7, 8, and 9 are involved with osmotic or salt stress. R. Bressan and co-workers discuss salt tolerance on the cellular level. One of their significant findings is a distinct difference in the occurrence of ionically bound proteins to the walls of osmotically adapted and unadapted cells. J. C. Cushman and co-workers, on the other hand, examine salt tolerance and gene expression in the whole halophytic plant. They find an increase, a decrease, and a transient expression of several genes involved in the tolerance reaction as a function of time. W. J. Hurkman reviews the use of two-dimensional gel electrophoresis to characterize the changes in gene expression during salt stress. In addition, he describes the application of this tool for his research on salt-tolerant barley.

Chapter 10 by P. J. Jackson *et al.* deals with the increasing presence of heavy and toxic metals in the growing environment caused by the practice of recycling sewage sludge into the soil for fertilizer. They discuss how plants cope with this stress by means of a unique molecular adaptation.

Chapter 11 by H. W. Gausman and J. E. Quisenberry reviews the detection of plant stress by remote sensing devices. This technique can distinguish between several kinds of environmental stress on plants in the field.

It will probably be apparent to the reader that most of the stress topics discussed in this publication (e.g., osmotic, chill, drought, and heat) appear to have a few of the induced proteins in common with one another. Future research should further clarify a similar set of physiological or biomolecular reactions or both, that are directly related to the respective shock protein induction and to a development of tolerance for a given stress by the plant.

Frank R. Katterman

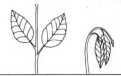

CHAPTER 1

Cold Acclimation and Freezing Injury from a Perspective of the Plasma Membrane

Peter L. Steponkus

Department of Agronomy
Cornell University
Ithaca, New York

I. INTRODUCTION

Cold acclimation is a complex developmental process that occurs in winter annuals, biennials, and perennials in which the ability to withstand freezing temperatures increases during the fall and winter months. There is a wide range in the increase in freezing tolerance that occurs, ranging from a few degrees in herbaceous species to tens of degrees in winter cereals to over 100 degrees in some extremely hardy deciduous species. Because of the impacts of freezing injury on a wide range of agricultural crop species, a considerable amount of attention has been devoted to plant cold hardiness and the process of cold acclimation.

During the cold-acclimation period, there is an orchestration of many events that are required to achieve maximum cold hardiness. These events include hormonal responses to environmental cues, altered gene activity and new gene products, and alterations in metabolism resulting in the accumulation of solutes and changes in lipid composition. Although seemingly disparate, these events ultimately contribute to the increased stability of cellular membranes, including the plasma membrane. To understand the role of each facet, it is necessary to first understand the collective effect of these changes on the primary site of freezing injury—the plasma membrane.

The plasma membrane is a primary site of freezing injury because of its central role in cellular behavior during a freeze–thaw cycle. As the principal barrier between the cytoplasm and the extracellular milieu, the semipermeable characteristics of the plasma membrane are of primary importance in allowing for the efflux/influx of water during a freeze–thaw cycle while restricting the efflux of intracellular solutes and, most important, precluding seeding of the cytosol by extracellular ice. Maintenance of these structural and functional characteristics during a freeze–thaw cycle is essential to survival. There are an increasing number of reports on the effect of freezing and cold acclimation on the plasma membrane because of its central importance. Although there is a general consensus that the plasma membrane is a primary site of freezing injury, there are divergent opinions on the nature of injury, its cause, and the effect of cold acclimation. In part, this is because different approaches have been taken to address these questions. Some of the studies have not dealt with the plasma membrane directly and are only inferential because manifestations of injury are measured long after the freeze–thaw event; other studies are more direct. Similarly, there are many divergent reports regarding alterations in the plasma membrane during cold acclimation.

To establish the nature and cause of injury during a freeze–thaw cycle, we have based our approach on cryomicroscopic studies of isolated protoplasts. These studies, which describe the phenomenology of destabilization of the plasma membrane during a freeze–thaw cycle, have been especially useful in providing a perspective from which to view the molecular and cellular aspects of both freezing injury and cold acclimation; they will be reviewed in this chapter. These studies have focused on winter rye (*Secale cereale* L. cv. Puma), which is among the most cold-hardy winter cereals. Following cold acclimation under artificial conditions (5°C and 12-hr daylength), the cold hardiness increases from -5 to -25°C over a period of 4 weeks. Protoplasts isolated from the leaves of the seedlings also reflect this increase in hardiness, even though the conditions during the freeze–thaw cycle are different from those experienced during freezing *in situ*.

II. THE FREEZING PROCESS

During cooling of a protoplast suspension, ice formation typically occurs first in the aqueous suspending medium. Ice formation occurs either as a result of heterogeneous nucleation or seeding by an ice crystal. During the subsequent growth of the ice crystals, solutes and gases are largely excluded from the ice matrix and accumulate in an unfrozen portion of the partially frozen mixture. During cooling to a given subzero temperature, ice formation will continue until the chemical potential of the unfrozen solution is in equilibrium with the chemical potential of the ice, which is a direct function of the subzero temperature. At equilibrium, the osmolality of the unfrozen solution will be equal to $(273 - T)/1.86$. Thus, when a solution is cooled and seeded at its freezing point, the osmolality of the unfrozen portion of the solution increases linearly as a function of the subzero temperature (e.g., 0.53 at $-1°C$, 2.69 at $-5°C$, 5.38 at $-10°C$, 10.75 at $-20°C$).

At any given subzero temperature, the osmolality of the unfrozen solution is independent of the initial osmolality of the solution (see Mazur, 1970). However, the proportion of the original solution that remains unfrozen at any given subzero temperature depends on the initial osmolality and the osmotic coefficient of the solute. The unfrozen portion is most accurately determined from the liquidus curve of the phase diagram for the solution, with the fraction (weight percent) of the unfrozen solution calculated as the ratio of the initial solute concentration (weight percent) to the solute concentration (weight percent) in the unfrozen portion at a given subzero temperature (see Rall *et al.*, 1983). For example, during freezing of a 0.53 osm sorbitol solution over the range of 0 to $-20°C$, approximately 28% of the solution remains unfrozen at $-5°C$, 18% at $-10°C$, and 14% at $-20°C$. If the initial osmolality of the solution is doubled to 1.06, the osmolality of the unfrozen solution at any given subzero temperature will be the same as the more dilute solution, but less solution will have to be frozen before the unfrozen solution is sufficiently concentrated to achieve the equilibrium osmolality (e.g., approximately 48% of the solution will remain unfrozen at $-5°C$, 32% at $-10°C$, and 24% at $-20°C$).

III. FREEZE-INDUCED CELL DEHYDRATION

Because of the freeze-induced concentration of the suspending medium, there will be a gradient in the chemical potential between the extracellular solution and the intracellular solution. As a result, the protoplasts will respond osmotically and will begin to dehydrate. The rate of water efflux will be a function of the magnitude of the gradient in chemical potential between the intracellular and the extracellular solution, the water permeability of the plasma membrane, and the area to volume ratio of the cell. The extent of dehydration will be a function of the external osmolality of the suspending medium, which is a direct function of the subzero temperature. At equilibrium, the extent of cell dehydration can be estimated from the Boyle van't Hoff relationship $V = V_b + x(\text{osm}^{-1})$. Because of the extremely high solute concentrations that are effected during the freezing of aqueous solutions, the extent of freeze-induced cell dehydration will be considerable. For example, at $-5°C$, the osmotic potential of the suspending medium is approximately -6 MPa and over 80% of the osmotically active water will be removed from the protoplasts. At $-10°C$, the osmotic potential is approximately -12 MPa and more than 90% of the osmotically active water will be removed. During warming of the suspension and melting of the suspending medium, the gradient in chemical potential will be reversed and, if the plasma membrane has not been destabilized, the cells will expand osmotically. Thus, during a freeze–thaw cycle, the plasma membrane must remain stable in the presence of extremely high solute concentrations and must be able to withstand the mechanical stresses incurred during the large osmotic excursions; otherwise survival of the cell is precluded. Accordingly, destabilization of the plasma membrane can occur as a result of different stresses in the cellular environment. In studies of protoplasts isolated from leaves of rye (*Secale cereale* L. cv. Puma), the conditions that lead to destabilization of the plasma membrane have been characterized for each of these possibilities (see Steponkus, 1984; Steponkus and Lynch, 1989a).

IV. THE ROLE OF THE PLASMA MEMBRANE IN INTRACELLULAR ICE FORMATION

Cell dehydration is only possible as long as ice formation does not occur in the cytosol. Two conditions are required for intracellular ice formation: (1) the cytosol must be supercooled, and (2) it must be either nucleated or seeded. Both conditions are influenced by characteristics of the plasma membrane.

Supercooling of the cytosol will occur during cooling, with the extent of supercooling a function of the rate of cooling (heat transfer) relative to the efflux of water (mass transfer). Water flux will be determined by the water permeability of the plasma membrane, the surface area to volume ratio of the protoplast, and the magnitude of the gradient in chemical potential of the extracellular solution and the cytosol. For a given set of conditions, the faster the rate of cooling, the greater the extent of supercooling. However, although there is a strong cooling rate dependence for intracellular ice formation, the probability of intracellular ice formation is not a simple function of the extent of supercooling (Dowgert and Steponkus, 1983). Instead, the probability of intracellular ice formation will increase only when the cells are supercooled at temperatures below some characteristic "ice nucleation temperature" (Mazur, 1970). A quantitative prediction of the temperature and cooling rate dependence of intracellular ice formation in isolated protoplasts requires information on the probability distribution of the temperature at which intracellular ice formation occurs, the dependence of intracellular ice formation on the extent of supercooling, and parameters, such as the water permeability of the plasma membrane and its temperature dependence, that determine volumetric behavior of the protoplasts during cooling (Pitt and Steponkus, 1989).

Intracellular ice crystallization can be effected either by seeding of the supercooled cytosol by extracellular ice or by nucleation (either homogeneous or heterogeneous) of the cytosol. Homogeneous nucleation depends on random density fluctuations that lead to the formation of a cluster of water molecules that can serve as an effective nucleus for the condensation of additional water molecules (see Franks, 1985). The probability of the formation of such a nucleus with a sufficiently long lifetime is a function of temperature and only occurs at relatively low temperatures. For example, the homogeneous nucleation of pure water occurs over the range of -38 to $-44°C$ (Mossop, 1955; Thomas and Slavely, 1952), depending on the volume of water (Wood and Walton, 1970). For solutions, the homogeneous nucleation temperature is decreased by about $1.8°C$ for every $1°C$ decrease in the freezing point (Rasmussen and MacKenzie, 1972). Heterogeneous nucleation, however, involves the condensation of water molecules on a nonaqueous impurity and can occur at much warmer temperatures. Although there is evidence for intracellular heterogeneous nucleating agents (Franks, 1985), most often they are only effective at temperatures a few degrees above the homogeneous nucleation temperature. The majority of studies of intracellular ice formation suggests that when cells are frozen in an aqueous medium, intracellular ice formation is a consequence of seeding by the extracellular ice (Mazur, 1977; Rasmussen *et al.*, 1975).

For isolated rye protoplasts, the critical cooling rate for intracellular ice formation is approximately $3°C$ per min, with the median temperature for intracellular ice at $-15°C$ (Dowgert and Steponkus, 1983). When cooled at rates greater than this, the incidence of intracellular ice formation increases greatly such that intracellular ice formation occurs in 100% of the protoplasts cooled at $10°C$ per min. Cold acclimation greatly decreases the incidence of intracellular ice formation. For protoplasts isolated from cold-acclimated rye leaves, the incidence of intracellular ice formation is low, over the range of 0 to $-30°C$, regardless of the cooling rate. However, at lower temperatures, the incidence increases at cooling rates greater than $3°C$ per min with a median intracellular ice formation temperature of $-42°C$. Regardless of whether the protoplasts are isolated from nonacclimated or cold-acclimated leaves, the temperature at which intracellular ice formation occurs is also influenced by the composition of the suspending medium and, under some conditions, by the cooling rate (Pitt and Steponkus, 1989). Nevertheless, the apparent decrease in the incidence of intracellular ice formation is a consequence of a decrease in the temperature at which seeding of the supercooled cytosol occurs rather than of a decrease in the extent of supercooling. Prior to these studies, it was long inferred that the lower incidence of intracellular ice formation in cold-acclimated tissues was a consequence of an increase in the hydraulic conductivity of the plasma membrane after cold acclimation (Levitt and Scarth, 1936). However, measurements of the hydraulic conductivity of the plasma membrane at subzero temperatures reveal that there is no difference in the hydraulic conductivity of rye protoplasts isolated from nonacclimated or cold-acclimated leaves (Dowgert and Steponkus, 1983). Thus, assuming that intracellular ice formation is a consequence of seeding by extracellular ice, the primary effect of cold acclimation is to decrease the temperature at which the plasma membrane is an effective barrier to extracellular ice.

Although the role of the plasma membrane in precluding seeding of the cytosol by extracellular ice was discussed as early as 1932 by Chambers and Hale; the reason its efficacy is diminished at low subzero temperatures has remained an enigma. Mazur (1965) has suggested that seeding is a consequence of the growth of ice crystals through channels in the membrane, however, Dowgert and Steponkus (1983) have observed mechanical failure of the plasma membrane preceding intracellular ice formation in isolated protoplasts. Resolution of the mechanism by which seeding occurs is of fundamental importance because under most conditions intracellular ice formation is lethal. However, the manner by which intracellular ice formation causes injury has never been resolved. Alternatively, if seeding is a consequence of the mechanical failure of the plasma membrane, then intracellular ice formation is a consequence rather than a cause of injury.

Destabilization of the plasma membrane during rapid cooling may be the result of freeze-induced electrical transients (Steponkus *et al.,* 1984, 1985). During the freezing of aqueous solutions of electrolytes, large electrical potentials may arise because of the differential exclusion of ions so that a charge separation occurs across the ice interface (Workman and Reynolds, 1950). These potentials, which may be in excess of tens of volts depending on the electrolyte, its concentration, and the velocity of ice crystal growth, are of sufficient magnitude to result in lysis of the plasma membrane. To date, only indirect evidence implicates freeze-induced electrical transients in destabilization of the plasma membrane. For example, in studies of the correlation between freeze-induced potentials and survival, rapid freezing of protoplast suspensions in sorbitol resulted in electrical potentials of nearly 4 V and fewer than 12% of the protoplasts survived (Steponkus *et al.,* 1985). In contrast, when the electrical transients were precluded by freezing at slower rates, survival was greater than 60%. If electrical transients result in destabilization of the plasma membrane, which causes seeding of the intracellular solution by extracellular ice, one would expect that cold acclimation would increase the stability of the plasma membrane to electrical fields. Such is the case. The critical transmembrane potential required for lysis of protoplasts isolated from nonacclimated rye leaves is 0.85 V, whereas the critical potential for protoplasts isolated from cold-acclimated leaves is 2.0 V (Steponkus *et al.*, 1985). Nevertheless, direct measurements of the transmembrane potential during ice formation in the suspending medium are required to establish the role of freeze-induced electrical transients in destabilization of the plasma membrane.

V. BEHAVIOR OF THE PLASMA MEMBRANE DURING OSMOTIC EXCURSIONS

During a freeze–thaw cycle, isolated protoplasts are subjected to extreme osmotic excursions; one form of freezing injury is a consequence of the behavior of the plasma membrane during osmotic excursions (see Steponkus, 1984; Steponkus and Lynch, 1989a). When suspended in isotonic solutions, the plasma membrane of isolated protoplasts is under a small tension of approximately 100 $\mu N \cdot m^{-1}$ (Wolfe and Steponkus, 1981, 1983). Only small reductions in the osmotic volume of the protoplast are sufficient to relax this tension to zero because of the relatively large area modulus of elasticity of the plasma membrane (≈ 200 $mN \cdot m^{-1}$). When the tension in the plasma membrane is reduced to zero in protoplasts that have been isolated from nonacclimated rye leaves (NA protoplasts), deletion of membrane material occurs via endocytotic vesiculation of the plasma membrane.

The process of endocytotic vesiculation is relatively slow in comparison to the efflux of water from the protoplast during osmotic contraction. For example, when protoplasts are transferred to a hypertonic solution, volumetric equilibration occurs within a few minutes, depending on the tonicity of the suspending medium and the temperature. During this time, the protoplasts appear flaccid and irregular in shape. However, after several minutes, they regain their spherical shape and are no longer flaccid. This is a consequence of the removal of the excess membrane material by endocytotic vesiculation. By using computer-enhanced video microscopy, endocytotic vesiculation is readily observed during osmotic contraction (Dowgert and Steponkus, 1984). The vesicles, which are observable within the cytoplasm, are 0.1–1.5 μm in diameter. That they are derived from the plasma membrane has been documented by labeling the plasma membrane with fluorescein-labeled Concanavalin A prior to osmotic contraction. Following transfer to a hypertonic solution, the fluorescent label is internalized. Freeze-fracture studies of the plasma membrane reveal that the intramembrane particle density on both fracture faces remains unchanged following osmotic contraction, which indicates that a unit membrane deletion has occurred (Gordon-Kamm and Steponkus, 1984a).

Endocytotic vesiculation per se is not injurious. However, sufficiently large area reductions are irreversible and the protoplasts lyse during subsequent osmotic expansion as occurs following thawing of the suspending medium. This form of injury is referred to as expansion-induced lysis. Lysis occurs when a critical tension in the plasma membrane is attained during volumetric expansion. For rye protoplasts, the critical tension is on the order of 4 to 6 $mN \cdot m^{-1}$ (Wolfe and Steponkus, 1981, 1983). Studies of the stress–strain relationship of the plasma membrane reveal that intrinsic or elastic expansion is limited to changes in area of approximately 2–3%. The larger changes in area (30–50%) that occur during osmotic expansion require that new membrane material be incorporated into the plasma membrane. The transfer of material into the membrane is also mediated by the tension in the plasma membrane (Wolfe *et al.*, 1985, 1986b). Somewhat paradoxically, the area expansion potential of the protoplasts appears to be independent of the rate of area expansion over time scales of tens of seconds to minutes, which are normally encountered during osmotic-induced expansion. Only during very rapid increases in area that are effected by aspiration with a micropipette does the extent of area expansion depend on the rate of expansion. Currently, the source of this membrane material is not known. However, the material deleted during osmotic contraction apparently is not readily reincorporated into the plane of the membrane because the expansion potential of rye protoplasts is independent of the extent of prior contraction. Thus, the area expansion potential of rye protoplasts is characterized by an incremental constant rather than by a constant maximum (Wiest and Steponkus, 1978).

Expansion-induced lysis is the predominant form of injury in NA protoplasts frozen to temperatures over the range of -2 to $-5°C$ (Dowgert and Steponkus, 1984). However, this form of injury is not observed in protoplasts isolated from leaves of cold-acclimated rye leaves (ACC protoplasts). This is because cold acclimation alters the behavior of the plasma membrane during osmotic contraction. Rather than resulting in the formation of endocytotic vesicles, osmotic contraction of protoplasts isolated from cold-acclimated leaves results in the formation of exocytotic extrusions of the plasma membrane (Dowgert and Steponkus, 1984). The extrusions appear as tethered spheres and have a densely osmiophilic interior (Gordon-Kamm and Steponkus, 1984b). Freeze-fracture studies indicate that there is an increase in the frequency of the intramembrane particles, suggesting that the densely osmiophilic cores are the result of preferential deletion of lipids from the bilayer. Nevertheless, both osmotic contraction and the formation of exocytotic extrusions are readily reversible in ACC protoplasts. Measurements of the tension in the plasma membrane during osmotic contraction–expansion excursions demonstrate that the tension does not increase until after the isotonic area is exceeded (Dowgert *et al.*, 1987).

Studies have provided compelling evidence that the differential behavior of the plasma membrane during osmotic contraction of NA and ACC protoplasts is a consequence of alterations in the lipid composition of the plasma membrane. Cryomicroscopic studies of large unilamellar vesicles (LUVs) prepared from plasma membrane lipid extracts of nonacclimated and cold-acclimated rye leaves also exhibit the differential behavior with respect to membrane flow during freeze-induced osmotic contraction (Steponkus and Lynch, 1989b). During osmotic contraction of LUVs prepared from plasma membrane lipids of nonacclimated rye leaves, the bilayer begins to flutter and undulate. Subsequently, numerous daughter vesicles are subduced from the parent bilayer and sequestered in the liposome interior. Following equilibration, there is good quantitative agreement between the decrease in the surface area of the liposome bilayer and the total surface area of the daughter vesicles. On thawing of the suspending medium, the LUVs begin to expand osmotically but lyse before there is an appreciable increase in volume. Lysis occurs either as a single lytic event without resealing of the bilayer or as a series of transient lytic events during which there is a partial release of the intraliposomal contents and resealing of the bilayer. In contrast, LUVs prepared from plasma membrane lipid extracts of cold-acclimated leaves form tubular or vesicular extrusions that remain continuous with the parent bilayer. During subsequent osmotic expansion, the extruded material is observed to be drawn back into the plane of the bilayer, and some of the LUVs regain their initial volume. However, the majority lyse during osmotic expansion but at an area that is greater than that attained by LUVs prepared from plasma membrane lipids of nonacclimated rye leaves. Nevertheless, the differential behavior during osmotic contraction parallels that observed in the plasma membrane of NA and ACC protoplasts and demonstrates that the differential behavior is a consequence of differences in the lipid composition of the plasma membrane following cold acclimation.

Establishing that the differential behavior of the plasma membrane during osmotic contraction is a consequence of alterations in its lipid composition is a significant finding. However, the changes in the lipid composition following cold acclimation are extremely complex. There are more than 100 different lipid molecular species in the plasma membrane of rye leaves, but none of the species is unique to the plasma membrane of either nonacclimated or cold-acclimated leaves (Lynch and Steponkus, 1987). Instead, cold acclimation alters the proportion of virtually every species. The major lipid classes are sterols, sterylglucosides, acylated sterylglucosides, glucocerebrosides, and phospholipids. Following cold acclimation, the proportions of free sterols and phospholipids are increased and the proportions of sterylglucosides, acylated sterylglucosides, and glucocerebrosides are decreased. The major free sterols are β-sitosterol and campesterol, with lesser amounts of stigmasterol and cholesterol. These are also the major steryl moieties in the sterylglucosides and acylated sterylglucosides, in which glucose is the sugar moiety. The major acyl groups of the acylated sterylglucosides are palmitic and linolenic acid. The glucocerebrosides consist of a single glucose moiety, a long chain base (the major bases are hydroxysphingenine and sphingadienine), and a long-chain hydroxy fatty acid (the major ones are hydroxynervonic, hydroxybehenic, and hydroxylignoceric acid) (Cahoon and Lynch, 1988). The major phospholipids are phosphatidylethanolamine and phosphatidylcholine, with lesser amounts of phosphatidylglycerol, phosphatidylserine, and phosphatidylinositol. Although there are only small differences in the proportions of the individual phospholipids following cold acclimation, there are substantial differences in the individual molecular species. The major di-unsaturated species (1-oleoyl-2-linoleoylphosphatidylcholine, dilinoleoylphosphatidylcholine, 1-linoleoyl-2-linoleoylphosphatidylcholine, and dilinolenoylphosphatidylcholine) increase twofold, with similar trends in the phosphatidylethanolamine species.

Given that the differential behavior of the plasma membrane during osmotic contraction of NA and ACC protoplasts is also observed in liposomes of the plasma membrane lipid extracts, and that there are no lipid species that are unique to either, it can be inferred that the differential behavior is a consequence of altered lipid–lipid interactions as a result of differences in the proportions of the individual molecular species. On the basis of this assumption, studies to establish structure and function relationships involved modification of the lipid composition of the plasma membrane. Selective enrichment of the plasma membrane lipid composition can be effected by using a protoplast-liposome fusion technique (Arvinte and Steponkus, 1988). With this procedure, the freezing tolerance of NA protoplasts is significantly increased following fusion with liposomes composed of the phospholipids isolated from the plasma membrane of cold-acclimated rye leaves (Steponkus *et al.,* 1988; Uemura and Steponkus, 1989). The increase in freezing tolerance can also be effected by fusion with liposomes composed of either mono-unsaturated (1-palmitoyl-2-linoleoylphosphatidylcholine or 1-palmitoyl-2-oleoylphosphatidylcholine) or di-unsaturated (dioleoylphosphatidylcholine, dilinoleoylphosphatidylcholine, or dilinolenoylphosphatidylcholine) species of phosphatidylcholine. However, fusion with di-saturated phosphatidylcholine species (dimyristoylphosphatidylcholine or dipalmitoylphosphatidylcholine) does not affect survival, either positively or negatively.

The increase in freezing tolerance occurs over the range of 0 to 5°C, the temperature range over which expansion-induced lysis is the predominant form of injury in protoplasts. Enrichment of the plasma membrane of protoplasts with dilinoleoylphosphatidylcholine results in 100% survival over this temperature range; survival only declines following a freeze–thaw cycle to $-7.5°C$ or lower. The increased freezing tolerance is a result of a transformation in the behavior of the plasma membrane during osmotic contraction. Following enrichment of the plasma membrane of nonacclimated protoplasts with mono- or di-unsaturated species of phosphatidylcholine, osmotic contraction results in the formation of exocytotic extrusions, which also occurs in ACC protoplasts. The morphology and frequency of the extrusions in NA protoplasts fused with dilinoleoylphosphatidylcholine are indistinguishable from those observed in ACC protoplasts. Further, studies of the expansion potential of the NA protoplasts fused with dilinoleoylphosphatidylcholine demonstrate that the protoplasts are able to undergo osmotic contraction–expansion excursions without a significant decrease in survival. Thus, alteration of the proportion of a single lipid species can dramatically alter both the functional and morphological characteristics of the plasma membrane during osmotic contraction.

VI. EFFECT OF SEVERE DEHYDRATION ON THE PLASMA MEMBRANE

When isolated protoplasts are frozen to temperatures below $-5°C$, they are subjected to extreme levels of dehydration. For example, at $-10°C$, the osmolality of the unfrozen portion of the suspending medium is 5.4 with an osmotic potential of approximately -12 MPa and more than 90% of the osmotically active water is removed from the protoplasts. Under these conditions, there is a loss of the semipermeable characteristics of the plasma membrane, and the protoplasts are osmotically unresponsive (see Steponkus, 1984). This form of injury is associated with several changes in the ultrastructure of the plasma membrane including the formation of large aparticulate domains, aparticulate lamellae subtending the plasma membrane, and lamellar-to-hexagonal$_{II}$ phase transitions in the plasma membrane and subtending lamellae (Gordon-Kamm and Steponkus, 1984c). Both the loss of osmotic responsiveness and the ultrastructural changes are a consequence of dehydration rather than temperature per se and can be effected by dehydration in a 5.4 osm sorbitol solution at 0°C. Under these conditions, neither the loss of osmotic responsiveness nor the ultrastructural modifications of the plasma membrane are observed in ACC protoplasts. Thus, cold acclimation also increases the stability of the plasma membrane to severe dehydration. And, as in the case of the differential behavior of the plasma membrane during osmotic contraction of NA and ACC protoplasts, the differential propensity for dehydration-induced lamellar-to-hexagonal$_{II}$ phase transitions is also observed in liposomes prepared from the plasma membrane lipid extracts of nonacclimated and cold-acclimated rye leaves (Cudd and Steponkus, 1988).

Although the mechanism responsible for the dehydration-induced lamellar-to-hexagonal$_{II}$ phase transitions in a complex biological membrane remains to be resolved, some insights are emerging. In freeze-fracture electron microscopy studies of the plasma membrane (Gordon-Kamm, 1985), aparticulate domains are observed in approximately 60% of the fracture faces of the plasma membrane following cooling of the protoplasts to $-10°C$ at 1°C/min prior to quenching in liquid propane. Under these conditions, hexagonal$_{II}$ phase is observed in less than 10% of the fracture faces. However, if the protoplasts are maintained at $-10°C$ for 15 min before quenching in liquid propane, there is little increase in the number of fracture faces that contain aparticulate domains, but hexagonal$_{II}$ phase is observed in nearly 50% of the fracture faces. These studies are consistent with the interpretation that lamellar-to-hexagonal$_{II}$ phase transitions in the complex mixture of the plasma membrane are preceded by dehydration-induced demixing of the membrane components. Thus, although the plasma membrane contains substantial amounts of nonbilayer-forming lipids such as phosphatidylethanolamine, they will be maintained in a bilayer configuration because of the influence of other bilayer-forming lipids such as phosphatidylcholine. If demixing occurs, there is the possibility that domains will become enriched with nonbilayer-forming lipids, which will then form the hexagonal$_{II}$ phase. For example, in mixtures of phosphatidylcholine and phosphatidylethanolamine, decreasing the proportion of phosphatidylcholine increases the tendency for a lamellar-to-hexagonal$_{II}$ phase transition (Cullis and de Kruijff, 1979).

Thus, it is necessary to understand how dehydration could result in demixing of the membrane components. One possibility suggested by Crowe and Crowe (1986) is that lateral phase separations result from dehydration-induced liquid crystalline-to-gel (L_α-to-L_β) phase transitions in lipid species such as phosphatidylcholine. This possibility is plausible because dehydration increases the L_α-to-L_β phase transition temperature T_m of phospholipids (Chapman *et al.*, 1967). In the case of phosphatidylcholine species, the increase in the T_m occurs at water contents of less than 20 wt % water and can be as great as 70°C, depending on the extent of dehydration. However, to invoke such a mechanism to explain the demixing observed during freeze-induced dehydration, it is necessary for the decreased water content to be achieved at the water potential that occurs during freezing to $-10°C$ (i.e., -12 MPa).

Studies of the lyotropic phase behavior of the major unsaturated phosphatidylcholine species present in the plasma membrane of rye leaves (1-palmitoyl-2-oleoyl-phosphatidylcholine, 1-palmitoyl-2-linoleoylphosphatidylcholine, dilinoleoylphosphatidylcholine, and dilinolenoylphosphatidylcholine) show that this is not the case (Lynch and Steponkus, 1989). Although the T_m's of these phosphatidylcholine species increased by 30 to 40°C over the range of water contents from 20 wt % to 5 wt %, construction of the desorption isotherms for these lipids revealed that these water contents were only attained at water potentials that were well below -12 MPa. For example, at -16 MPa, the water content was an average of 17 wt %, and there was only a small increase in the T_m's of the lipid species. Water potentials of -160 MPa were required to decrease the water content to 4 wt %, with a resultant increase in the T_m's of 30 to 40°C. Thus, under the conditions of temperature and dehydration that occur at $-10°C$, these phosphatidylcholine species would remain in the liquid crystalline phase. Although it might be argued that an L_α-to-L_β phase transition occurs in other lipid components of the plasma membrane, differential scanning calorimetry studies of the total plasma membrane lipid extract reveal no evidence for this at water potentials above -16 MPa. Only under conditions of severe dehydration (-160 MPa) is there evidence of a lyotropic phase transition.

An alternative mechanism to account for the dehydration-induced demixing of the membrane components derives from the studies of hydration forces by Parsegian, Rand, and co-workers (Parsegian *et al.*, 1979, 1986; Rand, 1981; Rand and Parsegian, 1984, 1986). Interbilayer interactions are subject to both long- and short-range forces, including long-range van der Waals forces of attraction and electrostatic repulsive forces. However, at distances of less than 2 to 3 µm, the interbilayer forces are dominated by strongly repulsive hydration forces (LeNeveu *et al.*, 1976, 1977; Cowley *et al.*, 1978; Lis *et al.*, 1982), which increase exponentially with a characteristic length of 0.25 to 0.35 nm (Parsegian *et al.*, 1979). Hydration repulsion, which is the result of the affinity of the polar head groups for water, presents a large energy barrier to the close approach of two bilayers. The pressures required to overcome the hydration forces are on the order of tens of megapascals (MPa). While these pressures are large, they occur during the freezing of aqueous solutions (e.g., the osmotic potential increases from -12 MPa at $-10°C$ to -25 MPa at $-20°C$).

Several changes in bilayer structure have been predicted to occur when bilayers are forced closer together (see Rand, 1981; Rand and Parsegian,

1984, 1986). Among the predicted changes are (1) demixing of the bilayer components and segregation of lipids into coexisting lamellar phases, (2) L_α-to-L_β phase transitions, and (3) bilayer-to-nonbilayer phase transitions. Whereas the characteristic increase in the T_m's of lipids occurs at water contents of less than 20 wt %, hydration forces begin to dominate the interbilayer interactions at water contents of less than 70% (Bryant and Wolfe, 1989). Recently, Wolfe and his colleagues (Wolfe *et al.*, 1986a; Wolfe, 1987; Bryant and Wolfe, 1989) have developed a model that predicts phase separation of lipid mixtures at hydration levels above those that result in L_α-to-L_β phase transitions. This model is based on the fact that the magnitude of the hydration forces is different for different lipid species (Marra and Israelachvili, 1985). For example, the hydration repulsion of dipalmitoylphosphatidylcholine is several orders of magnitude greater than that of dipalmitoylphosphatidylethanolamine at separations of less than 2 *nm*. Thus, during the close approach of two bilayers consisting of a multicomponent mixture, lateral segregation of lipids is expected. Proteins contained within a bilayer are also subject to the same hydration forces and would be expected to exhibit lateral segregation on the close approach of two bilayers. From this perspective, we view the ultrastructural modifications in the plasma membrane and suggest that the aparticulate regions observed during dehydration need not be a consequence of L_α-to-L_β phase transitions.

VII. SUMMARY

Destabilization of the plasma membrane is a primary cause of freezing injury; however, destabilization can be effected by different stresses that occur during a freeze–thaw cycle. Accordingly, the manifestations of injury will also be different. Under conditions that preclude intracellular ice formation, injury in protoplasts isolated from nonacclimated rye leaves is a consequence of (1) osmotic excursions incurred during a freeze–thaw cycle, and (2) dehydration-induced demixing of the plasma membrane components during severe dehydration. Cold acclimation increases the tolerance of the plasma membrane to both osmotic excursions and severe freeze-induced dehydration. In both cases, the increased tolerance in large part, is, a consequence of alterations in the lipid composition of the plasma membrane.

ACKNOWLEDGMENTS

Portions of this work were supported by the U.S. Department of Energy (Grant No. DE-FG02-84ER13214) and the U.S. Department of Agriculture Competitive Research Grant Program (Grant No. 5-CRCR-1-1651 and Grant No. 88-37264-3988).

REFERENCES

Arvinte, T. A., and Steponkus, P. L. (1988). *Biochemistry* **27**, 5671–5677.
Bryant, G., and Wolfe, J. (1989). *Eur. J. Biophys.* **16**, 369–374.
Cahoon, E. B., and Lynch, D. V. (1988). *Plant Physiol.* **86**, S–53.
Chambers, R., and Hale, H. P. (1932). *Proc. R. Soc. London, Ser. B* **110**, 336–352.
Chapman, D., Williams, R. M., and Ladbrooke, B. D. (1967). *Chem. Phys. Lipids* **1**, 445–475.
Cowley, A. C., Fuller, N., Rand, R. P., and Parsegian, V. A. (1978). *Biochemistry* **17**, 3163–3168.
Crowe, J. H., and Crowe, L. M. (1986). *In* "Membranes, Metabolism, and Dry Organisms" (A. C. Leopold, ed.), pp. 188–209. Cornell Univ. Press (Comstock), Ithaca, New York.
Cudd, A., and Steponkus, P. L. (1988). *Biochim. Biophys. Acta* **941**, 278–286.
Cullis, P. R., and de Kruijff, B. (1979). *Biochim. Biophys. Acta* **559**, 399–420.
Dowgert, M. F., and Steponkus, P. L. (1983). *Plant Physiol.* **72**, 978–988.
Dowgert, M. F., and Steponkus, P. L. (1984). *Plant Physiol.* **75**, 1139–1151.
Dowgert, M. F., Wolfe, J., and Steponkus, P. L. (1987). *Plant Physiol.* **83**, 1001–1007.
Franks, F. (1985). "Biophysics and Biochemistry at Low Temperatures." Cambridge Univ. Press, London and New York.
Gordon-Kamm, W. J. (1985). Ph.D. Thesis, Cornell University, Ithaca, New York.
Gordon-Kamm, W. J., and Steponkus, P. L. (1984a). *Protoplasma* **123**, 83–94.
Gordon-Kamm, W. J., and Steponkus, P. L. (1984b). *Protoplasma* **123**, 161–173.
Gordon-Kamm, W. J., and Steponkus, P. L. (1984c). *Proc. Natl. Acad. Sci. U.S.A.* **81**, 6373–6377.
LeNeveu, D. M., Rand, R. P., and Parsegian, V. A. (1976). *Nature (London)* **259**, 601–603.
LeNeveu, D. M., Rand, R. P., Gingell, D., and Parsegian, V. A. (1977). *Biophys. J.* **18**, 209–230.
Levitt, J., and Scarth, G. W. (1936). *Can. J. Res., Sect. C* **14**, 285–305.
Lis, L. J., McAlister, M., Fuller, N., Rand, R. P., and Parsegian, V. A. (1982). *Biophys. J.* **37**, 657–666.
Lynch, D. V., and Steponkus, P. L. (1987). *Plant Physiol.* **83**, 761–767.
Lynch, D. V., and Steponkus, P. L., (1989). *Biochim. Biophys. Acta* **984**, 267–272.
Marra, J., and Israelachvili, J. (1985). *Biochemistry* **24**, 4608–4618.
Mazur, P. (1965). *Ann. N.Y. Acad. Sci.* **125**, 658–676.
Mazur, P. (1970). *Science* **168**, 939–940.
Mazur, P. (1977). *Cryobiology* **14**, 251–272.
Mossop, S. C. (1955). *Proc. Phys. Soc. London, Sect. B* **68**, 193–208.
Parsegian, V. A., Fuller, N., and Rand, R. P. (1979). *Proc. Natl. Acad. Sci. U.S.A.* **76**, 2750–2754.
Parsegian, A., Rau, D., and Zimmerberg, J. (1986). *In* "Membranes, Metabolism, and Dry Organisms" (A. C. Leopold, ed.), pp. 306–317. Cornell Univ. Press (Comstock), Ithaca, New York.
Pitt, R. E., and Steponkus, P. L. (1989). *Cryobiology* **26**, 44–63.
Rall, W. F., Mazur, P., and McGrath, J. J. (1983). *Biophys. J.* **41**, 1–12.
Rand, R. P. (1981). *Annu. Rev. Biophys. Bioeng.* **10**, 277–314.
Rand, R. P., and Parsegian, V. A. (1984). *Can. J. Biochem. Cell Biol.* **62**, 752–759.
Rand, R. P., and Parsegian, V. A. (1986). *Annu. Rev. Physiol.* **48**, 201–212.
Rasmussen, D. H., and MacKenzie, A. P. (1972). *In* "Water Structure at the Water-Polymer Interface" (H. H. G. Jellinek, ed.), pp. 126–135. Plenum, New York.

Rasmussen, D. H., Macauley, M. N., and MacKenzie, A. P. (1975). *Cryobiology* 12, 328–329.

Steponkus, P. L. (1984). *Annu. Rev. Plant Physiol.* 35, 543–584.

Steponkus, P. L., and Lynch, D. V. (1989a). *J. Bioenerg. Biomembr.* 21, 21–41.

Steponkus, P. L., and Lynch, D. V. (1989b). *Cryo-Lett.* 10, 43–50.

Steponkus, P. L. Stout, D. G., Wolfe, J., and Lovelace, R. V. E. (1984). *Cryo-Lett.* 5, 343–348.

Steponkus, P. L., Stout, D. G., Wolfe, J., and Lovelace, R. V. E. (1985). *J. Membr. Biol.* 85, 191–198.

Steponkus, P. L., Uemura, M., Balsamo, R. A., Arvinte, T. A., and Lynch, D. V. (1988). *Proc. Natl. Acad. Sci. U.S.A.* 85, 9026–9030.

Thomas, D. G., and Slavely, L. A. M. (1952). *J. Chem. Soc.*, pp. 4569–4577.

Uemura, M., and Steponkus, P. L. (1989). *Plant Physiol.* 91, 1131–1137.

Wiest, S. C., and Steponkus, P. L. (1978). *Plant Physiol.* 62, 699–705.

Wolfe, J. (1987). *Aust. J. Plant Physiol.* 14, 311–318.

Wolfe, J., and Steponkus, P. L. (1981). *Biochim. Biophys. Acta* 643, 663–668.

Wolfe, J., and Steponkus, P. L. (1983). *Plant Physiol.* 71, 276–285.

Wolfe, J., Dowgert, M. F., and Steponkus, P. L. (1985). *J. Membr. Biol.* 86, 127–138.

Wolfe, J., Dowgert, M. F., Maier, B., and Steponkus, P. L. (1986a). *In* "Membranes, Metabolism, and Dry Organisms" (A. C. Leopold, ed), pp. 286–305. Cornell Univ. Press (Comstock), Ithaca, New York.

Wolfe, J., Dowgert, M. F., and Steponkus, P. L. (1986b). *J. Membr. Biol.* 93, 63–74.

Wood, R. G., and Walton, A. G. (1970). *J. Appl. Phys.* 41, 3027–3036.

Workman, E. J., and Reynolds, S. E. (1950). *Phys. Rev.* 78, 254–259.

CHAPTER 2

Chilling Injury in Plants: The Relevance of Membrane Lipids

Daniel V. Lynch

Department of Biology
Williams College
Williamstown, Massachusetts

I. INTRODUCTION

Many plants native to warm habitats are sensitive to low nonfreezing temperatures, displaying abrupt reductions in the rates of physiological processes and exhibiting signs of injury following exposure to temperatures in the range of 0 to 15°C. Many crop species including corn, rice, tomato, cucumber, soybeans, sorghum, and cotton fall into this category of chilling-sensitive plants. Several crops of temperate origin, including potatoes, asparagus, and apples, also exhibit chilling sensitivity, albeit at lower temperatures (0–5°C) (Bramlage, 1982). As such, there is a great potential benefit to increasing resistance to chilling temperatures in order to extend the effective growing season and the geographical (climatic) range in which sensitive crop species can be cultivated. Reducing chilling sensitivity of fruits and vegetables may serve to standardize and improve long-term storage (Bramlage, 1982; Couey, 1982).

The manifestations of chilling injury are numerous. Imbibitional chilling injury is associated with germination of seeds (Wolk and Herner, 1982). Expansive growth of cotton, corn, soybeans, and rice is inhibited by low temperatures, and wilting, chlorosis, and/or bleaching are common symptoms of chilling injury in seedlings (Wang, 1982; Wilson, 1976; Wolk and Herner, 1982). In some species, reproductive development is extremely sensitive to low temperatures. For example, exposure of rice plants to temperatures below 17°C during anthesis results in sterility (Satake, 1974). Storage disorders of many fruits and vegetables are attributable to temperatures between 0 and 5°C (Bramlage, 1982).

Chilling injury depends not only on the species and tissue type, but also on the severity (i.e., the temperature below the critical temperature for chilling injury) and duration of exposure to low temperature. Other environmental variables contribute to the extent of injury manifested as a consequence of chilling. High humidity partially ameliorates the effects of wilting that occurs as a consequence of increased stomatal water loss from the leaves and diminished hydraulic conductivity of the roots at low temperature (Bagnall *et al.*, 1983; Wilson, 1976). Exposure to chilling temperatures in the light is more injurious than exposure to the same temperatures in the dark (Garber, 1977; Hodgson *et al.*, 1987; Kee *et al.*, 1986; Powles *et al.*, 1983; Powles, 1984). Tomato seedlings exhibit differences in diurnal chilling sensitivity. Maximum sensitivity at the end of the dark period was found to coincide with carbohydrate depletion (King *et al.*, 1982, 1988), although a mechanistic explanation is lacking.

The obvious morphological manifestations or symptoms of chilling injury are numerous and appear to reflect a range of metabolic dysfunctions. At a more fundamental level, chilling injury is reflected in a diverse array of biochemical and cellular changes (see Graham and Patterson, 1982; Levitt, 1972; Lyons, 1973; Minorsky, 1985). In fact, few, if any, cellular–biochemical processes remain unaltered during the progression of events following exposure to chilling temperatures. The most prominent chill-induced metabolic and structural changes reported include the cessation of protoplasmic streaming (Woods *et al.*, 1984a), altered rates of respiration (Lyons, 1973), impaired photosynthesis (Peeler and Naylor, 1988a,b,c), alterations in protein synthesis (Cooper and Ort, 1988), and changes in membrane permeability (Minorsky, 1985).

The diversity of responses to chilling injury and the many types of biochemical–metabolic dysfunctions suggest that either (1) many cellular–molecular sites are sensitive to low temperature in chilling-sensitive plants or (2) one primary site or lesion initiates a cascade of events, leading to cell injury and/or death. The most widely accepted hypothesis for the general phenomenon of chilling injury, the "membrane hypothesis" proposed by Lyons, Raison, and co-workers (Lyons, 1973), falls into the latter category, assuming, for conceptual simplicity, that all cellular membrane types are considered a single site. The hypothesis initially proposed that a reversible change in the membrane's physical state is the primary response to low temperature. This change occurs at a temperature characteristic for a given plant species or tissue. As a consequence of alterations in membrane properties, physiological, metabolic, and biochemical dysfunctions occur. These include changes in membrane-associated enzyme activities and membrane permeability and, consequently, decreases in ATP levels, ion–solute leakage, loss of compartmentation, and metabolic imbalances. Whereas the primary response is reversible, the secondary events occur in a time-dependent fashion and are not reversible in all cases. In fact, many secondary processes are accelerated if elevated temperatures are encountered following a chilling stress.

The origins of this proposal have been described elsewhere (Lyons, 1973; Steponkus, 1981) and will not be reiterated here. The membrane hypothesis has continued to evolve in response to additional evidence, much of it supporting the hypothesis (in spirit if not in letter) and some of it apparently contradicting the basic tenets and assumptions on which the hypothesis is based (Martin, 1986; Minorsky, 1985; Steponkus, 1981). The hypothesis and the proposed role of lipids in chilling sensitivity has received renewed interest since Murata and colleagues (Murata *et al.*, 1982; Murata, 1983) proposed that chilling sensitivity is correlated to the molecular species composition of phosphatidylglycerol. Rather than attempt a comprehensive presentation of the subject and to avoid redundancy with many excellent reviews of chilling injury (Graham and Patterson, 1982; Levitt, 1972; Lyons, 1973; Quinn and Williams, 1978), this review will concentrate on more recent work, referring to earlier literature only to supply required background information. This review focuses on several points that are relevant to the membrane hypothesis and the role of lipids in chilling injury. Emphasis is placed on recent studies implicating high-melting-point molecular species of phosphatidylglycerol in conferring chilling sensitivity. Although the role of lipids in chilling sensitivity is the centerpiece of this review, I do not wish to imply that the membrane hypothesis and/or Murata's "phosphatidylglycerol hypothesis" discount or exclude alternate hypotheses proposed for chilling injury of plant tissues (Graham and Patterson, 1982; Martin, 1986; Minorsky, 1985).

II.   LIPID PHYSICAL PROPERTIES AND MEMBRANE FUNCTION

The thermotropic change in membrane lipids induced by low temperatures in chilling-sensitive plants was first suggested to be a liquid crystalline-to-gel phase transition in the bulk lipids (Lyons, 1973). Subsequent studies of chilling-sensitive plants employing electron spin resonance and demonstrating two "breaks" (changes in slope) in Arrhenius plots (log [spin label motion parameter] *versus* the reciprocal of absolute temperature) were interpreted as reflecting changes in the proportions of gel-phase and fluid-phase lipid, the higher breakpoint representing the temperature at which some gel-phase lipid appears and the lower breakpoint representing the completion of phase separation (Raison *et al.*, 1971; Raison and Chapman, 1976). More recent investigations suggest that breaks inferred from electron spin resonance studies do not reflect major thermotropic phase transitions but rather a phase transition involving a small subpopulation of lipids (Raison and Wright, 1983; Raison and Orr, 1986). Furthermore, studies of the thermotropic behavior of thylakoid lipids demonstrate that the results obtained using calorimetry, electron spin resonance, and fluorescence techniques are consistent and that each is an appropriate technique for detecting phase separations (Raison and Orr, 1986). Although many studies have demonstrated phase separations in lipid extracts (or, more frequently, polar lipid extracts), there is little evidence of these separations occurring in intact membranes. This is attributable, in part, to the limitations of the methods used. However, studies of lipid dispersions do not allow for protein–lipid interactions, and studies of phospholipid or polar lipid fractions obviate possible lipid–lipid interactions. As such, the occurrence of and role of membrane phase separations per se in chilling-sensitive plants are still equivocal.

One important but ill-defined role of membrane lipids is their modulation of enzyme/protein function via protein–lipid interactions (Sandermann, 1978). The respective break temperatures observed for mitochondria or chloroplasts using spin probes coincided with break temperatures in Arrhenius plots of respiratory activities (Raison and Chapman, 1976) and photosynthetic activities (Shneyour *et al.*, 1973), suggesting that abrupt thermotropic changes in physiological and biochemical activities are related to changes in the molecular ordering of membrane lipids. However, there is still much debate concerning the use and interpretation of Arrhenius plots. Discontinuities in Arrhenius plots of enzyme activity at temperatures corresponding to those at which changes in the molecular ordering of membrane lipids are observed has prompted the view that breaks reflect changes in the conformation of one or more enzymes that are attributable to alterations in the surrounding lipid microenvironment. It has been questioned whether (1) data points are most appropriately fit with straight lines of differing slope or a smooth curve, and (2) breaks are indicative of a change in the activation energy of enzymatic reactions as a consequence of a phase transition (Bagnall and Wolfe, 1978, 1982; Martin, 1986; Steponkus, 1981; Wolfe and Bagnall, 1980). In addition, there are numerous examples of breaks in Arrhenius plots that are the result of inadequate or inappropriate experimental protocols (see Martin, 1986). Although it can be demonstrated in certain cases that changing the lipid environment does not alter the breakpoints in Arrhenius plots, the actual enzyme activity is affected (Maeshima *et al.*, 1980).

Breaks in Arrhenius plots were not observed in early studies of chilling-resistant plants; however, subsequent studies did find such breaks in plots of enzyme activities and electron spin resonance data (Quinn and Williams, 1978), casting doubt on the apparent distinction between chilling-sensitive and chilling-resistant plants. Nolan and Smillie (1976, 1977) suggested that the difference is not in the presence or absence of breaks, but in the magnitude of the associated change in slope of the Arrhenius plots.

An alternative means of determining the influence of lipid fluidity on the structure and function of membrane enzymes is to use a physical approach to assess changes in the properties of the proteins. Complementary biophysical techniques have provided evidence that temperature-induced changes in lipid fluidity affect protein conformation in smooth microsomal membranes of bean cotyledons (Lynch *et al.*, 1987). Lipid fluidity was measured by fluorescence polarization after labeling the membranes with diphenylhexatriene (DPH), and alterations in protein conformation were determined by labeling the membranes with the paramagnetic sulfhydryl reagent, 3-maleimido proxyl, and following changes in the ratio of weakly immobilized to strongly immobilized (w/s) spin label. Simple plots of w/s as a function of temperature featured characteristic breakpoints. Corresponding breakpoints at virtually identical temperatures were also observed in plots of DPH polarization versus temperature for membranes and liposomes prepared from lipid extracts of membranes. The breaks did not correspond to liquid crystalline-to-gel phase transitions in the lipid. These results indicate that subtle changes in the molecular ordering of lipid bilayers alter the relative proportions of weakly and strongly immobilized sulfhydryl groups in the membrane proteins, which is interpreted as reflecting changes in protein conformation.

More detailed biophysical characterization of barley root plasma membrane–enriched microsomes using protein spin labels and lipid spin probes (Caldwell and Whitman, 1987) and intrinsic protein (tryptophan) fluorescence (Caldwell, 1987) also revealed abrupt changes in lipid and protein mobilities and lateral diffusion within the membrane at 12 to 14°C that coincided with changes in the K_m of plasma membrane Mg^{++}-dependent ATPase. These changes were attributed to the formation of quasi-crystalline lipid clusters that would be expected to influence the activity of membrane-bound enzymes. The identity and nature of the lipid species participating in clustering were not identified. Lipid clustering of this type would not necessarily be detectable by calorimetric techniques. This is consistent with previous studies of the effects of temperature on spin probe motion and enzyme activity, which suggest that changes in membrane structure occur in the absence of a phase transition (Quinn and Williams, 1978; Sandermann, 1978; Wolfe, 1978).

These studies provide direct physical evidence of the influence of lipids on protein conformation and mobility without resorting to the use of Arrhenius plots. Whereas these two studies have examined the behavior of bulk lipids and membrane proteins, increasing evidence of specific protein–lipid interactions leads to the possibility that certain lipid species may influence specific protein-mediated processes. Such a role is proposed for phosphatidylglycerol (see the following section).

III.   LIPID COMPOSITION AND CHILLING SENSITIVITY

One major tenet of the original membrane hypothesis is that the lipid (fatty acid) compositions of chilling-sensitive and chilling-resistant plants are different. Early analyses revealed that mitochondrial membranes of chilling-sensitive plants contained a higher proportion of saturated (high-melting-point) fatty acids than chilling-resistant plants (Lyons *et al.*, 1964), and it was demonstrated that small changes in fatty acid composition can dramatically influence the temperature at which a mixture solidifies (Lyons and Asmundson, 1965).

Subsequent analyses of the membrane fatty-acid composition of tissues (e.g., leaves) from chilling-sensitive and chilling-resistant plants did not necessarily support the contention that chilling sensitivity was related to fatty acid unsaturation (Wilson and Crawford, 1974; Quinn and Williams, 1978). The constituent acyl lipids of leaf tissues are monogalactosyldiacylglycerol, digalactosyldiacylglycerol, sulfoquinovosyldiacylglycerol, phosphatidylglycerol, phosphatidylcholine, and lesser amounts of other phospholipids. The major components—the galactolipids—contain unsaturated acyl chains almost exclusively and, as such, have a phase transition temperature well below 0°C. The phospholipid fraction also typically contains over 70% unsaturated acyl chains. Early physical studies (Patterson *et al.*, 1978; Wade *et al.*, 1974) of polar lipid extracts of leaves or subcellular membranes were interpreted as indicating a change in the physical state of a portion of the lipids extracted from chilling-sensitive species, but not in those from chilling-resistant species. In spite of these differences in physical properties, the degree of unsaturation of lipids was not obviously correlated to chilling sensitivity. Thus, it became apparent that bulk fatty-acid compositional analyses of tissues were not appropriate.

This lack of correlation can be explained, however, if one considers that the pairing of acyl chains at the *sn*-1 and *sn*-2 positions of glycerolipids can influence the thermotropic phase behavior. The importance of the molecular species composition (i.e., the intermolecular and intramolecular arrangement of acyl chains), rather than acyl chain composition per se, in determining lipid phase behavior was demonstrated by Phillips *et al.* (1972): A dispersion of 1-palmitoyl-2-oleoylphosphatidylcholine exhibits a single gel-to-liquid crystalline phase transition at $-3°C$, whereas an equimolar mixture of dipalmitoylphosphatidylcholine and dioleoylphosphatidylcholine exhibits transitions at $-20°C$ (attributable to the unsaturated species) and $41°C$ (attributable to the saturated species). Thus, these two lipid preparations—both consisting of a single phospholipid class, phosphatidylcholine, and two acyl chains, palmitate and oleate, in a 1 : 1 ratio—exhibit vastly different phase behavior. This example demonstrates that the critical feature of lipid composition that accounts for differences in the thermotropic behavior of membrane lipid physical properties between chilling-sensitive and chilling-resistant plants may be subtle differences in the molecular species of a population of the glycerolipids.

Murata and co-workers (1982; Murata, 1983) suggested that chilling sensitivity of higher plants is related to the proportion of molecular species of phosphatidylglycerol consisting of two saturated acyl chains in the *sn*-1 and *sn*-2 positions of the glycerol backbone (dipalmitoyl or 1-stearoyl-2-palmitoylphosphatidylglycerol) or one saturated acyl chain and *trans*-3-hexadecenoic acid, which has physical properties similar to those of a saturated acyl chain (see below). Phosphatidylglycerol is the predominant phospholipid in chloroplasts, and *trans*-3-hexadecenoyl-containing species of phosphatidylglycerol are localized in chloroplasts (see Dubacq and Tremolières, 1983).

Initially, Murata *et al.* (1982) examined the compositions and positional distributions of acyl chains of phosphatidylcholine, phosphatidylethanolamine, and phosphatidylglycerol from the leaf tissue of 9 chilling-sensitive plants and 11 chilling-resistant plants. The sum of the contents of palmitic acid and *trans*-3-hexadecenoic acid of phosphatidylglycerol isolated from 8 chilling-sensitive species (including sweet potato, castor bean, tobacco, squash, and corn) ranged from 60 to 78% of the total fatty acids. The sum of the contents of the two fatty acids ranged from 50 to 57% in chilling-resistant species (including pea, spinach, oat, and wheat). Tomato, a chilling-sensitive plant, was an apparent exception, having 54% of the acyl chains comprised of the two C_{16} acyl chains. Analysis of the positional distribution of the acyl chains suggested that high proportions of dipalmitoyl and 1-palmitoyl-2-(*trans*-3-hexadecenoyl)phosphatidylglycerol are correlated to chilling sensitivity. In contrast, the composition and positional distribution of fatty acids of phosphatidylcholine and phosphatidylethanolamine were similar for chilling-sensitive and chilling-resistant plants.

This work was subsequently extended to performing actual molecular species analysis of phosphatidylglycerols from chilling-sensitive and chilling-resistant plants (Murata, 1983). The sum of the contents of dipalmitoyl and 1-palmitoyl-2-(*trans*-3-hexadecenoyl) species ranged from 26 to 65% in chilling-sensitive plants, but only from 3 to 19% in chilling-resistant plants, suggesting that these two species of phosphatidylglycerol are correlated to chilling sensitivity.

In support of this contention, Murata and co-workers have reported that (1) phosphatidylglycerol is the only major plant membrane glycerolipid class (with the possible exception of sulfoquinovosyldiacylglycerol) that undergoes a gel-to-liquid crystalline phase transition above 0°C, and (2) phosphatidylglycerol fractions isolated from chilling-sensitive plants and having high levels of palmitate and *trans*-3-hexadecenoate display consistently higher phase separation temperatures than do phosphatidylglycerol fractions from chilling-resistant plants (Murata and Yamaya, 1984).

The apparent simplicity of the hypothesis and the facility of testing it (i.e., by analyzing the composition of phosphatidylglycerol from a variety of plants) have led other workers to expand on Murata's work (Norman *et al.*, 1984; Roughan, 1985; Kenrick and Bishop, 1986; Bishop, 1986; Li *et al.*, 1987). With the exception of Normon *et al.* (1984) and Li *et al.* (1987), these investigators did not directly determine the molecular species of phosphatidylglycerol. Instead, the molecular species compositions were inferred from the fatty acid compositions of phosphatidylglycerol. This is possible because phosphatidylglycerol is comprised of C_{18}/C_{16} and C_{16}/C_{16} species and lacks C_{18}/C_{18} species. In contrast, all other glycerolipids contain species having two C_{18} acyl chains. This difference in molecular species profiles of phosphatidylglycerol and other phospholipids and galactolipids is a consequence of the biosynthetic pathways involved (Andrews and Mudd, 1985; Cronan and Roughan, 1987; Roughan, 1986; Roughan and Slack, 1984). The majority of subsequent reports frequently did not even consider the molecular species compositions, but simply expressed phosphatidylglycerol composition in terms of the proportion of "high-melting-point fatty acids," which include palmitate (16 : 0), *trans*-3-hexadecenoate (16 : 1[13]), and stearate (18 : 0) (Bishop, 1986; Kenrick and Bishop, 1986; Roughan, 1985).

The results of these subsequent investigations were not entirely consistent with the predictions of Murata's hypothesis. Kenrick and Bishop (1986) examined the fatty acid composition of phosphatidylglycerol isolated from the leaves of 27 species of higher plants. The content of high-melting-point fatty acids varied from 50% in *Hordeum vulgare* to 80% in *Passiflora ligularis*. However, they concluded that the content of high-melting-point fatty acids in phosphatidylglycerol varies little between members of the same plant family, even those containing species differing in chilling sensitivity.

Roughan (1985) examined the fatty acid composition (and the inferred molecular species composition) of phosphatidylglycerol in over 70 plant species, and concluded that, within a genus, chilling sensitivity appeared correlated to the molecular species of phosphatidylglycerol, with dipalmitoyl and 1-palmitoyl-2-(*trans*-3)hexadecenoyl species accounting for less than 25% of the phosphatidylglycerol in chilling-resistant plants and 50–60% in most chilling-sensitive plants. However, solanaceous and other "16 : 3"-plants and C_4 grasses proved to be apparent exceptions to the rule.

These studies may not represent rigorous testing of the relationship between phosphatidylglycerol composition and chilling-sensitivity because (1) analyses of the actual molecular species of PG phosphatidylglycerol were not undertaken, (2) only plants having similar genotypes and overall lipid compositions and differing in chilling sensitivity are appropriate for comparison, and (3) characterization of chilling sensitivity was based on some general horticultural or agronomic assessment rather than on uniform quantitative assessments of chilling sensitivity using plants grown under identical conditions. Norman *et al.* (1984) compared phosphatidyglycerol compositions of a chilling-sensitive and chilling-resistant population of black mangrove *Avicennia germinans* (L.) L. (Markley *et al.*, 1982) grown under identical conditions. The dipalmitoyl and 1-palmitoyl-2-(*trans*-3-hexadecenoyl)phosphatidylglycerol content of the leaves was higher in chilling-sensitive species, although the sum of the contents of palmitate and *trans*-3-hexadecenoate was similar. This difference in molecular species composition was not observed for root phosphatidylglycerol extracts.

Investigation of the molecular species compositions of phosphatidylglycerol from thylakoid preparations obtained from 17 varieties of rice differing in chilling sensitivity was undertaken (Li *et al.*, 1987). The rice plants were grown under identical conditions and chilling sensitivity was determined for each variety. This allowed a simple statistical analysis of the relationship between chilling sensitivity and parameters of phosphatidylglycerol composition. A correlation analysis of the relationship between the extent of chilling injury and the sum of dipalmitoyl plus 1-palmitoyl-2-(*trans*-3-hexadecenoyl) species yielded a correlation coefficient $r = 0.7307$ (significant at $P = 0.01$) and a coefficient of determination $r^2 = 0.534$ (i.e., over half of the variation in chilling sensitivity may be related to the variation in the relative content of these two species). Some varieties of rice contained appreciable amounts of 1-stearoyl-2-palmitoyl and 1-stearoyl-2-(*trans*-3-hexadecenoyl) species. These species would be expected to have phase transition temperatures similar to the other two species. Performing another correlation analysis but employing the sums of the proportions of these four high-melting-point molecular species yielded a slightly higher correlation coefficient $r = 0.7512$ and a coefficient of determination $r^2 = 0.564$. The values obtained for the sum of the proportions of high-melting-point fatty acids displayed less variation among varieties of rice differing in chilling sensitivity. A correlation analysis of chilling injury versus content of high-melting-point fatty acids yielded a correlation coefficient $r = 0.6182$ and a coefficient of determination $r^2 = 0.382$. These results indicate a better correlation between chilling sensitivity and molecular species composition than fatty acid composition. This may be attributed to the lack of mechanistic and quantitative significance of fatty acid analyses.

Expressing phosphatidylglycerol composition in terms of the content of high-melting-point fatty acids is not necessarily appropriate from a mechanistic perspective because the pairing of acyl chains on the phosphatidylglycerol molecule is an important determinant of phase behavior: Dipalmitoylphosphatidylglycerol has a phase transition temperature (T_m) of 41°C (Raison and Wright, 1983), whereas 1-palmitoyl-2-oleoylphosphatidylglycerol has a T_m of -5°C (Borle and Seelig, 1985). Thus, only when two high-melting-point fatty acids are paired does the parent molecule exhibit a phase transition above 30°C. On the other hand, if the fatty acid is paired with an unsaturated C_{18} fatty acid, the T_m of the phosphatidylglycerol species is below 0°C, and so the potentially deleterious influence of the high-melting-point fatty acid is diminished.

In addition, consideration of phosphatidylglycerol composition in terms of high-melting-point fatty acids is imprecise and less sensitive than actual molecular species analyses (Li *et al.*, 1987): The possible range of values for high-melting-point fatty acids is limited as a consequence of the specific pairing of acyl chains of phosphatidylglycerol. Because the *sn*-2 position of phosphatidylglycerol is occupied by palmitic acid or *trans*-hexadecenoic acid (Murata *et al.*, 1982; Roughan, 1985), these two fatty acids must comprise a minimum of 50% of the total fatty acids of phosphatidylglycerol. Further, any proportion of palmitate in excess of 50% of the total C_{16} fatty acids must occupy the *sn*-1 position (since *trans*-3-hexadecenoate is restricted to the *sn*-2 position) and, hence, be paired with another C_{16} acyl chain (Kenrick and Bishop, 1986; Roughan, 1985). A similar argument applies to stearic acid, which also must be localized at the *sn*-1 position, at least in phosphatidylglycerol of chloroplast origin. As such, the proportion of species comprised of palmitic acid in both the *sn*-1 and *sn*-2 positions, or palmitic acid (or stearic acid) and *trans*-3-hexadecenoic acid can be approximated, where the percentage of high-melting-point species equals $2[(\Sigma 16{:}0 + 16{:}1^{t3} + 18{:}0) - 50]$. Therefore, if the sum of palmitate plus *trans*-3-hexadecenoate is 50%, the two major high-melting-point species of phosphatidylglycerol will be absent; if, however, the sum of high-melting-point fatty acids is 60%, 70%, or 80% of the total fatty acids, disaturated plus *trans*-3-hexadecenoyl-containing species will comprise 20%, 40% or 60%, respectively, of the total phosphatidylglycerol species. As a result, minor differences in high-melting-point fatty acids reflect more pronounced differences in phosphatidylglycerol molecular species. Given these limitations, a strong correlation between high-melting-point fatty acids and chilling sensitivity should not be expected.

Analysis of high-melting-point fatty acids has been used to infer the molecular species composition using the rationale described earlier (Kenrick and Bishop, 1986; Roughan, 1985). However, the precise composition of phosphatidylglycerol molecular species comprising the high-melting-point component is not obtained from these calculations. Variations in the levels of dipalmitoyl and 1-palmitoyl-2-(*trans*-3-hexadecenoyl) species relative to each other were observed among different rice varieties (Li *et al.*, 1987). These compositional differences may have subtle implications for chilling sensitivity if the physical properties of *trans*-3-hexadecenoyl species exhibit phase behavior different from saturated species.

In the absence of divalent cations, the temperature of the onset of phase separation in dispersions of phosphatidylglycerol varied from 29 to 33°C for chilling-sensitive plants and from less than 5 to 14°C for chilling-resistant plants (Murata and Yamaya, 1984). These values were 6–10°C higher in the presence of Mg^{++} salts. Raison and Wright (1983) have demonstrated that as little as 1 to 2% dipalmitoylphosphatidylglycerol in plant polar lipid extracts can induce a transition above 0°C. Limited information is available on the phase behavior of individual phosphatidylglycerol species. One assumption of Murata's hypothesis is that *trans*-3-hexadecenoic acid (because its double bond is the *trans* configuration and is located near the carboxyl terminus of the acyl chain) has thermal properties similar to palmitic acid; likewise, species containing *trans*-3-hexadecanoic acid are similar in behavior to those containing palmitic acid. Bishop and Kenrick (1987) reported that *trans*-3-hexadecenoic acid has a melting point of 54 to 55°C, approximately 10°C lower than that of palmitic acid. The T_m of 1-palmitoyl-2-(*trans*-3-hexadecenoyl)phosphatidylglycerol (Na^+ salt) was determined to be 32°C, compared to 41°C for dipalmitoylphosphatidylglycerol. Comparison of the T_m of 1-stearoyl-2-(*trans*-3-hexadecenoyl)phosphatidylglycerol (46°C) with the palmitoyl-containing analogue (D. V. Lynch, unpublished) also indicates an 8–10°C difference in the phase transition temperature as a consequence of the *trans*-3 double bond. These findings suggest caution when attempting to correlate chilling sensitivity with the levels of high-melting-point fatty acids of phosphatidylglycerol without specific information about the molecular species composition. The influence of divalent cations on lipid phase behavior and membrane properties also merits consideration.

Although phosphatidylglycerol contains high-melting-point molecular species, sulfoquinovosyldiacylglycerols isolated from the leaves of paw paw, corn, and rice have a high proportion of palmitate and stearate (Murata and Hoshi, 1984; Kenrick and Bishop, 1986). The molecular species composition of sulfoquinovosyldiacylglycerol cannot be accurately determined from fatty acid analyses, however, the presence of high melting species of sulfolipid in paw paw may be inferred: Since palmitic and stearic acids make up 62% of the total fatty acids, *at least* 29% of the species must contain two saturated acyl chains. Using argentation thin-layer chromatography to subfractionate sulfoquinovosyldiacylglycerols, Murata and Hoshi (1984) determined that this chloroplast lipid contains 15% and 20% saturated species in rice and maize, respectively. Other chilling-sensitive and chilling-resistant plants examined contained lesser amounts (less than 8%) of saturated sulfoquinovosyldiacylglycerol species. Although it was concluded that saturated sulfoquinovosyldiacylglycerol is not the primary lipid component correlated to chilling sensitivity, it may contribute to thermotropic membrane changes in the range of chilling temperatures and, hence, play a role in chilling injury.

Implicit in these previous studies is the notion that phosphatidylglycerol influences chilling sensitivity by altering the phase properties of bulk lipids. Given the high concentration of phosphatidylglycerol within the chloroplast, it is inferred that this primary event is associated with the thylakoid membranes. Although the observed influence of saturated phosphatidylglycerol species on the phase behavior of plant lipid extracts is consistent with this (Raison and Wright, 1983), calorimetric studies of thylakoids from spinach, pea, tomato, kidney bean, and soybean failed to show a relationship between chilling sensitivity of photosynthesis and the presence of a phase transition of bulk membrane lipids (Low *et al.*, 1984). An alternative hypothesis, based on several lines of evidence, takes into consideration the role of phosphatidylglycerol as a structural component of the photosynthetic apparatus and the mechanism of chill-induced photoinhibition. Chilling in the presence of light results in inhibition of photosynthesis in many chilling-sensitive plants (see Oquist, 1983), a consequence of damage to photosystem II (PSII) and changes in energy distribution between PSII and photosystem I (PSI) (Baker *et al.*, 1983; Powles and Björkman, 1983). For example, leaves of *Zea mays* chilled for 6 hr at 5°C in the light exhibit a 45% decrease in the primary photochemical quantum yield of photosynthetic O_2 evolution and a 20–25% decrease in the primary photochemical quantum yield of PSII following a return to 20°C (Baker *et al.*, 1983). Chilling-induced photoinhibition results in loss of the photosynthetic capacity that persists after chilling and can lead to subsequent photo-oxidation (Wise and Naylor, 1987) and, ultimately, death.

Chill-induced photoinhibition in chilling-sensitive species has been associated with inactivation of the reaction center and turnover of the 32-kDa herbicide binding protein (Q_B protein) associated with PSII, a consequence of the inability to regulate the distribution of light energy between PSII and PSI at chilling temperatures. It has been established that the distribution of light energy between the two photosystems (state 1 to state 2 transition) is regulated by phosphorylation of light harvesting chlorophyll protein (Allen *et al.*, 1981; Allen, 1983). Recently, it was demonstrated that chilling temperatures inhibited thylakoid protein phosphorylation in rice but not in barley, and state 1 to state 2 transitions did not occur in rice but did occur in barley at low temperature (Moll and Steinback, 1986). *In vitro* phosphorylation of thylakoid proteins in rice was inhibited below 20°C; however, phosphorylation of histones by thylakoid-bound kinases was not altered by temperature (Moll *et al.*, 1987). This suggests that low temperatures affect the accessibility of the substrate to the kinase as a result of changes in the conformation of the polypeptides of the light harvesting complex (and other phosphoproteins), or in decreased mobility of the kinase, and/or protein substrates so that they cannot diffuse laterally in the plane of the thylakoid membrane and effect a close approach (Moll *et al.*, 1987). Such changes in protein mobility and/or conformation may be mediated by changes in the thermotropic behavior of associated lipids specifically associated with the protein complexes.

There is widespread evidence that *trans*-3-hexadecenoyl-containing phosphatidylglycerol is associated with the oligomeric structure of the light-harvesting chlorophyll protein complex of PSII (LHCII). Purified LHCII is enriched in phosphatidylglycerol (Krupa *et al.*, 1987). Phospholipase A_2 treatment of thylakoids results in a decrease in the oligomeric form of LHCII (LHCII[1]) and concomitant increase in the monomeric form LHCII[3]. Conversely, mixing LHCII[3] with liposomes of phosphatidylglycerol leads to reconstitution of the oligomeric form (Remy *et al.*, 1982). Analyses of isolated subfractions indicated that one phosphatidylglycerol molecule is present for each protein molecule in the light-harvesting chlorophyll protein complex (Dubacq and Tremolières, 1983).

Taken together, the findings suggest that phosphatidylglycerol does not play a typical role in forming the lipid matrix of thylakoid membranes, but rather it plays a direct role in relation to light harvesting. The physical basis for this unique protein–lipid interaction remains to be elucidated. Nevertheless, it is conceivable that a thermotropic change in the physical state of phosphatidylglycerol associated with LHCII may lead to a conformational change in the associated protein, masking the phosphorylation site. Alternatively, a change in the ordering of the lipid may inhibit diffusional processes, decreasing the likelihood of kinase–LHCII interactions. Either of these processes could lead to a disruption in the regulation of light energy.

The mechanism involving phosphatidylglycerol may be proposed to account for chilling injury of photosynthetic tissues in the light. However, as indicated in the introduction to this chapter, many processes are inhibited by chilling temperatures, and chilling injury certainly is not restricted to photosynthetic tissues nor does it occur only in the light. This raises the question of whether phosphatidylglycerol or other lipid constituents may be involved in altering the associated properties (molecular ordering, permeability, enzymatic activities) of membranes and result, ultimately, in chilling injury. Several studies have provided evidence suggesting a possible role in chilling injury for lipids localized in other cellular membranes.

The most sensitive stage of the life cycle of rice plants is during pollen formation (Satake, 1974). Toriyama *et al.* (1987) have analyzed the fatty acid compositions of individual lipids isolated from rice anthers and compared them to those from leaf tissue. Phosphatidylglycerol isolated from anthers contained 67% saturated fatty acids, or 34% (calculated) saturated molecular species. This was similar to the proportions found in leaf tissue, although the stearic acid content was high and *trans*-3-hexadecenoic acid was absent from phosphatidylglycerol of achlorophyllous anther tissue.

Information on the phosphatidylglycerol composition of organelles and membranes other than the chloroplast is limited. Phosphatidylglycerol comprises less than 5% of the phospholipids of the inner mitochondrial membranes (Mazliak, 1980; McCarty *et al.*, 1973). There is little information on acyl chain composition and nothing is available on the molecular species profiles of chilling-sensitive and chilling-resistant plants. Cardiolipin is localized to the inner mitochondrial membranes and comprises 15–20% of the membrane phospholipid. Since it is derived from phosphatidylglycerol and CDP-diacylglycerol (Moore, 1982), cardiolipin may contain species having high phase transition temperatures, although data on the acyl chain composition of cardiolipin from potato tubers (Mazliak, 1980; McCarty *et al.*, 1973) suggest otherwise. At present, it is not clear which, if any, lipid species in mitochondrial membranes account for the thermotropic behavior exhibited by chilling-sensitive species. To resolve this will require detailed lipid analyses of mitochondria from chilling-sensitive and chilling-resistant plants.

The plasma membrane and tonoplast have received increased attention in recent years, and both membranes have been implicated in chilling injury, especially in relation to the proposed involvement of Ca^{++} (Woods *et al.*, 1984a,b; Minorsky, 1985; Joyce *et al.*, 1988). Yoshida and co-workers (Yoshida and Uemura, 1986; Yoshida *et al.*, 1988) have examined the lipid composition of the plasma membrane and tonoplast of chilling-sensitive mung bean and the thermotropic properties of the respective lipid extracts. In both the plasma membrane and tonoplast, phosphatidylglycerol contained high proportions of saturated fatty acids (73% and 83%, respectively). Although phosphatidylglycerol accounted for less than 3% of the lipid content in plasma membrane and tonoplast, the (calculated) proportion of saturated phosphatidylglycerol molecular species (46% and 66%, respectively) is such that these saturated species would account for 1–2% of the total lipid, a proportion sufficient to induce a phase separation above 0°C (Raison and Wright, 1983). Using differential scanning calorimetry, Yoshida *et al.* (1988) have demonstrated phase separations between 3 and 11°C in dispersions of phospholipids and total lipid extracts of plasma membrane and tonoplast.

This study also found that in addition to highly saturated phospholipids, glucocerebroside is also involved in manifesting phase separations in the total lipid extracts: Addition of glucocerebrosides to the total phospholipid fraction in proportions approximating that of the respective membranes resulted in increases in the phase transition temperatures of 4 to 5°C. Cerebrosides are derived from sphingoid long-chain bases and exhibit high transition temperatures and poor miscibility with phospholipids (Curatolo, 1987). Purified glucocerebrosides from mung bean tonoplast and plasma membrane exhibited phase transitions at 35°C and 38°C, respectively. These temperatures are considerably lower than those reported for other natural and synthetic cerebrosides, even those from rye that contain high proportions of monoenoic fatty acids and exhibit a transition at 56°C (Lynch and Steponkus, 1987a). These low phase transition temperatures exhibited by mung bean glucocerebrosides may be a consequence of trace impurities or of supercooling, since only cooling scans were demonstrated.

These studies concluded that a low temperature—induced phase transition or phase separation of glucocerebroside and/or saturated phospholipid in plasma membranes and tonoplast may be involved in chilling injury. However, glucocerebrosides are also present in the plasma membrane of rye (Lynch and Steponkus, 1987b) and barley (Rochester *et al.*, 1987), two chilling-resistant species. This indicates that their presence in the membrane may not be directly related to chilling sensitivity per se. Other lipid components (e.g. sterols) may modulate the phase behavior of cerebrosides in the tonoplast and plasma membrane. As outlined in Section II, the validity of extrapolating membrane behavior from the thermotropic phase behavior of lipid dispersions not containing the full complement of lipid species must be questioned. Further detailed lipid analyses and studies of lipid–lipid interactions are required to resolve these apparent problems and establish the role, if any, of tonoplast and plasma membrane lipids in the mechanism of chilling injury.

IV. CONCLUSION

There is a great deal of correlative and/or circumstantial evidence suggesting that membranes and constituent lipids are involved in chilling injury. Furthermore, differences in lipid composition between chilling-sensitive and chilling-resistant plants appear consistent with the membrane hypothesis. However, it is still not clear if phase separation of a small subpopulation of the membrane lipid occurs in biological membranes, or if the thermotropic changes in membrane physical properties reflect more subtle interactions between different membrane components. Future efforts at defining the role of lipids should focus on the mechanism of chilling injury, the specific sub-cellular membranes involved, and considerations of lipid–lipid and lipid–protein interactions.

REFERENCES

Allen, J. F. (1983). *Trends Biochem. Sci.* **8**, 369–373.

Allen, J. F., Bennett, J., Steinback, K. E., and Arntzen, C. J. (1981). *Nature (London)* **291**, 1–5.

Andrews, J., and Mudd, J. B. (1985). *Plant Physiol.* **79**, 259–265.

Bagnall, D. J., and Wolfe, J. (1978). *J. Exp. Bot.* **29**, 1231–1242.

Bagnall, D. J., and Wolfe, J. (1982). *Cryo-Lett.* **3**, 7–16.

Bagnall, D. J., Wolfe, J., and King, R. W. (1983). *Plant, Cell Environ.* **6**, 457–464.

Baker, N. R., East, T. M., and Long, S. P. (1983). *J. Exp. Bot.* **34**, 189–197.

Bishop, D. G. (1986). *Plant, Cell Environ.* **9**, 613–616.

Bishop, D. G., and Kenrick, J. R. (1987). *Phytochemistry* **26**, 3065–3067.

Borle, F., and Seelig, J. (1985). *Chem. Phys. Lipids* **36**, 263–283.

Bramlage, W. J. (1982). *HortScience* **17**, 165–168.

Caldwell, C. R. (1987). *Plant Physiol.* **84**, 924–924.

Caldwell. C. R., and Whitman, C. E. (1987). *Plant Physiol.* **84**, 918–923.

Cooper, P., and Ort, D. R. (1988). *Plant Physiol.* **88**, 454–461.

Couey, H. M. (1982). *HortScience* **17**, 162–165.

Cronan, J. E., Jr., and Roughan, P. G. (1987). *Plant Physiol.* **83**, 676–680.

Curatolo, W. (1987). *Biochim. Biophys. Acta* **906**, 111–136.

Dubacq, J.-P., and Tremolières, A. (1983). *Physiol. Veg.* **21**, 293–312.

Garber, M. P. (1977). *Plant Physiol.* **59**, 981–985.

Graham, D., and Patterson, B. D. (1982). *Annu. Rev. Plant Physiol.* **33**, 347–372.

Hodgson, R. A. J., Orr, G. R., and Raison, J. K. (1987). *Plant Sci.* **49**, 75–79.

Joyce, D. C., Cramer, G. R., Reid. M. S., and Bennett, A. B. (1988). *Plant Physiol.* **88**, 1097–1103.

Kee, S. C., Martin, B., and Ort, D. (1986). *Photosynth. Res.* **8**, 41–51.

Kenrick, J. R., and Bishop, D. G. (1986). *Plant Physiol.* **81**, 946–949.

King, A. I., Reid, M. S., and Patterson, B. D. (1982). *Plant Physiol.* **70**, 211–214.

King, A. I., Joyce, D. C., and Reid, M. S. (1988). *Plant Physiol.* **86**, 764–768.

Krupa, Z., Huner, N. P. A., Williams J. P., Maissan, E., and James, D. R. (1987). *Plant Physiol.* **84**, 19–24.

Levitt, J. (1972). "Responses of Plants to Environmental Stresses." Academic Press, New York.

Li, T., Lynch, D. V., and Steponkus, P. L. (1987). *Cryo-Lett.* **8**, 314–321.

Low, P. S., Ort, D. R., Cramer, W. A., Whitmarsh, J., and Martin, B. (1984). *Arch. Biochem. Biophys.* **231**, 336–344.

Lynch, D. V., and Steponkus, P. L. (1987a). *Cryobiology* **24**, 555–556.

Lynch, D. V., and Steponkus, P. L. (1987b). *Plant Physiol.* **83**, 761–767.

Lynch, D. V., Lepock, J. R., and Thompson, J. E. (1987). *Plant Cell Physiol.* **28**, 787–797.

Lyons, J. M. (1973). *Annu. Rev. Plant Physiol.* **24**, 445–466.

Lyons, J. M., and Asmundson, C. M. (1965). *J. Am. Oil Chem. Soc.* **42**, 1056–1058.

Lyons, J. M., Wheaton, T. A., and Pratt, H. K. (1964). *Plant Physiol.* **39**, 262–268.

Maeshima, M., Asahi, T., and Uritani, I. (1980). *Agric. Biol. Chem.* **44**, 2351–2356.

Markley, J. L., McMillan, C., and Thompson, G. A., Jr. (1982). *Can. J. Bot.* **60**, 2704–2715.

Martin, B. (1986). *Plant, Cell Environ.* **9**, 323–331.

Mazliak, P. (1980). *Prog. Phytochem.* **6**, 49–102.

McCarty R. E., Douce, R., and Benson, A. A. (1973). *Biochim. Biophys. Acta* **316**, 266–270.

Minorsky, P. V. (1985). *Plant, Cell Environ.* **8**, 75–94.

Moll, B. A., and Steinback, K. E. (1986). *Plant Physiol.* **80**, 420–423.

Moll, B. A., Eilmann, M., and Steinback, K. E. (1987). *Plant Physiol.* **83**, 428–433.

Moore, T. S. (1982). *Annu. Rev. Plant Physiol.* **33**, 235–259.

Murata, N. (1983). *Plant Cell Physiol.* **24**, 81–86.

Murata, N., and Hoshi, H. (1984). *Plant Cell Physiol.* **25**, 1241–1245.

Murata, N., and Yamaya, J. (1984). *Plant Physiol.* **74**, 1016–1024.

Murata, N., Sato, N., Takahashi, N., and Hamazaki, T. (1982). *Plant Cell Physiol.* **23**, 1071–1079.

Nolan, W. G., and Smillie, R. M. (1976). *Biochim. Biophys. Acta* **440**, 461–475.

Nolan, W. G., and Smillie, R. M. (1977). *Plant Physiol.* **59**, 1141–1145.

Norman, H. A., McMillan, C., and Thompson, G. A., Jr. (1984). *Plant Cell Physiol.* **25**, 1437–1444.

Öquist, G. (1983). *Plant, Cell Environ.* **6**, 281–300.

Patterson, B. D., Kenrick, J. R., and Raison, J. K. (1978). *Phytochemistry* **17**, 1089–1092.

Peeler, T. C., and Naylor, A. W. (1988a). *Plant Physiol.* **86**, 143–146.

Peeler, T. C., and Naylor, A. W. (1988b). *Plant Physiol.* **86**, 147–151.

Peeler, T. C., and Naylor, A. W. (1988c). *Plant Physiol.* **86**, 152–154.

Phillips, N. C., Hauser, H., and Paltauf, F. (1972). *Chem. Phys. Lipids* **8**, 127–133.

Powles, S. B. (1984). *Annu. Rev. Plant Physiol.* **35**, 15–44.

Powles, S. B., and Björkman, O. (1983). *Planta* **156**, 97–101.

Powles, S. B., Berry, J. A., and Björkman, O. (1983). *Plant, Cell Environ.* **6**, 117–123.

Quinn, P. J., and Williams, W. P. (1978). *Prog. Biophys. Mol. Biol.* **34**, 109–173.

Raison, J. K., and Chapman, E. A. (1976). *Aust. J. Plant Physiol.* **3**, 291–299.

Raison, J. K., and Wright, L. C. (1983). *Biochim. Biophys. Acta* **1731**, 69–78.

Raison, J. K., and Orr, G. R. (1986). *Plant Physiol.* **80**, 638–645.

Raison, J. K., Lyons, J. M., Melhorn, R. J., and Keith, A. D. (1971). *J. Biol. Chem.* **246**, 4036–4040.

Remy, R., Tremolières, A., Duval, J.-C., Ambard-Bretteville, F., and Dubacq, J.-P. (1982). *FEBS Lett.* **137**, 271–275.

Rochester, C. P., Kjellbom, P., Andersson, B., and Larsson, C. (1987). *Arch. Biochem. Biophys.* **255**, 385–391.

Roughan, P. G. (1985). *Plant Physiol.* **77**, 740–746.

Roughan, P. G. (1986). *Biochim. Biophys. Acta* **878**, 371–379.

Roughan, P. G., and Slack, C. R. (1984). *Trends Biochem. Sci.* **9**, 383–386.

Sandermann, H. (1978). *Biochim. Biophys. Acta* **515**, 209–237.

Satake, T. (1974). *Proc. Crop. Sci. Soc. Jpn.* **43**, 31–35.

Shneyour, A., Raison, J. K., and Smillie, R. M. (1973). *Biochim. Biophys. Acta* **292**, 152–161.

Steponkus, P. L. (1981). *Encycl. Plant Physiol. New Ser.* **12A**, 371–402.

Toriyama, S., Hinata, K., Nishida, I., and Murata, N. (1987). *In* "The Metabolism, Structure and Function of Plant Lipids" (P. K. Stumpf, J. B. Mudd, and W. D. Nes, eds.), pp. 345–347. Plenum, New York.

Wade, N. L., Breidenbach, R. W., and Lyons, J. M. (1974). *Plant Physiol.* **54**, 320–323.

Wang, C. Y. (1982). HortScience **17**, 173–186.

Wilson, J. M. (1976). *New Phytol.* **76**, 257–270

Wilson, J. M., and Crawford, R. M. M. (1974). *J. Exp. Bot.* **25**, 121–131

Wise, R. R., and Naylor, A. W. (1987). *Plant Physiol.* **83**, 278–282.

Wolfe, J. (1978). *Plant, Cell Environ.* **1**, 241–247.

Wolfe, J., and Bagnall, D. J. (1980). *Ann. Bot. (London)* [N.S.] **45**, 485–488.

Wolk, W. D., and Herner, R. C. (1982). *HortScience* **17**, 169–173.

Woods, C. M., Reid, M. S., and Patterson, B. D. (1984a). *Protoplasma* **121**, 8–16.

Woods, C. M., Polito, V. S., and Reid, M. S. (1984b). *Protoplasma* **121**, 17–24.

Yoshida, S., and Uemura, M. (1986). *Plant Physiol.* **82**, 807–812.

Yoshida, S., Washio, K., Kenrick, J., and Orr, G. (1988). *Plant Cell Physiol.* **29**, 1411–1416.

Molecular Mechanisms of Cold Acclimation

Charles Guy

Ornamental Horticulture Department
Institute of Food and Agricultural Science
University of Florida
Gainesville, Florida

I. INTRODUCTION

A. Ice Formation in Plants

The endless wax and wane of harsh winter conditions in temperate and alpine regions of the world are accompanied by the potential for the temperature of exposed plants to fall below the freezing point of water. Crystallization of cellular water into ice as the temperature drops below the freezing point can impose tremendous stresses and strains on the cells of a plant, ultimately causing injury (Levitt, 1980). If the injury is severe enough, death of the cell, tissue, and plant will ensue. For crop plants, this usually means a reduction of yield or, worse, loss of the entire crop. Most of the world's major crop-growing regions are subject to varying degrees of risk caused by freezing temperatures.

In plant tissues, ice formation can occur anywhere inside the cell, and when this happens, freezing is termed intracellular. The formation of intracellular ice in nature is thought to be universally lethal to the affected cell. If intracellular freezing is rampant in a plant, it will most certainly be lethally injured upon thawing. In contrast, in the vast majority of plants adapted to regions where freezing conditions are common, ice does not commonly form inside the cells; instead it forms outside in the intercellular space where the solute concentration of the water is decidedly lower (Siminovitch and Scarth, 1938).

The formation of ice in the intercellular spaces is termed extracellular freezing (Levitt, 1980). If intercellular ice formation begins at high subzero temperatures, then as the tissue cools, liquid water will be removed from the cell and will coalesce with the growing ice crystals outside the cell. This occurs because of the chemical potential difference of supercooled water and ice at the same temperature. A vapor pressure gradient forms owing to the lower vapor pressure of ice when compared to liquid water at the same temperature. As the tissue temperature declines during equilibrium freezing, more and more cellular water is withdrawn to the extracellular ice and the cells become increasingly dehydrated. Figure 3.1 illustrates, on a macroscopic scale, an example of extracellular freezing in a castor bean petiole. The petiole in this plant is characterized by a large central lumen running its entire length. In the unfrozen plant, the lumen is simply free air space saturated with water vapor; in the frozen plant, however, the petiole is filled with ice. Clearly, cellular water from the tissues surrounding the lumen has accumulated as a result of extracellular ice formation. In situations where

Figure 3.1. Extracellular freezing of a castor bean leaf petiole. The petiole on the left shows a common water-soaked appearance and accumulation of ice in the petiolar lumen; the water for the ice buildup was drawn from the tissues surrounding the lumen. The petiole on the right shows a normal appearance and no accretion of ice.

intracellular ice formation is excluded, injury to plants undergoing extracellular freezing can largely be attributed to the deleterious effects of extreme cellular dehydration and the physical stresses and strains of large amounts of cellular water becoming ice in the extracellular spaces of the tissue.

B. Cold Acclimation

For a vast number of species, tolerance to freezing stress is not static, but can change seasonally or when the temperature and other environmental conditions are varied. A large part of the seasonal dynamics of freezing tolerance variation in plants is related to the process of cold acclimation. Many plant species are able to undergo cold acclimation during exposure to low nonfreezing temperatures (0–10°C) and increase their tolerance to the stresses of extracellular freezing (Levitt, 1980). Strictly speaking, cold acclimation is the process, requiring low-temperature exposure, by which plants are able to alter the freezing tolerance of their cells. It is abundantly clear from the extensive literature spanning almost 300 years of research that cold acclimation is a highly active process resulting from a combination of physiological and metabolic alterations in the plant in response to low temperatures (Levitt, 1980; Graham and Patterson, 1982; Steponkus, 1984). This view of cold acclimation is exactly the opposite of what might be expected in a plant that has become dormant and ceased to grow. While a great deal is known about the cold acclimation process, a definitive understanding of the mechanisms that lead to increases in freezing tolerance is lacking. For the most part, it has not been possible to distinguish the specific biochemical alterations that play a central role in freezing tolerance mechanisms from the metabolic and physiological adjustments to low nonfreezing temperatures. Therefore, an understanding of all plant responses to low nonfreezing temperatures may be required before we can fully comprehend the mechanism of freezing stress tolerance induced during low-temperature exposure.

By now it should not be surprising that the ability to cold-acclimate and increase freezing tolerance varies greatly in nature (Levitt, 1980). Plants of tropical origin have evolved under conditions of constantly warm temperatures and never need to be freezing tolerant, whereas temperate and alpine species have evolved under climatic conditions that require freezing tolerance capabilities in the absence of effective freezing avoidance mechanisms (Levitt, 1980). Chen and Gusta (1983) showed that freezing sensitive species in cell culture cannot be induced by abscisic acid (ABA) treatment to become more freezing tolerant, whereas cold-hardy species can. This fact and the heritability of freezing tolerance suggest that the ability to tolerate freezing has evolved in temperate species and may be completely lacking in tropical species. The large diversity of temperate species and the great range of climatic and ecological conditions in which they have evolved is likely to have led to an array of cold acclimation mechanisms and responses. This is reflected, in part, by the fact that many temperate perennial species have developed mechanisms that sense daylength changes as a more reliable means to signal the initiation of cold acclimation (Smithberg and Weiser, 1968), while in many other species freezing tolerance induction can be controlled strictly by temperature (Guy and Carter, 1984). Such adaptive diversity may also explain, in part, the nearly infinite array of physiological and metabolic adjustments now recorded in the literature.

Present knowledge about the metabolic and physiological changes in plant cells during cold acclimation has been well documented in several reviews and books, as well as elsewhere in this volume (Weiser, 1970; Li and Sakai, 1978; 1982; Lyons *et al.*, 1979; Levitt, 1980; Graham and Patterson, 1982; Steponkus, 1984). Much work has concentrated on physiologically relevant changes in soluble sugars, amino acids (including proline), proteins, and membranes (Levitt, 1980). More recent emphasis has focused on plasma membrane composition (Yoshida and Uemura, 1984; Yoshida, 1984; Uemura and Yoshida, 1984; Lynch and Steponkus, 1987) and stability during freezing and thawing (Steponkus *et al.*, 1983; Gordon-Kamm and Steponkus, 1984; Dowgert *et al.*, 1987). Also, there has been a growing interest in the last four years in the molecular-genetic mechanisms involved in cold acclimation (Guy *et al.*, 1985; Guy and Haskell, 1987; Johnson-Flanagan and Singh, 1987; Laroche and Hopkins, 1987; Marmioli *et al.*, 1986; Mohapatra *et al.*, 1987, 1988; Meza-Basso *et al.*, 1986; Ougham, 1987; Robertson and Gusta, 1986; Robertson *et al.*, 1987).

C. Genetics of Cold Tolerance

Mounting evidence points to the involvement of a genetic component associated with the ability to cold-acclimate and the induction of freezing tolerance. The most well-established evidence is the inheritance pattern of hardiness in progeny of parents with different tolerance levels, but it also includes (1) accumulation of soluble proteins, (2) changes in protein electrophoretic patterns, (3) alterations in isoenzyme composition, (4) appearance of new proteins, (5) accumulation of rRNA and soluble RNAs, (6) changes in RNA base composition, (7) altered mRNA content, and (8) induction of freezing tolerance by ABA.

Most plant cryobiologists consider freezing tolerance and the ability of plants to cold-acclimate to be a quantitative trait (Bouwkamp and Honma, 1969; Rehfeldt, 1977; Hummel *et al.*, 1982; Norell *et al.*, 1986). Normally, progeny will have hardiness capabilities intermediate of parents having different hardiness levels (Auld *et al.*, 1983). However, few studies are directed at a determination of the inheritance of cold hardiness and they are almost always concerned with cultivar or variety improvement (Quamme, 1978). Almost no information is available on the mechanism of inheritance or the number of genes involved in freezing tolerance, although there is some evidence to suggest that a maternal inheritance influence is operable in certain apple cultivars (Harris, 1965; Wilner, 1964). Clearly, additional and more extensive studies concerning the genetics of freezing tolerance are warranted and will be best facilitated in species that exhibit a diversity in freezing tolerance and sexual compatibility. Whether plants with different cold acclimation mechanisms and freezing tolerance capabilities have common cold-acclimation-specific genes or completely different genes remains unknown. The scant information regarding the evolutionary relationships of freezing tolerance, which can be inferred from the cold-hardiness literature, gives no hint whether the basic tolerance mechanism was derived from an ancient progenitor and is somewhat conserved or whether the capability to resist injury from freezing stress arose independently and exists as many variations on numerous themes.

In recent years, it has become apparent that plants respond to adverse environmental stress conditions through alterations in protein synthesis patterns (Sachs *et al.*, 1980; Key *et al.*, 1981; Heikkila *et al.*, 1984; Mason *et al.*, 1988; Hurkman and Tanaka, 1988). In general, when the environmental conditions become unsuitable for optimal growth and development of the plant, new stress-related proteins are synthesized. The newly synthesized proteins appear to be more or less specific to a given environmental stress condition, i.e., anaerobic proteins for anoxia and anaerobiosis (Sachs *et al.*, 1980), heat-shock proteins for supraoptimal temperatures (Key *et al.*, 1981), water-stress proteins for water deficits (Heikkila *et al.*, 1984; Mason *et al.*, 1988), and salt-stress proteins for highly saline conditions (Hurkman and Tanaka, 1988). Although the functional identity of most stress-induced proteins remains undetermined, it seems certain that some will be adaptive and function to enhance plant survival (Sachs and Ho, 1986).

Based on the models for the molecular responses of plants to anaerobiosis and heat shock (Sachs *et al.*, 1980; Key *et al.*, 1981), researchers are searching for proteins that might be linked to tolerance mechanisms for a host of other environmental stresses (Sachs and Ho, 1986). This is also true for cold tolerance. Prior to the advances in understanding anaerobic and heat shock responses, there was considerable evidence to suggest that cold-tolerant or cold-acclimated plants contained unique proteins (Levitt, 1980; Briggs and Siminovitch, 1949; Craker *et al.*, 1969; Rochat and Therrien, 1975a). As early as 1970, Weiser proposed that altered gene expression and the synthesis of new proteins during cold acclimation is necessary for the induction of maximum freezing tolerance in temperate perennials (Fig. 3.2). Nevertheless the impetus provided by the heat shock and anaerobiosis phenomena, coupled with the development of powerful analytical and molecular-genetic techniques, has led to much of the effort now directed at understanding the molecular responses of plants to low temperature. This new approach offers the potential to provide an understanding of the cold acclimation process from a molecular-genetic perspective. Such an approach may help to resolve which metabolic and physiological adjustments occurring at low temperature are important in freezing tolerance mechanisms.

II. HYPOTHESIS

It must be considered that the sum total of freezing tolerance for any given plant is a multigenic characteristic derived from many factors such as cell size, cell morphology, tissue structure, development stage, dormancy requirements, and physiological makeup. With the quantitative nature of cold hardiness in mind, it is clear that each component of the overall equation that goes into determining a given plant's tolerance to freezing stress must be examined separately. Our approach has been to focus only on the ability to increase tolerance to freezing stress through cold acclimation. This approach is based on the belief that a small number of genes could control any one of the major components of freezing tolerance and, in our case, the physiological process of cold acclimation that leads to greater cellular freezing tolerance. This is a somewhat radical concept that is not embraced by most plant cryobiologists. However, it is not a new hypothesis. Weiser (1970) hypothesized that altered gene expression is involved in the development of freezing tolerance in plants during cold acclimation (Fig. 3.2). What is different in our view of cold acclimation and the scheme proposed by Weiser is that few genes control cold acclimation, and their major role is to specifically direct some of physiological and biochemical alterations that occur at low temperature. If this is borne out, then identifying cold-acclimation-specific genes that function at low nonfreezing temperatures to increase freezing tolerance seems possible.

We recognize that this hypothesis, which is based on the heat shock and anaerobiosis models, is at odds with the quantitative nature of cold tolerance, but can an understanding at the molecular level of plant cold hardiness be approached from the standpoint of a widely multigenic nature? It should be made clear that almost all genetic studies of cold hardiness have actually been based on overwintering ability (Liesenfeld *et al.*, 1986) not on cold acclimation (Limin and Fowler, 1988). While quantitative traits can now be dissected in a Mendelian fashion using restriction fragment map-

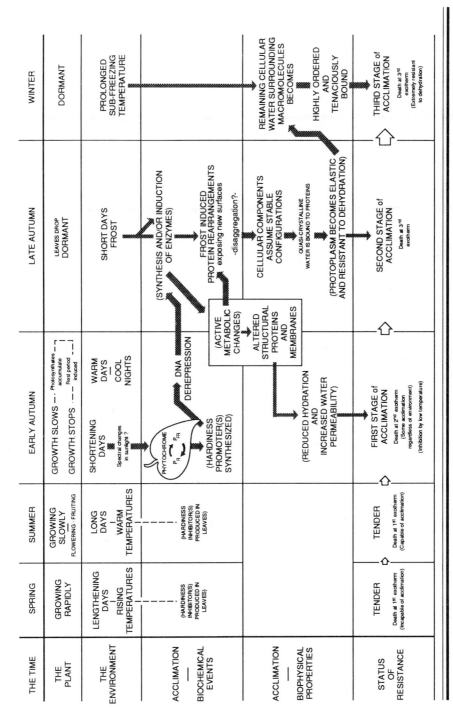

Figure 3.2. A hypothetical mechanism for the cold acclimation of woody perennials involving changes in gene expression. (Reprinted from Weiser, 1970, with permission.)

ping techniques (Paterson *et al.*, 1988), other more direct approaches may also be feasible. Although it is true that overall hardiness (including cold acclimation potential, ultimate freezing tolerance, timing of acclimation, deacclimation, and morphological and physiological processes) is multi-genic, we propose that only one aspect of the overall hardiness—cold acclimation—is controlled by a small number of genes. Cold acclimation genes, by definition, would be inducible and could be studied via standard molecular and recombinant DNA techniques. Freezing tolerance derived from cold acclimation is, without question, an inducible trait that depends on alterations of cellular physiology (Fig. 3.3). The study of genes associated with cold acclimation processes, including those associated with metabolic adjustment to low nonfreezing temperatures and freezing tolerance, can offer a second class of temperature-stress genes to contrast with the well-known heat-shock genes. Identification of low-temperature-regulated genes will also provide insight into the processes that are truly important in freezing tolerance mechanisms.

A. Nucleic Acid Changes in Cold-Hardy Plants

Information regarding nucleic acid changes during cold acclimation remains extremely limited. This is reflected in the fact that characterization of cold acclimation at the molecular level is still in the most preliminary stages of investigation. Several reports have indicated an accumulation of RNA during cold acclimation (Li and Weiser, 1967, 1969; Siminovitch *et al.*, 1968; Devay and Paldi, 1977). The most significant findings are those of Simino-vitch *et al.* (1968) demonstrating increased RNA content of black locust bark cells prior to and during the induction of freezing tolerance in the fall (Fig. 3.4). The accumulation of RNA in black locust paralleled and preceded a similar increase in soluble protein content, while DNA content remained constant. The dramatic increase in the levels of total RNA and protein closely matched the rapid increase of freezing tolerance in the fall (the

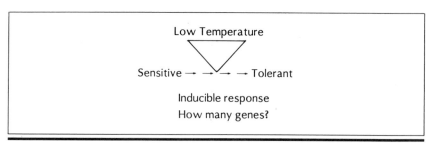

Figure 3.3.   Induction of freezing tolerance in spinach at low temperature.

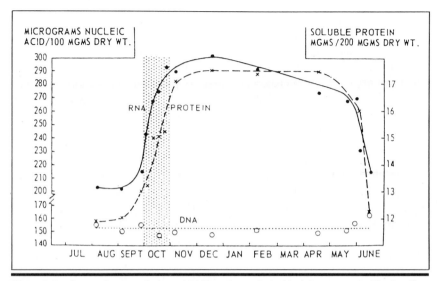

Figure 3.4. Seasonal accumulation of RNA and protein in black locust cortical bark cells. (Reprinted from Siminovitch *et al.*, 1968, with permission.)

stippled area in Fig. 3.4). Particularly noteworthy was the fact the RNA and protein levels remained high until early summer, when the cells reached their least-hardy state. This study measured total RNA content and no effort to further characterize the various classes of RNA was made. Apple and dogwood also show increased RNA content during cold acclimation (Li and Weiser, 1967, 1969), with both soluble RNA (low molecular weight) and total RNA levels again paralleling changes in freezing tolerance during cold acclimation and deacclimation (Li and Weiser, 1969). The increase in total RNA may be due to increased rRNA content, or lower RNase activity in cold acclimated plants (Brown and Bixby, 1973a). Ribosomal RNA synthesis in wheat appeared to increase rapidly during cold acclimation, apparently without an amplification of rRNA genes (Devay and Paldi, 1977). The mechanism of the increase in ribosomal RNA synthesis in cold-acclimated plant tissue remains unknown. Brown (1972) isolated polyribosomes from tissues subjected to cold acclimation and found that they remained intact during the hardening process. More recent evidence has shown that not only do polysomes remain intact, but that cold-acclimated tissues contain 2–3 times more polysomes than nonacclimated tissues (Laroche and Hopkins, 1987).

This increase in rRNA, ribosomes, and polysomes in plants acclimated to low temperatures can be rationalized from a kinetic perspective. At low temperature, enzymatic processes including protein synthesis are significantly reduced, therefore increased protein-synthetic capacity is needed to offset the reduced rate of synthesis. When cold-tolerant winter wheat was subjected to cold acclimation, both rRNA and soluble RNA were increased; the less-hardy spring wheat, subjected to the same conditions, showed smaller increases in rRNA and soluble RNA (Sarhan and D'Aoust, 1975). Increases in rRNA can probably be explained on the basis of more ribosomes being associated in polysomes in cold-acclimated tissues. More interesting, however, was the finding that the base composition of the soluble RNA was altered during cold acclimation (Sarhan and D'Aoust, 1975). Their isolated soluble RNA fraction no doubt contained both mRNA and low-molecular-weight rRNAs; therefore, it is not clear whether the change in base ratios $(G + C/A + U)$ occurred in the rRNA or in the mRNA portion or both. Nevertheless, Sarhan and D'Aoust's data did provide the first hint that different RNAs may characterize cold-acclimated tissues. Measurement of chromatin DNA-dependent RNA polymerase activity in cold-acclimated wheat has revealed an increase in polymerase I activity when compared to nonacclimated tissue (Sarhan and Chevrier, 1985). The increase in polymerase I activity appeared to be associated with increased synthesis of rRNA, which may be needed for additional protein-synthetic capacity at low temperature. Since plants at low temperature contain more rRNA and polysomes, it is clear that protein synthesis must continue. The question arises, what mRNAs are the polysomes translating? The answer to this question may provide the key that unlocks part of the mystery of freezing tolerance.

Recently, two reports have appeared describing the molecular cloning of low-temperature-induced genes (Schaffer and Fischer, 1988; Mohapatra *et al.*, 1988). Three low-temperature-specific cDNAs were isolated from poly(A) RNA obtained from 4°C-incubated tomato fruit. One cDNA had a region homologus to the plant thiol proteases actinidin and papain (Schaffer and Fischer, 1988). From this finding, it was proposed that low temperature may cause some proteins to denature, which may in turn stimulate the expression of the protease. Three cold-acclimation-specific cDNAs were also isolated for cold-acclimated alfalfa (Mohapatra *et al.*, 1988). While low temperature resulted in elevated steady-state levels of the three RNAs, heat shock, water stress, ABA, or wounding did not alter RNA levels when compared to nonacclimated plants.

B. Proteins in Cold-Hardy Plants

Siminovitch and Briggs (1949) were the first to associate the accumulation of soluble proteins with the induction of freezing tolerance. They found that soluble proteins in black locust (*Robinia pseudoacacia*) bark cells increased in the fall and declined in the spring and that these changes closely paralleled changes in tissue freezing tolerance (Fig. 3.4). Accumulation of soluble protein in cold-acclimating tissues, while not universal, is generally accepted as an important step in the development of freezing tolerance (Levitt, 1980) since many researchers have reported similar responses during cold acclimation (Pomeroy *et al.,* 1970; Brown and Bixby, 1973b; Levitt, 1980). Gel electrophoresis of proteins extracted from nonacclimated and cold-acclimated plants has revealed many apparent changes that could be associated with freezing tolerance. Alterations in the electrophoretogram patterns of ribosomal proteins (Bixby and Brown, 1975) were associated with cold acclimation of black locust. Like the rRNA alterations mentioned earlier, the changes in ribosomal protein content occurred at the same time the melting profile of the ribosomes changed and suggested that shifts in ribosomal protein content may be implicated in altered ribosomal function and stability. Similar shifts in the composition of ribosomal proteins following cold acclimation were not found in winter rye, but altered melting points and increased levels of polysomes were observed (Laroche and Hopkins, 1987). Also in black locust the appearance of a glycoprotein complex (Brown and Bixby, 1975) during cold acclimation has been associated with freezing tolerance. The appearance of this glycoprotein component in the soluble protein fraction during the later stages of freezing tolerance induction was considered a possible factor in resistance to ice formation inside the cell via a putative high water-binding capacity (Brown and Bixby, 1975). It now appears that the accumulated protein in the bark tissue of black locust is a lectin (Peumans *et al.,* 1986). The postulated function of this lectin is that of a storage form of organic reserves from the senescent leaves. However, the lectin may have additional biological functions besides that of a storage protein and could play a role in freezing tolerance mechanisms.

Pulse-labeling studies of cold-tolerant and cold-sensitive wheat failed to reveal changes in chloroplast or membrane proteins (Rochat and Therrien, 1975b), but during cold acclimation cold-tolerant Kharkov wheat did synthesize two soluble proteins that the cold-sensitive Selkirk wheat did not (Rochat and Therrien, 1975a). The two proteins had very high molecular weights and appeared to be hydrophilic. Cloutier (1983) observed quantitative changes in proteins but did not detect the appearance of any new polypeptides resulting from cold acclimation of wheat. Other protein-labeling studies of plants subjected to low temperatures have provided additional evidence of low-temperature-induced changes in protein content (Guy *et al.,* 1985; Marmiroli *et al.,*1986; Meza-Basso *et al.,* 1986; Robertson *et al.,* 1987; Yacoob and Filion, 1986; Ougham, 1987; Gilmour *et al.,* 1988; Kurkela *et al.,* 1988; Perras and Sarhan, 1989). In all of these studies, exposure to low temperature caused a change in the protein synthesis pattern when compared to the pattern of synthesis at warm temperature. The changes included the appearance of new polypeptide bands and either diminished concentration or disappearance of a few bands present in warm-grown plants. In parallel, chilling-sensitive tomato also showed alterations in protein synthesis on exposure to 4°C (Cooper and Ort, 1988). The synthesis of a 27-kDa chlorophyll *a/b* binding protein was found to be greatly decreased, and a 35-kDa protein was induced. Unlike cold-tolerant plants, these changes in tomato may be symptomatic of stress or injury instead of being adaptive. When taken together, all of the reports on altered protein synthesis and content in plants subjected to low temperature present no clear picture other than the appearance and disappearance of bands. In no case are the changes identical, and most are not even reasonably similar. The reason for this lack of uniformity in response is unclear; it may result from several factors including the use of different species, different temperature conditions, different experimental approaches, and possibly the different quality of the electrophoretic separations.

Analysis of plasma membrane proteins of plants subjected to low temperatures has shown that cold acclimation causes the appearance of several new polypeptides (Yoshida and Uemura, 1984; Uemura and Yoshida, 1984). Some polypeptides present in membranes of nonacclimated tissue disappeared or decreased in concentration, while others remained unchanged (Uemura and Yoshida, 1984). Changes were also observed in the glycoprotein and lipid composition of the plasma membranes of cold-acclimated tissues. However, similar changes in polypeptide composition were not observed after cold acclimation in chloroplast thylakoid membranes (Rochat and Therrien, 1975b; Griffith *et al.*, 1982). The lack of protein changes in thylakoid membranes during cold acclimation could mean that alterations in the membrane protein composition are not general in the cell, but are specific to each type of membrane.

In cold-acclimated freezing-tolerant spinach leaves, a group of soluble proteins capable of protecting membranes against freezing-stress damage were reported (Volger and Heber, 1975). These cryoprotective proteins were partially purified and reported to be a thousand times more effective on a molar basis than sucrose in preventing cryo-injury. They ranged in molecular weight from 10,000 to 20,000 and were very heat stable. The cryoprotective proteins were not present in nonacclimated leaf tissue, but appeared to be induced during cold acclimation and were believed to be involved in freezing resistance. How this cryoprotective protein protects the thylakoid membrane against freezing stress remains unknown. Work in our laboratory on spinach protein synthesis during cold acclimation has not confirmed the synthesis of small-molecular-weight proteins in cold-acclimated tissue that could be the putative cryoprotective proteins (Guy and Haskell, 1987). However, recent work on isolated chloroplasts from nonacclimated and acclimated tissues has revealed a unique protein in chloroplast from acclimated tissue (Guy and Haskell, 1989). This unique protein is soluble, presumably stromal, and has a molecular weight similar to the cryo-protective protein (Volger and Heber, 1975).

Enzymes from cold-acclimated plants also show significant changes when compared to nonacclimated plants. Cold acclimation of *Dianthus* resulted in altered amylase and lactate dehydrogenase activities. The isoenzyme composition of acid phosphatase and esterase also changed and the synthesis of new peroxidase isoenzymes was found (McCown *et al.*, 1969a). Higher peroxidase activity and new isoenzymes appeared in four unrelated genera during hardening (McCown *et al.*, 1969b), while cold acclimation of winter wheat changed the relative amounts of various molecular weight forms of invertase. Many of the changes in invertase were correlated with increased freezing tolerance (Roberts, 1975). Exposure of alfalfa to low temperature resulted in quantitative increases in several dehydrogenases, as well as the appearance of new isoenzymes for isocitrate, lactate, and glu-cose-6-phosphate dehydrogenase (Krasnuk *et al.*, 1976). Ribonuclease activity of mimosa epicotyl and hypocotyl tissue was altered during cold acclimation (Brown and Bixby, 1973a), while glutathione reductase from spinach leaf tissue exhibited both increased activity and altered isoenzyme composition (Guy and Carter, 1984). Examination of ribulose-bisphos-phate-carboxylase-oxygenase (RUBPCase) from nonacclimated and cold-acclimated "Puma" rye has demonstrated changes in substrate binding, reaction kinetics, sensitivity to freezing, net charge, and conformation (Huner and MacDowall, 1978; Huner and Carter, 1982; Huner and Hayden, 1982). Shomer-Ilan and Waisel (1975) detected hydrophobicity changes in fraction-1-protein (another name for RUBPCase). Glutathione reductase isolated from nonacclimated and cold-acclimated spinach also exhibited changes in substrate affinity, reaction kinetics, electrophoretic mobility, and freezing stability (Guy and Carter, 1984). These changes in the structure, function, activity, and stability of enzymes during cold acclimation have been consistently demonstrated, but the responsible mechanism is not yet understood. However, nearly all of the above changes in enzymes from cold-acclimated tissue can be reconciled by a differential gene expression hypothesis. Furthermore, these changes can provide a conceptual frame-work for the well-documented change in protein synthesis during cold acclimation.

III. SPINACH COLD HARDINESS

A. Experimental System

Several model systems have been developed to facilitate the study of cold acclimation (Chen *et al.*, 1979; Chen and Gusta, 1983; Gordon-Kamm and Steponkus, 1984; Orr *et al.*, 1986; Siminovitch and Cloutier, 1982). Most of these systems consist of callus cultures, cell suspensions, or protoplasts. These cellular-based systems have been most useful in the study of membranes (Gordon-Kamm and Steponkus, 1984; Lynch and Steponkus, 1987) and the role of ABA in freezing tolerance induction (Chen and Gusta, 1983; Orr *et al.*, 1986). However, a major disadvantage is that these cellular systems may not show the same responses to environmental change or have the physiological capabilities possessed by an intact plant. To overcome some of these disadvantages, we have developed a model system using intact spinach seedlings, cultured *in vitro*. This *in vitro* system is particularly suited for long-term labeling studies of the metabolism associated with cold acclimation. Earlier we were able to demonstrate that young soil-grown spinach seedlings, 1 week of age, cold-acclimate when subjected to a constant 5°C, and plants with true leaves, 3 weeks post-germination, acclimate to the same levels of freezing tolerance as older seedlings (Fennell and Li, 1985; Guy *et al.*, 1985). Plants cultured *in vitro*, 1 week post-germination, having only cotyledonary leaves can also cold-acclimate at 5°C (Guy *et al.*, 1987). Furthermore, true leaves and cotyledons from *in vitro*-cultured seedlings show freezing tolerance induction kinetics similar to that of older soil-grown plants (Fennell and Li, 1985; Guy *et al.*, 1985). Clearly, hardiness levels for cotyledons and true leaves from soil and *in vitro*-grown seedlings are comparable to those observed by others using leaf tissue from much older spinach. Since *in vitro*-cultured seedlings could cold-acclimate, this made it possible to conduct radiolabeling experiments. Radiolabeled methionine was readily absorbed into the tissue of *in vitro*-cultured seedlings and incorporated into protein. Using this system, we radiolabeled leaf tissue proteins to high specific activities over a reasonable interval without resorting to wounding or fear microbial uptake and metabolism of the applied label.

B. Protein Synthesis Associated with Cold Acclimation

The first direct evidence to support an alteration in gene expression associated with the cold acclimation process was the observation that newly translatable mRNAs were induced in spinach leaf tissues exposed to 5°C (Guy *et al.*, 1985). The low-temperature-induced mRNAs were found to be present at the same time freezing tolerance was increasing and these particular mRNAs continued to be present throughout a 16-day 5°C treatment. In more recent work, we have demonstrated that exposure of *in vitro*-cultured spinach seedlings to 5°C results in the synthesis of several high-molecular-weight polypeptides in leaf tissue (Guy and Haskell, 1987). Figure 3.5 shows a two-dimensional gel analysis of the proteins synthesized in spinach at 25°C and 5°C. The induction or increased synthesis of four high-molecu-

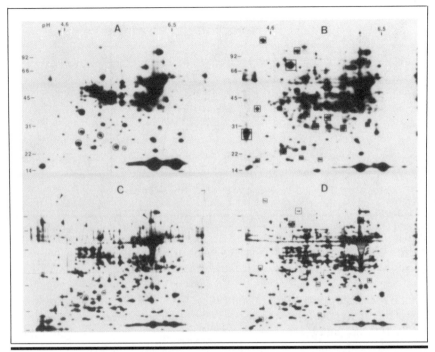

Figure 3.5. Protein synthesis and silver-staining patterns of spinach leaf tissues exposed to 25°C and 5°C. Fluorograms of *in vivo*-labeled spinach leaf proteins separated by two-dimensional gel electrophoresis are shown in panels A and B, while silver staining of leaf proteins is shown in panels C and D. A and C are proteins extracted from tissue grown at 25°C and B and D are proteins from tissue grown at 5°C. The major differences in B and D are indicated by squares. (Reprinted from Guy and Haskell, 1988, with permission.)

lar-weight proteins (160, 117, 85, and 79 kDa) during cold acclimation was highly correlated with the induction of freezing tolerance (Fig. 3.6). Increased synthesis of these polypeptides was observed within 24 hr of exposure to low temperatures and synthesis remained high throughout a 14-day 5°C treatment as freezing tolerance peaked. During deacclimation, the synthesis of the three cold acclimation proteins (CAPs) (160, 117, and 85) was greatly diminished within 24 hr of exposure to 25°C and the cessation of synthesis of the CAPs coincided with the decline in freezing tolerance. CAP 79 was synthesized in nonacclimated tissues, but at a lower level than in cold-acclimated leaf tissue. CAP 79 could also be detected by silver staining in extracts from nonacclimated tissues. Three polypeptides—CAPs 160, 85, and 79—were present in both leaves and hypocotyls of spinach seedlings following cold acclimation, but they were apparently lacking in

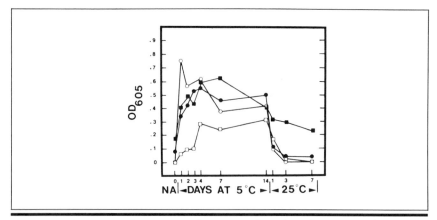

Figure 3.6.   Quantitative densitometry of protein synthesis from fluorograms of two-dimensional gel separations of *in vivo*-labeled leaf proteins. Symbols are (●) CAP 160, (□) CAP 117, (○) CAP 85, and (■) CAP 79. (Reprinted from Guy and Haskell, 1987.)

similarly treated roots (Guy *et al.,* 1988). CAP 117 does not appear to be detectable in hypocotyl or root tissue. The absence of increased amounts of CAPs 160, 85, and 79 in root tissues also suggests that they could play a role in freezing tolerance, since roots were unable to increase in freezing tolerance on exposure to 5°C. It should be noted that in nonacclimated tissue, CAPs 160, 117, and 85 could occasionally be detected by silver staining. More often than not, we were unable to detect CAPs 160, 117, and 85 in nonacclimated leaves. In view of this result, we suspect that either one or all three polypeptides may be constitutively expressed at much lower levels in nonacclimated leaves. In contrast, CAP 79 was always found in nonacclimated leaf, hypocotyl, and root tissue. However, the synthesis and accumulation of all four polypeptides are favored or increased during cold acclimation. In agreement with our findings, synthesis of high-molecular-weight polypeptides during cold acclimation has been reported for *Arabidopsis,* barley, potato, and winter wheat (Gilmour *et al.,* 1988; Kurkela *et al.,* 1988 Marmiroli *et al.,* 1986; Rochat and Therrien, 1975a; Sarhan and Perras, 1987; Perras and Sarhan, 1989; Tseng and Li, 1987). Unfortunately, it is impossible at this point to assign a common identity to any of these polypeptides. However, several studies from our laboratory show that both spinach and citrus may accumulate the same 160-kDa polypeptide (Fig 3.7) (Guy *et al.,* 1988) and that petunia contains detectable quantities of a 160-kDa polypeptide. A low-temperature-induced polypeptide with a molecular weight of 160 kDa and isoelectric point of 4.5 was detected in *Arabidopsis*

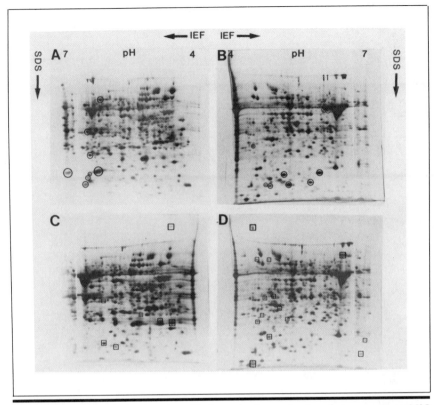

Figure 3.7.   Pattern of silver staining of leaf proteins from spinach and citrus grown at 25°C and 5°C, separated by two-dimensional gel electrophoresis. Citrus leaf proteins are from tissue grown at (A) 25°C and (C) 5°C; spinach leaf proteins are from tissue grown at (B) 25°C and (D) 5°C. (Reprinted from Guy *et al.,* 1988, with permission).

callus by *in vivo* labeling (Gilmour *et al.,* 1988). To our knowledge, this represents the first evidence that unrelated plants could synthesize a similar protein in response to low-temperature treatment. Table 3.1 shows the species and tissues that contain a 160-kDa protein that seems to be increased following cold acclimation. The data in this table are based only on electrophoretic mobilities and therefore represent only indirect evidence for the existence of similar proteins in cold-acclimated tissues of different species. However, due to the unusual electrophoretic characteristics of the 160-kDa protein, the probability of relatedness of this protein across species seems more likely.

Table 3.1. Detection of Increased Amounts of High-Molecular-Weight Proteins in Cold-Acclimated Tisues

Species	Tissue	Nonac-climated	Cold Acclimated	Killing Point LT_{50}°C	CAP 160	CAP 85	CAP 79
S. oleracea	Leaf	+		-4	+[a]/-[b]	+/-	+
S. oleracea	Leaf		+	-9	+++[c]	+++	+++
S. oleracea	Hypocotyl	+		-4	++	++	++
S. oleracea	Hypocotyl		+	-9	++++	+++	+++
S. oleracea	Root	+		-4	-	-	-
S. oleracea	Root		+	-4	-	-	-
P. hybrida	Leaf	+		-4	+	ND[d]	ND
P. hybrida	Leaf		+	-5	+	ND	ND
C. sinensis	Leaf	+		-4	-	ND	ND
C. sinensis	Leaf		+	-7	++	ND	ND
P. trifoliata[e]	Leaf mRNA	+		-8	+[f]	ND	ND
P. trifoliata	Leaf mRNA		+	-18	++++	ND	ND
P. trifoliata	Stem mRNA	+		-8	+	ND	ND
P. trifoliata	Stem mRNA		+	-18	++++	ND	ND
A. thaliana[g]	Leaf mRNA	+		-4	-	ND	ND
A. thaliana	Leaf mRNA		+	-9	+++	ND	ND

[a] + = trace detected
[b] - = trace not detected
[c] +++ = protein present in high quantity
[d] ND = not determined
[e] R. Durham, personal communication.
[f] The isoelectric point of this protein is about 5.3.
[g] Gilmour et al., 1988.

C. Messenger RNA Content

Not surprisingly, the observed changes in the *in vivo* synthesis of the high-molecular-weight polypeptides during cold acclimation and deacclimation are largely paralleled by similar changes in the levels of the translatable mRNAs for these polypeptides (Fig 3.8). Within 24 hr of 5°C exposure, mRNAs for the 160-, 85-, and 79-kDA polypeptides are present in significant quantities. While this would appear to be a rather slow response in comparison to heat shock (Key *et al.*, 1981) or anaerobiosis (Sachs *et al.*, 1980), synthesis of new proteins at low temperature within 1 day is much faster than previously thought (Weiser, 1970). How quickly these mRNAs appear following transfer of spinach seedlings to low temperature has not been determined. It is likely they could appear sooner than 24 hr. In *Arabi-*

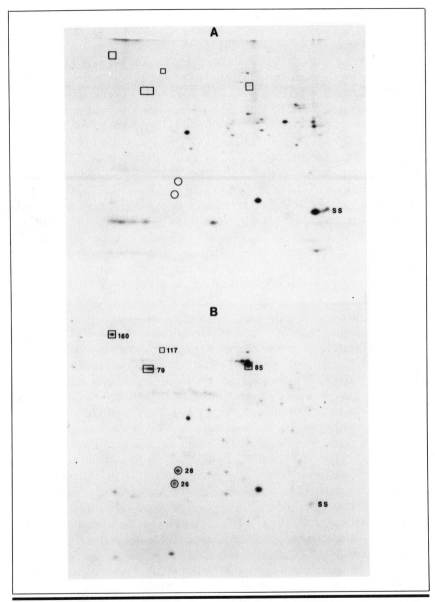

Figure 3.8. Fluorograms of two dimensional gel separations of in vitro translation products from total RNA isolated from leaf tissue grown at (A) 25°C and (B) 5°C.

dopsis, the appearance of an mRNA for a 160-kDa protein was observed within 12 hr of exposure to 4°C (Gilmour *et al.,* 1988). It is reasonable to consider, at least for some species, that low-temperature induction of new mRNAs may be as early as a few hours. In spinach, the rapid increase of the mRNAs for the high-molecular-weight proteins on transfer of the plant to 5°C is matched by an equally rapid disappearance during deacclimation at 25°C (Guy and Haskell, 1988). This observation is significant because the translatability of these mRNAs closely corresponds to the previously documented *in vivo* protein synthesis patterns (Guy and Haskell, 1987). Furthermore, the kinetics of the changes in translatable mRNA content exactly match the changes in freezing tolerance during both cold acclimation and deacclimation (Guy *et al.,* 1985; Guy and Haskell, 1987). Our two-dimensional analyses of *in vitro* translation assays largely confirm previous findings (Guy *et al.,* 1985) and unambiguously demonstrate that the changes in protein synthesis are reflected by similar changes in the mRNA content (Guy and Haskell, 1988). Although this is not direct evidence for a casual relationship with cold acclimation nor is it indicative of an actual role in the induction of freezing tolerance, the data provide a strong justification for further exploration of the functions of these low-temperature-regulated genes.

That the high-molecular-weight *in vitro* translation products are indeed the same as those observed in the *in vivo* labeling studies was confirmed by their co-migration in two-dimensional gels (Fig 3.9). The fact that the 160-, 117-, 85-, and 79-KDa polypeptides co-migrated as single spots on two-dimensional gels when the *in vivo* and *in vitro* translation products were mixed and run together is evidence that they are not synthesized as precursors that are later proteolytically processed. While we would not want to rule out all possible posttranslational modifications on the basis of this experiment, we believe that significant modifications that would greatly alter the electrophoretic mobilities (i.e., signal peptides) can be excluded. In the absence of apparent signal peptides, we would not expect the high-molecular-weight polypeptides induced by low temperature to be targeted to locations within the mitochondria or chloroplasts. However, the subcellular location of these four polypeptides during cold acclimation remains in question.

In addition to the appearance of mRNAs for high-molecular-weight translation products in spinach leaf tissue in response to low temperature and cold acclimation, a number of new lower-molecular-weight translation products also appear. Most prominent are the mRNAs for polypeptides of 28 and 26 kDa (Fig. 3.8). In contrast to the high-molecular-weight *in vitro* translation products, these two polypeptides have not been observed by *in vivo* labeling. Possible explanations why such abundant mRNAs would not

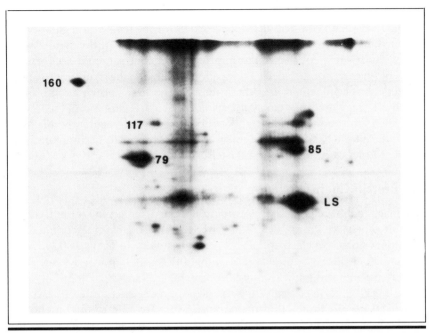

Figure 3.9. Co-electrophoresis of the *in vivo* and *in vitro* translation products for the high-molecular-weight CAPs. (Reprinted from Guy and Haskell, 1988, with permission.)

result in detectable polypeptides *in vivo* are that either these two mRNAs are not translated *in vivo* at high efficiency or the translation products are processed to disguise their presence on the gels of *in vivo*-labeled polypeptides. Unlike the response of plants to heat shock (Key *et al.*, 1981) or anaerobiosis (Sachs *et al.*, 1980), where the synthesis of most proteins present prior to the imposition of the stress is halted, virtually all proteins present in spinach leaf tissue prior to low-temperature exposure continue to be synthesized (Fig. 3.5) (Guy and Haskell, 1987; Guy *et al.*, 1988; Meza-Basso *et al.*, 1986; Mohapatra *et al.*, 1987). In addition to being synthesized, the overall relative abundance of most housekeeping proteins appears to remain largely unchanged as a result of low-temperature treatment (Fig. 3.7). This is indicated by the close similarity of synthesis and protein patterns of silver-stained gels for the various temperature treatments and is in keeping with the idea that plants at low temperature need to continue normal metabolism. Furthermore, this close similarity of the silver-staining pattern holds true regardless of whether low-temperature treatments are for short periods, on the order of hours, or extend up to 2 weeks or more. This clearly contrasts with observations that stress-induced protein responses can be characterized by the synthesis of "early" and/or "late" proteins (Sachs *et al.*, 1980). To date, no compelling evidence exists for the temporal regulation of protein synthesis in spinach, which would be involved in the synthesis of "early" and "late" cold-acclimation proteins. In contrast, cold acclimation of a cold-hardy potato species appears to promote the synthesis of new proteins at specific times and these may be the first known examples of temporally regulated protein synthesis during cold acclimation (Tseng and Li, 1987). Additional evidence to support temporal regulation includes cold-induced mRNAs in barley coleoptiles that appear to accumulate with different kinetics (Cattivelli and Bartels, 1989).

D.   Protein Characterization

In spinach, cold-acclimated hypocotyl tissue can also become more freezing tolerant when grown at 5°C. When hypocotyl proteins are separated by two-dimensional gel electrophoresis, proteins similar to CAPs 160, 85, and 79 can be detected by Coomassie staining. In fact, in hypocotyl tissue CAP 85 may be one of the most abundant proteins (Guy *et al.*, 1988; Guy and Haskell, 1989). By electroblotting hypocotyl proteins separated by two-dimensional gels to Immobilon membranes, we were able to obtain enough of each protein in pure form to determine the amino acid composition for CAPs 160, 85, and 79. Given the fact that the total amino acid composition by itself does not provide a great deal of definitive information, we are able to draw three conclusions from the data. First, CAP 160 is rich in threonine and serine, both common residues for phosphorylation. We point out these amino acids because CAP 160 appears to be a phosphoprotein (Guy and Haskell, 1989). CAP 160 also contains trace amounts of tyrosine; tyrosine can also be a site for phosphorylation, although this type of phosphorylation has rarely been shown to occur in plants (Ranjeva and Boudet, 1987). Second, CAP 160 is rich in glycine 14.8, glutamic/glutamine 15.3, and aspartic/asparagine at 12.4 mol %. Third, from previous studies we could not rule out the remote possibility that CAPs 79 and 85 were cleavage products of CAP 160, since they could add together to yield a molecular weight of approximately 160 kDa. However, the amino acid compositions of CAPs 79 and 85 were so divergent that it would be virtually impossible for them to be cleavage products of CAP 160. For example, CAP 160 lacked phenylalanine, while CAP 85 had 0.6 mol % phenylalanine. Coupled with the evidence from *in vitro* translations of mRNAs (Guy and Haskell, 1988), it is highly plausible that CAPs 160, 85, and 79 represent products of separate genes. Fourth, from the compositional analyses we were able to determine the relative abundance of CAPs 160, 85, and 79 to one another and in the total hypocotyl protein. All were between 0.1 and 1.0% of the total protein.

CAP 160 was lowest, followed by CAP 79; CAP 85 was highest. The estimated mole ratio for the three CAPs $85 : 79 : 160$ is $3 : 2 : 1$.

E. Abscisic Acid and Freezing Tolerance

The trigger or mechanism that regulates the induction or increased synthesis of certain mRNAs at low temperature is unknown in spinach. Evidence from cell culture systems of other species suggests that ABA could play an important role in the induction of freezing tolerance and control the expression of genes responsible for increased hardiness (Chen *et al.*, 1983; Chen and Gusta, 1983; Heikkila *et al.*, 1984; Johnson-Flanagan and Singh, 1987; Robertson *et al.*, 1987). It has been clearly demonstrated that ABA alone can provoke greater freezing tolerance in cell cultures (Chen *et al.*, 1983; Chen and Gusta, 1983; Johnson-Flanagan and Singh, 1987; Robertson *et al.*, 1987). Since ABA also influences the expression of a number of genes in a temporal fashion that is physiologically relevant, it could play a direct role in the induction of freezing tolerance (Tseng and Li, 1987; Johnson-Flanagan and Singh, 1987; Robertson *et al.*, 1987). Evidence to support an ABA role in controlling expression of genes associated with cold tolerance has recently been reported (Lang *et al.*, 1989.) In our spinach system, the ABA response in leaf tissue exposed to low temperature (Fig. 3.10) suggests a possible role as a triggering mechanism for influencing gene expression and increased hardiness exactly as Chen *et al.* (1983) proposed. However, we have not attempted to resolve for spinach, which changes in protein synthesis resulting from the direct effect of low temperature and which are under the control of ABA. As cDNA clones and monospecific antibodies for the high-molecular-weight proteins become available, studies directed at establishing ABA's role in controlling cold acclimation and freezing tolerance will begin.

IV. FUTURE PROSPECTS

Molecular studies of cold acclimation and freezing tolerance in plants are at a preliminary stage. Coupled with the lack of good quantitative genetic data, the question of the degree and importance of alterations in gene expression in these processes is open. However, what is known about the metabolism of plant cells at low temperature clearly indicates that significant adjustments must occur, if for no other reason than to accommodate the continuance of life's processes under what would be adverse growth conditions for most plants. Understanding the molecular basis for these adjustments of metabolism to low temperature brings into focus an important

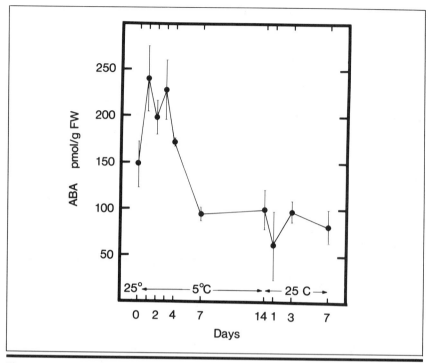

Figure 3.10.   Effect of low-temperature exposure on the free ABA content of spinach leaf tissue. (Reprinted from Guy and Haskell, 1988, with permission.)

question that may provide new insights into why a significant portion of the world's plant species are chilling sensitive. Beyond adjustment of metabolism to low temperature, molecular-genetic approaches offer the potential to detect the action of genes that play a central role in cold-acclimation-directed freezing tolerance. If the diversity of freezing tolerance in cold-acclimated plants can be partially explained by the expression of a small number of genes, then the prospect that it may be possible to alter either the onset or intensity of their expression and achieve an incremental increase in freezing tolerance is even more exciting.

REFERENCES

Auld, D. L., Adams, K. J., Swensen, J. B., and Murray, G. A. (1983). *Crop Sci.* **23**, 763–766.
Bixby, J. A., and Brown, G. N. (1975). *Plant Physiol.* **56**, 617–621.
Bouwkamp, J. C., and Honma, S. (1969). *Euphytica* **18**, 395–397.
Briggs, D. R., and Siminovitch, D. (1949). *Arch. Biochem.* **23**, 18–28.

Brown, G. N. (1972). *Plant Cell Physiol.* **13**, 345–351.
Brown, G. N., and Bixby, J. A. (1973a). *Cryobiology* **10**, 152–156.
Brown, G. N., and Bixby, J. A. (1973b). *Cryobiology* **10**, 529–530.
Brown, G. N., and Bixby, J. A. (1975). *Physiol. Plant.* **34**, 187–191.
Chen, H. H., and Gusta, L. V. (1983). *Plant Physiol.* **73**, 71–75.
Chen, H. H., Li, P. H., and Brenner, M. L. (1983). Plant Physiol. **71**, 362–365.
Chen, H. H., Gavinlertvatana, P., and Li, P. H. (1979). *Bot. Gaz. (Chicago)* **140**, 142–147.
Cloutier, Y. (1983). *Plant Physiol.* **71**, 400–403.
Cooper, P., and Ort, D. R. (1988). *Plant Physiol.* **88**, 454–461.
Craker, L. E., Gusta, L. V., and Weiser, C. J. (1969). *Can. J. Plant Sci.* **49**, 279–286.
Devay, M., and Paldi, E. (1977). *Plant Sci. Lett.* **8**, 191–195.
Dowgert, M. F., Wolfe, J., and Steponkus, P. L. (1987). *Plant Physiol.* **83**, 1001–1007
Fennell, A., and Li, P. H. (1985). *Acta Hortic,* **168**, 179–183.
Gilmour, S. J., Hajela, R. K., and Thomashow, M. F. (1988). *Plant Physiol.* **87**, 745–750.
Gordon-Kamm, W. J., and Steponkus, P. L. (1984). *Proc. Natl. Acad. Sci. U.S.A.* **79**, 6373–6377.
Graham, D., and Patterson, B. D. (1982). *Annu. Rev. Plant Physiol.* **33**, 347–372.
Griffith, M., Brown, G. N., and Huner, N. P. H. (1982). *Plant Physiol.* **70**, 418–423.
Guy, C. L., and Carter, J. V. (1984). *Cryobiology* **21**, 454–464.
Guy, C. L., and Haskell, D. (1987). *Plant Physiol.* **84**, 872–878.
Guy, C. L., and Haskell, D. (1988). *Electrophoresis* **9**, 787–796.
Guy, C. L., and Haskell, D. (1989). *Plant Physiol. Biochem.* **27**, 777–784.
Guy, C. L., Niemi, K. J., and Brambl, R. (1985). *Proc. Natl. Acad. Sci. U.S.A.***82**, 3673–3677.
Guy, C. L., Hummel, R. L., and Haskell, D. (1987). *Plant Physiol.* **84**, 868–871.
Guy, C. L., Haskell, D., and Yelenosky, G. (1988). *Cryobiology* **25**, 264–271.
Harris, R. E. (1965). *Can. J. Plant Sci.* **45**, 159.
Heikkila, J. J., Papp, J. E. T., Schultz, G. A., and Bewley, J. D. (1984). *Plant Physiol.* **76**, 270–274.
Hummel, R. L., Ascher, P. D., and Pellett, H. M. (1982). *Theor. Appl. Genet.* **62**, 385–394.
Huner, N. P. A., and Carter, J. V. (1982). *Z. Pflanzenphysiol.* **106**, 179–184.
Huner, N. P. A., and Hayden, D. B. (1982). *Can. J. Biochem.* **60** 897–903.
Huner, N. P. A., and MacDowall, F. D. H. (1978). Can. J. Biochem. **56**, 1154–1161.
Hurkman, W. J., and Tanaka, C. K. (1988). *Electrophoresis* **9**, 781–786.
Johnson-Flanagan, A. M., and Singh, J. (1987). *Plant Physiol.* **85**, 699–705.
Key, J. L., Lin, C. Y., and Chen, Y. M. (1981). *Proc. Natl. Acad. Sci. U.S.A.* **78**, 3526–3530.
Krasnuk, M., Jung, G. A., and Witham, F. H. (1976). *Cryobiology* **13**, 375–393.
Kurkela, S., Franck, M., Heino, R., Lang, V., and Palva, E. T. (1988). *Plant Cell Rep.* **7**, 495–498.
Lang, V., Heino, P. and Palva, E. T. (1989). *Theor. Appl. Genet.* **77**, 729–734.
Laroche, A., and Hopkins, W. G. (1987). *Plant Physiol.* **85**, 648–654.
Levitt, J. (1980). "Responses of Plants to Environmental Stresses," 2nd ed. Vol. 1. Academic Press, New York.
L., P. H., and Sakai, A., eds. (1982). "Plant Cold Hardiness and Freezing Stress: Mechanisms and Crop Implications," Vol 2. Academic Press, New York.
Li, P. H., and Sakai, A., eds. (1978). "Plant Cold Hardiness and Freezing Stress: Mechanisms and Crop Implications," Vol. 1. Academic Press, New York.
Li, P. H., and Weiser, C. J. (1967). *Proc. Am. Soc. Hortic. Sci.* **91**, 716–727.
Li, P. H., and Weiser, C. J. (1969). *Plant Cell Physiol.* **10**, 21–30.
Liesenfeld, D. R., Auld, D. L., Murray, G. A., and Swensen, J. B. (1986). *Crop Sci.* **26**, 49–54.
Limin, A. E., and Fowler, D. B. (1988). *Genome* **30**, 361–365.

Lynch, D., and Steponkus, P. L. (1987). *Plant Physiol.* **83**, 761–767.

Lyons, J. M., Graham, D., and Raison, J. K., eds. (1979). "Low Temperature Stress in Crop Plants." Academic Press, New York.

Marmioli, N., Terzi, V., Odoardi Stanca, M., Lorenzoni, C., and Stanca, A. M. (1986). *Theor. Appl. Genet.* **73**, 190–196.

Mason, H. S., Mullet, J. E., and Boyer, J. S. (1988). *Plant Physiol.* **86**, 725–733.

McCown, B. H., Hall, T. C., and Beck, G. E. (1969a). *Plant Physiol.* **44**, 210–216.

McCown, B. H., McLeester, R. C., Beck, G. E., and Hall, T. C. (1969b). *Cryobiology* **5**, 410–412.

Meza-Basso, L., Alberdi, M., Raynal, M., Ferrero-Cadinanos, M. L., and Delseny, M. (1986). *Plant Physiol.* **82**, 733–738.

Mohapatra, S. S., Poole, R. J., and Dhindsa, R. S. (1987). *Plant Physiol.* **84**, 1172–1176.

Mohapatra, S. S., Wolfraim, L., Poole, R. J., and Dhindsa, R. S. (1988). *Plant Physiol.* **89**, 375–380

Norell, L., Eriksson, G., Ekberg, I., and Dormling, I. (1986). *Theor. Appl. Genet.* **72**, 440–448.

Orr, W., Keller, W., and Singh, J. (1986). *J. Plant Physiol.* **126**, 23–32.

Ougham, H. J. (1987). *Physiol. Plant.* **70**, 479–484.

Paterson, A. H., Lander, E. S., Hewitt, J. D., Peterson, S., Lincoln, S. E., and Tanksley, S. D. (1988). *Nature (London)* **335**, 721–726.

Perras, M., and Sarhan, F. (1989). *Plant Physiol.* **89**, 577–585.

Peumans, W. J., Nsimba-Lubaki, M., Broekaert, W. F., and Van Damme, E. J. M. (1986). *In* "Molecular Biology of Seed Storage Proteins and Lectins" (L. M. Shannon and M. J. Crispeels, eds.) pp. 53–63. Waverly Press, Baltimore, Maryland.

Pomeroy, M. K., Siminovitch, D., and Wrightman, F. (1970). *Can. J. Bot.* **48**, 953–967.

Quamme, H. (1978). *In* "Plant Cold Hardiness and Freezing Stress: Mechanisms and Crop Implications" (P. H. Li and A. Sakai, eds.), pp. 313–332. Academic Press, New York.

Ranjeva, R., and Boudet, A. M. (1987). *Annu. Rev. Plant Physiol.* **38**, 73–93.

Rehfeldt, G. E. (1977). *Theor. Appl. Genet.* **50**, 3–15.

Roberts, D. W. A. (1975). *Can. J. Bot.* **53**, 1333–1337.

Robertson, A. J., and Gusta, L. V. (1986). *Can. J. Bot.* **64**, 2758–2763.

Robertson, A. J., Gusta, L. V., Reaney, M. J. T., and Ishikawa, M. (1987). *Plant Physiol.* **84**, 1331–1336.

Rochat, E., and Therrien, H. P. (1975a). *Can. J. Bot.* **53**, 2411–2416.

Rochat, E., and Therrien, H. P. (1975b). *Can. J. Bot.* **53**, 2417–2424.

Sachs, M. M., and Ho, T.-H. D. (1986). *Annu. Rev. Plant Physiol.* **37**, 363–376.

Sachs, M. M., Freeling, M., and Okimoto, R. (1980). *Cell (Cambridge, Mass.)* **20**, 761–767.

Sarhan, F., and Chevrier, N. (1985). *Plant Physiol.* **78**, 250–255.

Sarhan, F., and D'Aoust, M. J. (1975). *Physiol. Plant.* **35**, 62–65.

Sarhan, F., and Perras, M. (1987). *Plant Cell Physiol,* **28**, 1173–1179.

Schaffer, M. A., and Fischer, R. L. (1988). *Plant Physiol.* **87**, 431–436.

Shomer-Ilan, A., and Waisel, Y. (1975). *Physiol. Plant.* **34**, 90–96.

Siminovitch, D., and Briggs, D. R. (1949). *Arch. Biochem. Biophys.* **23**, 8–17.

Siminovitch, D., and Cloutier, Y. (1982). *Plant Physiol.* **69**, 250–255.

Siminovitch, D., and Scarth, G. W. (1938). *Can. J. Res., Sect. C* **16**, 467–481.

Siminovitch, D., and Rheaume, B., Pomeroy, K., and Lepage, M. (1968). *Cryobiology* **5**, 202–225.

Smithberg, M., and Weiser, C. J. (1968). *Ecology* **49**, 495–505.

Steponkus, P. L., Dowgert, M. F., Gordon-Kamm, W. J. (1983). *Cryobiology* **20**, 448–465.

Steponkus, P. L. (1984). *Annu. Rev. Plant Physiol.* **35**, 543–584.

Tseng, M. J., and Li, P. H. (1987). *In* "Plant Cold Hardiness" (P. H. Li and A. Sakai, eds.), pp. 1–27. Liss, New York.

Uemura, M., and Yoshida, S. (1984). *Plant Physiol.* **75,** 818–826.

Volger, H. G., and Heber, U. (1975). *Biochim. Biophys. Acta* **412,** 335–349.

Weiser, C. J. (1970). *Science* **169,** 1269–1278.

Wilner, J. (1964). *Can. J. Plant Sci.* **45,** 67.

Yacoob, R. K., and Filion, W. G. (1986). *Biochem. Cell Biol.* **65,** 112–119.

Yoshida, S. (1984). *Plant Physiol.* **75,** 38–42.

Yoshida, S., and Uemura, M. (1984). *Plant Physiol.* **75,** 31–37.

CHAPTER 4

Anatomy: A Key Factor Regulating Plant Tissue Response to Water Stress

Kaoru Matsuda*
Ahmed Rayan**

*Department of Molecular and Cellular
Biology and Department of Plant Sciences
University of Arizona
Tucson, Arizona

**Department of Molecular and Cellular
Biology
University of Arizona
Tucson, Arizona

I. INTRODUCTION

Plant productivity in many regions of the world is limited primarily by water deficits (Boyer, 1985), and field (e.g., Hanks, 1982) and laboratory studies (Sands and Correll, 1976; Mason and Matsuda, 1985) have shown that the growth rates of several glycophytic plants are directly proportional to the availability of water. Water deficits are also known to alter a variety

Environmental Injury to Plants

of biochemical and physiological processes ranging from photosynthesis to protein synthesis and solute accumulation (Hsiao, 1973; Hsiao *et al.*, 1976), but despite several decades of intense effort, little is known about how stress regulates growth and other plant functions. The lack of progress is due in part to the complexity of the problem, but now there is reason to believe that rapid advances will occur if stress responses are studied in terms of a plant's anatomy as well as its physiology and biochemistry. This optimism about the potential for progress and concern for the role that anatomy plays in regulating plant responses to stress has come largely from results of experiments with young Arivat barley (*Hordeum vulgare* L.) seedlings that were stressed by exposing their roots to different concentrations of osmotic solutions. These studies demonstrated that the basally located growing region of the leaf is much more responsive to stress than the expanded blade, and response differences appear to be due primarily to anatomical features that regulate water movement in tissues.

In this report, we first show how various levels of osmotic stress affect barley leaf growth and some biochemical and physiological processes in growing and expanded regions of the leaf. Subsequently, we present anatomical views and physiological studies to suggest how transpirational water moves in plants and why tissues at different stages of development can vary in their response to water deficits.

II. GROWTH RESPONSES OF BARLEY LEAVES TO OSMOTIC STRESS

Young Arivat barley seedlings grown at $25 \pm 2°C$ and 13 hr daily light of 200 μ mol m^{-2} s^{-1} photosynthetically active radiation (PAR) are used routinely in our studies. Seeds are germinated in vermiculite, and the seedlings are transferred on the fourth day to aerated Hoagland's medium and held for 24 hr before studies are initiated. Plants are stressed by exposing their roots to a nutrient solution containing either polyethelene glycol (PEG) 8000 or NaCl, and growth rates are determined by positioning the leaf tips against a precision 12.7-mm-wide ruler and measuring elongation with time-lapse photography (Matsuda and Riazi, 1981). The usual growth rate of unstressed plants is 1.5 mm/hr.

Intact barley seedlings exposed to osmotic solutions containing either NaCl or PEG yield similar growth responses (Matsuda and Riazi, 1981). Figure 4.1 provides a representative view of how leaf expansion is affected by different levels of stress. As in corn (Acevedo *et al.*, 1971), exposure of roots to osmotic solutions causes an almost immediate cessation of leaf elongation. Under continuous stress, barley leaves will resume growth after

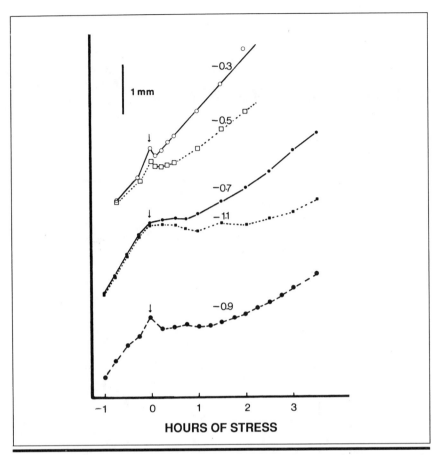

Figure 4.1. Rapid leaf growth responses of 5-day-old barley seedlings. Plants were grown in normal Hoagland's solution and replaced (at times indicated by the arrows) with nutrient solutions containing NaCl of the indicated ψ (MPa). Values are the means of three observations. (Reproduced with permission from Matsuda and Riazi, 1981; Copyright 1981, The American Society of Plant Physiology.)

a lag time of a few minutes for mildly stressed plants (e.g., −0.3 MPa) and after about 2 hr for more severely stressed seedlings (−1.1 MPa). Growth rates are proportional to the water potential (ψ) of the nutrient medium.

The occurrence of rapid stress-caused growth cessation indicates that all or nearly all cells involved in leaf elongation respond almost immediately to decreased water availability around the distally located roots. This sudden growth stoppage, although dramatic and useful in helping to understand how plant cells respond to stress, represents only part of the overall growth response. The subsequent lag period and regrowth suggests that cells undergo one or more adjustment steps to resume expansion at rates that are controlled in some way by the ψ around the roots. Longer-duration stress studies (Fig. 4.2b) demonstrate that the relationship of growth to solution ψ is maintained for several days. The physiological and biochemical studies described below were initiated to gain an understanding of the basis for sudden growth cessation and subsequent resumption of leaf elongation.

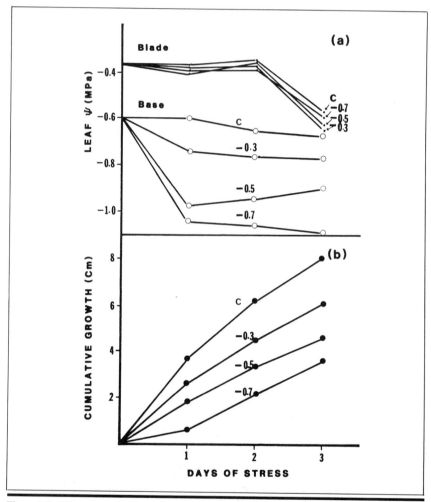

Figure 4.2.   Long-term effects of osmotic solutions on leaf water status and growth. Nutrient solutions containing PEG 8000 of indicated ψ (MPa) were applied on the fifth day. Blade and basal tissue ψ are (a) the means of four observations and cumulative growth and (b) the means from 20 plants. (Reproduced with permission from Matsuda and Riazi, 1981; Copyright 1981, The American Society of Plant Physiology.)

III.   STRESS RESPONSES OF GROWING AND EXPANDED AREAS OF LEAVES

Elongation of cereal leaves occurs as a result of the expansion of cells generated by an intercalary meristem located at the base of the leaf. Because there is no *a priori* reason for believing that cells in growing regions should behave like those in the blade, we specifically compared the stress responsiveness of the growing region (basal 10 mm closest to the seed) with segments of the same length obtained from the midblade and, in some cases, from intermediate regions.

A.   Stress-Induced Water Status Changes

In the leaves of unstressed young barley seedlings, the ψ of the growing region is usually between -0.6 to -0.8 MPa, which is 0.2–0.3 MPa lower than the -0.4 to -0.5 MPa found for the expanded blade (Figs. 4.2a, 4.3a, and 4.3b). Osmotic stress increases the difference in ψ between the growing region and the expanded blade because it lowers the ψ of the growing region in proportion to the degree of applied stress (Fig. 4.2a), while only slightly affecting the ψ of the expanded blade. Results of this type indicate that cells in the expanded blade are largely insulated from the stress signal that causes osmotic adjustment to occur in the growing region.

   Insight into the basis for the stress-caused reduction in tissue ψ of the growing region was obtained by summarizing kinetic data such as those in Fig 4.3a:

(1) Although significant reductions in ψ are not detected for 0.5 to 1.0 hr, its lowering in the growing region is initiated rapidly (Fig. 4.3a). This suggests that decreases in ψ are initiated at or very close to the time that stress-caused growth cessation occurs.

(2) Throughout the stress period, ψ and osmotic potential (Ψ_π) are reduced by equal amounts so turgor (Ψ_ρ) remains constant in the growing region (Fig. 4.3a), even though growth first stops and then resumes after a lag period (Fig. 4.1).

(3) When plants are stressed with different concentrations of osmotic solutions for periods sufficient to allow for stabilization (12 hr or more), reductions in Ψ and Ψ_π (osmotic potential) of the growing region approximate the stress that is applied in the external medium (e.g., Figs. 4.2a and 4.3a).

(4) Plants that are severely stressed by exposing their roots to solutions with Ψ lower than that present in the growing region of unstressed seedlings will not resume growth until the Ψ of the growing tissue drops to that of

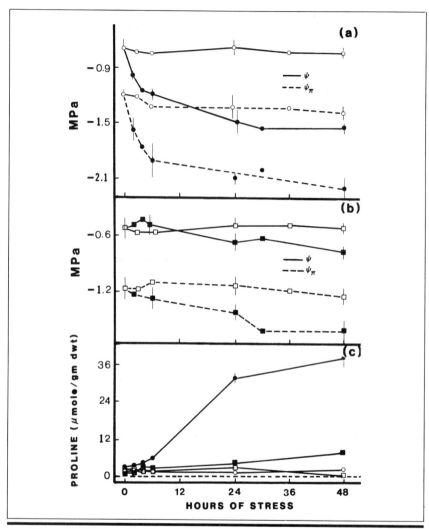

Figure 4.3.   Water status and proline values of leaf basal and midblade segments from unstressed and stressed (-0.8 MPa, PEG) barley seedlings. Plants were 5 days old at the start. Circles and squares refer to data from basal and midblade regions, respectively. Open symbols are from unstressed plants; closed symbols are from stressed plants. Values ($\pm$SD) are the means of three observations for proline and four observations for water status. (From Riazi *et al.*, 1985.)

the external solution. These data suggest that net water uptake by growing cells is one requirement for expansion.

Osmotic adjustment with turgor maintenance occurs commonly in many stressed plant tissues, which implies that while a reduction in Ψ_π is the driving force for water entry into cells, water entry into growing cells is somewhat restricted. This phenomenon of limited water entry into cells differs significantly from that found in guard cells, where solute accumulation leads to increased turgidity due to relatively free initial water entry.

Because there is a long-held belief that cell growth in plants is regulated by turgor, the accuracy of Ψ values obtained pyschrometrically for cut sections of the growing regions of barley leaves was questioned (Cosgrove *et al.*, 1984). These workers reasoned that because of the wall relaxation that occurs during the required equilibration period, Ψ (and therefore Ψ_p) values obtained in this way for cut sections of growing tissues would be artifactually low. However, subsequent studies have shown that ψ values obtained psychrometrically using excised sections from the growing region of corn leaves are virtually identical to those obtained with *in situ* methods (Westgate and Boyer, 1984), and also under conditions where steps were taken to minimize the possibility of wall relaxation in growing areas of corn (Hsiao and Jing, 1984), barley, and wheat leaves and in the hypocotyls and epicotyls, respectively, of dark-grown squash and pea seedlings (Mason and Matsuda, 1985). Wall relaxation by excised sections of growing tissues may generally not occur because of an elimination of water (Nonami and Boyer, 1987; Matyssek *et al.*, 1988) or other components (e.g., wall-loosening factors) required for growth.

Stress-caused growth reduction and turgor maintenance occur not only in the growing regions of barley leaves (Fig. 4.3; Matsuda and Riazi, 1981; Mason and Matsuda, 1985), but also in the growing regions of corn (Michlena and Boyer, 1982; Westgate and Boyer, 1984) and wheat leaves (Mason and Matsuda, 1985; Barlow, 1986), in the hypocotyls of dark-grown soybean (Meyer and Boyer, 1981; Cavalieri and Boyer, 1982; Nonami and Boyer, 1989), squash (Mason and Matsuda, 1985), and mung bean seedlings (Passos, 1989), and in the epicotyls of dark-grown pea seedlings (Mason and Matsuda, 1985). In contrast to the turgor maintenance that occurs in growing regions of stressed organs, the expanded regions of wheat leaves (Barlow, 1986) and soybean hypocotyls (Nonami and Boyer, 1989) were found to undergo a stress-caused turgor loss.

B. The Nature of Solutes Involved in Osmotic Adjustment

Calculations based on available data of dry weight gains of excised sections, their water content, and the extent of osmotic adjustment (Riazi *et al.*, 1985) suggest that molecules of about 200 daltons may account for the early stress-caused osmotic adjustment that occurs in growing regions of barley leaves. Possibly, hexoses (mol wt 180) or their derivatives can participate in the adjustment processes and in short-duration kinetic studies; stress-caused increases in glucose have been found to closely parallel reductions in Ψ_π (e.g, Fig. 4). However, glucose changes account for about one-third of the osmotic adjustment, and in Fig. 4.4b, its level increased as sucrose decreased. In other studies, sucrose was found to remain constant and even increase during stress (data are not shown). Because of this variability and because all osmotic components have not been identified, we can only hypothesize that osmotic adjustment is initiated because sucrose

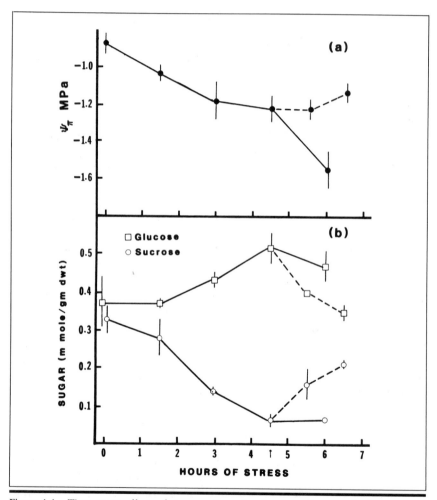

Figure 4.4. Time-course effects of stress on tissue ψ_π, glucose, and sucrose in the growing regions of barley leaves. Five-day-old seedlings were stressed (PEG, -0.7 MPa); after 4.5 hr, some plants were relieved of stress (dashed lines). Values ($\pm$SD) are the means of three replications for sugar and four replications for ψ_π. (From Riazi *et al.,* 1985.)

accumulation by stressed growing cells continues, whereas incorporation of carbohydrate precursors into cell walls is reduced. These precursors are likely metabolized to other osmotically active forms and accumulated.

The effect of stress on leaf proline content was also studied. As was found for osmotic adjustment (Figs. 4.3a and 4.3b), stress-caused increases in proline were much more pronounced in the growing region than in the expanded blade (Fig. 4.3c). After 24 hr, proline was calculated to account for about 5% of the total osmotic adjustment (Riazi *et al.*, 1985).

Despite the clear differences in proline accumulation by growing-region and blade tissues, it appears likely that the mechanisms causing its increase in both parts of the leaf are the same. This is suggested by the fact that proline contents of the growing-region, expanded blade, and intervening tissues from seedlings stressed for 24 hr or more were related inversely to the Ψ of the sampled tissue (Riazi *et al.*, 1985). On the other hand, since proline increases did not occur until 4 hr following continuous stress, the mechanism responsible for its increase in stressed tissues differs from that responsible for rapid osmotic adjustment.

C. Stress Effects on Polyribosome Proportions

Since Hsiao's (1970) early demonstration of rapid water stress-caused reduction in the proportion of ribosomes present as polyribosomes (polyribosome proportion) in corn seedlings, water deficits have been shown to reduce polyribosome proportions in pumpkin cotyledons and pea seedlings (Rhodes and Matsuda, 1976), black locust seedlings (Brandle *et al.*, 1977), wheat shoot apicies (Barlow *et al.*, 1977), and squash fruits (Cocucci *et al.*, 1976). Further, in pumpkin and pea seedlings (Rhodes and Matsuda, 1976) and in squash fruits (Cocucci *et al.*, 1976), stress-caused growth reductions were correlated highly with reductions in polyribosome proportions.

The suggestion of a close relationship of growth to polyribosome proportion was reinforced by more recent studies (Mason and Matsuda, 1985) with several plant species. Extensive kinetic studies performed with barley seedlings stressed with PEG (-0.8 MPa) demonstrated that reductions in polyribosome proportions occur in the growing region of leaves within 15 min, and reductions continue until stabilization is reached in about 4 hr (Fig. 4.5b). In contrast, polyribosome proportions in the expanded blade are not altered for at least 2 hr.

In addition to conducting studies with barley plants, wheat and dark-grown squash and pea seedlings were exposed to osmotic solutions of various concentrations for 12 hr or more, and then the solution water status, organ growth, and several properties of the growing region (Mason and Matsuda, 1985) were compared. Growth rates of barley and wheat leaves, squash hypocotyls, and pea epicotyls are related directly to the polyribo-

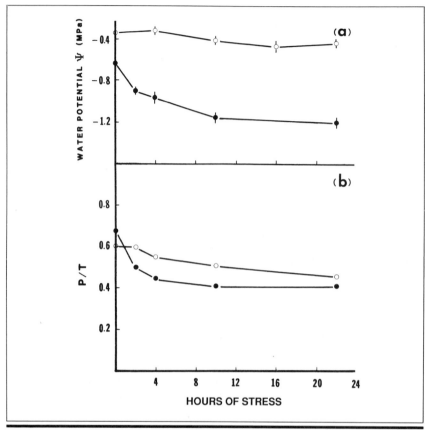

Figure 4.5.   Stress-caused changes in growing and expanded regions' tissue ψ and polyribosome proportions of leaves from barley seedlings. Open and closed circles refer to expanded and growing regions, respectively; (a) ψ values are the means of five replications and (b) P/T values are the means of three replications. (Adapted from Mason and Matsuda, 1985.)

some proportion in the growing region (Fig. 4.6a). Additionally, when results are expressed as the percentage of the values found in unstressed control plants, data from all plants fit a single curve that shows that a 50% reduction in the polyribosome proportions will result in a 90% inhibition of growth (Fig. 4.6b).

D.   A Hypothesis for Explaining Rapid Stress Responses

The results of the physiological studies summarized above have provided several independent lines of evidence that the growing region of young bar-

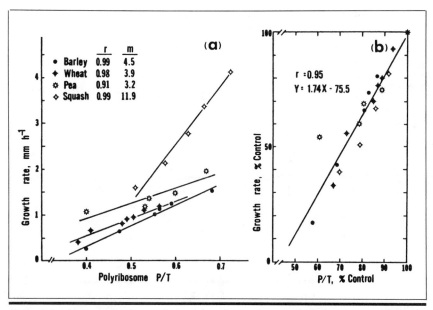

Figure 4.6. Linear regression analysis of growth rate versus polyribosome status of the growing region of various plants. Growth rates and proportions of total ribosomes present as polyribosomes percentages (PT) (a) for unstressed plants and plants stressed with various concentrations of PEG. In (b) results are expressed as percentages of values found in unstressed control plants. (From Mason and Matsuda, 1985.)

ley leaves is much more responsive to osmotic stress than the expanded blade. They have shown also that while stress causes leaf growth to stop before there is any change in the Ψ or Ψ_π of the growing region, it nevertheless rapidly initiates osmotic adjustment and reductions of polyribosome proportions and causes proline levels to begin to rise after 4 hr of continuous stress. Also, when plants are stressed for extended periods in osmotic solutions of different concentrations, polyribosome proportions, Ψ and Ψ_π of the growing region are, like leaf elongation, directly related to the solution Ψ whereas proline levels are inversely related to water availability. These data suggest that cell elongation, osmotic adjustment, and polyribosome proportions in the growing region may be controlled by the same stress signal, or changes in some processes are extremely rapid responses to a primary stress signal. On the other hand, proline increase clearly is regulated in another way.

Because the xylem represents a small fraction of the total water present in tissues and since pressure changes in water transmitted at the speed of sound, it was hypothesized that sudden reduced water availability plus ongoing transpiration would cause an immediate reduction in the hydrostatic pressure and therefore Ψ of the transpiration stream. Such Ψ changes would not be detectable with our present procedures, but cells in the growing zone are likely to respond to the changes in Ψ of the apoplasm around them (Matsuda and Riazi, 1981). It was proposed also that the expanded blade is relatively insensitive to stress because transpirational water movement from the xylem to stomata largely bypasses most mesophyll cells—a view supported by studies that showed that when [³H]water was supplied to roots of intact plants, radioactivity first appeared as transpiration before it was detected in the blade (Matsuda and Riazi, 1981).

An examination of the anatomy of the growing-region, expanded-blade, and intermediate tissues gave clear suggestions about how transpirational water might move through leaves, how stress effects are transmitted to most growing cells, and why cells in the expanded blade are relatively insensitive to stress. The anatomical views also led to tests to determine experimentally how water moves through plants and if stress will cause xylem hydrostatic pressure to drop.

IV.   THE RELATION OF ANATOMY TO STRESS RESPONSE

Fresh-cut and fixed cross sections from the intercalary meristem, the growing region 5 mm above the point of seed attachment, the expanded midblade, and, in some cases, intermediate region tissues were examined to infer how water will likely move in different parts of the leaf. Except for slight changes in the dimensions of cells and intercellular spaces, tissues prepared in both ways yielded similar data. Comparison of cross-sectional views from the intercalary meristem (Fig. 4.7a) and the more differentiated midblade (Fig. 4.7b) provided suggestions for explaining how transpirational water will move through the basal region and the blade of a young barley leaf.

Mature (open) vessel elements capable of transporting water are present in the intercalary meristem (Fig. 4.7a) and in the expanded blade (Fig. 4.7b). The summarized data in Table 4.1 show that all mature vessels are clustered in five functional vascular bundles (FVB) throughout the 5-day-old barley leaf. Furthermore, FVB are separated from each other by approximately 20 closely packed mesophyll cells and one to three immature vascular bundles (IVB), which have no open vessels.

The occurrence of mature vessels that extend throughout the leaf is analogous to the situation found in corn (Esau, 1943). Their presence in the growing region of barley leaves supports the view (Westgate and Boyer, 1984) that transpirational water will move in well-developed xylem

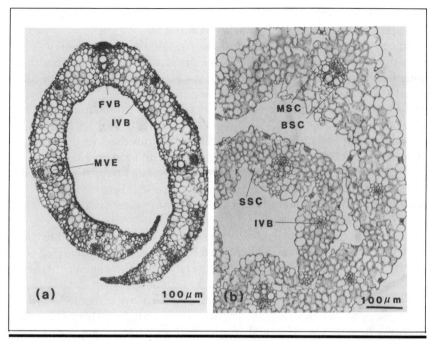

Figure 4.7. Micrographs from 5-day-old "Arivat" barley leaves. Cross-section views of (a) the intercalary meristem, which contains functional vascular bundles (FVB) with mature vessel elements (MVE) and immature vascular bundles (IVB), and (b) the expanded blade, which shows the association of stomata and substomatal cavities (SSC) with FVB, IVB, bundle sheath cells (BSC), and mestome sheath cells (MSC). These sections were obtained 1 and 70 mm from the seed, respectively. (Reproduced with permission from Rayan and Matsuda, 1988; Copyright 1988, The American Society of Plant Physiology.)

Table 4.1. Mesophyll and Vessel Areas, and Numbers of Functional Vascular Bundles (FVB) and Immature Vascular Bundles (IVB) in Different Areas of Young Barley Leaves

Distance from seed (mm)	Mesophyll area (mm²)	Vessel area (mm²)	Numbers of vascular bundles	
			FVB	IVB
1	0.40 ± 0.013	0.0029 ± 0.0002	5.5 ± 0.5	3.5 ± 0.5
5	0.43 ± 0.010	0.0036 ± 0.0004	5.2 ± 0.4	8.3 ± 0.7
15	0.55 ± 0.029	0.0041 ± 0.0001	5.1 ± 0.5	8.6 ± 0.5
25	0.76 ± 0.017	0.0044 ± 0.0002	5.1 ± 0.3	9.5 ± 0.7
45	0.98 ± 0.027	0.0048 ± 0.0003	5.0 ± 0.0	10.7 ± 0.7

Data (±SD) are the means of five fresh-cut cross sections or more and are reproduced with permission from Rayan and Matsuda (1988); Copyright 1988, The American Society of Plant Physiology.

through the growing region of cereal leaves rather than intercellularly as we (Matsuda and Riazi, 1981) had believed earlier.

Since sudden osmotic stress causes nearly instantaneous growth cessation, the anatomical arrangement in the growing zone suggests that there must be a rapid lateral transmission of a stress signal from the five functional vascular bundles to all intervening mesophyll cells. If the primary stress signal is the proposed reduction in xylem hydrostatic pressure (Matsuda and Riazi, 1981), pressure changes potentially can be transmitted through pits in vessel walls and into the continuous mass of water in the mesophyll cell walls. Apoplastic (cell wall + xylem) pressure changes, however, can occur only if the apoplasm is largely isolated from the symplasm, a possibility first presented (Matsuda and Riazi, 1981) because stress causes growth cessation with no apparent change in any water status measure of the growing region.

Except for the fact that guard cells are developed and functional, the cross-sectional anatomy of the expanded blade (Fig. 4.7b) is modified only slightly from that of the growing region (Fig. 4.7a). The numbers of mesophyll cells and IVB (Table 4.1) are increased somewhat in the expanded blade, but the cells remain tightly packed and there are still only five functional vascular bundles that can transfer water into the blade. The cross-sectional view in Fig. 4.7b and longitudinal views (data are not shown) show that guard cells are present on both abaxial and adaxial surfaces and are aligned parallel to both sides of the IVB as well as the FVB. In addition, their substomatal cavities form trenches that are two or three cells away from the vascular bundles. Because of this proximity, transpirational water brought into the blade will preferentially leave the leaf through the closest stomata rather than through more distal stomata that are eight cells or more away. Thus, cells immediately around the five functional vascular bundles may be responsive to stress, but most cells in the mesophyll of the expanded blade will be insulated from stress effects on the vessel Ψ.

V. THE PATHWAY OF TRANSPIRATIONAL WATER MOVEMENT

The anatomically derived suggestion that transpirational water will move through vessels in the growing zone was tested by heat-pulse-transport velocity experiments. In these experiments, the end of a sensitive thermocouple probe (Sensortek IT-23 linked to a BAT-12 meter) was placed in the fold of the leaf 45 mm from the point of seed attachment; heat was applied at points 20mm or more away from the probe (Rayan and Matsuda, 1988). In preliminary studies, a 0.2-sec pulse sufficient to raise the temperature about 10°C at the source was applied 20 mm below the probe. No temperature increases were noted within 2 min in plants held in darkness and prevented from transpiring (data are not shown), but rapid and apparently transpiration-rate-dependent heat transfer did occur in plants held in light (Fig. 4.8). Because 0.1°C is the measured temperature closest to the "heat

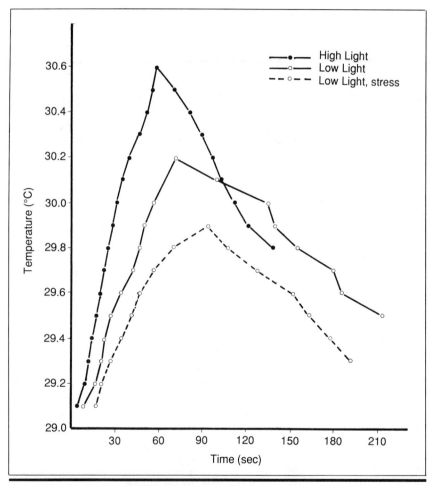

Figure 4.8. Heat-pulse transfer in the expanded lower leaf region of intact barley seedlings. The thermocouple probe was held 45 mm away from the point of attachment to the seed and a heat pulse sufficient to raise the temperature 10°C was applied 20 mm below the probe. Studies were performed under high (400 μmol m^{-2} s^{-1}) and low (200 μmol m^{-2} s^{-1}) light intensities, and under low light and NaCl (-0.8 MPa). Values are the means of four observations or more.

front," this temperature increase was used to measure transport velocities in subsequent studies.

Although transpiration rates were not determined for the plants used in Fig. 4.8, plants receiving the same high light, low light, and low light plus salt transpired at rates of 0.022, 0.011, and 0.0056 mm^3 water plant^{-1}-sec^{-1}, respectively. Heat-pulse-transport studies showed that the velocity of transport throughout the basal part of the leaf was the same as that expected for movement through the lumens of vessels (Rayan and Matsuda, 1988). These data, together with visual results obtained when dyes were supplied through cut roots of otherwise intact wheat seedlings (Barlow, 1986), provide experimental evidence that supports the anatomical suggestion that water moves in xylem through the growing area of cereal leaves.

Heat-pulse-transport velocity studies also tested the hypothesis that sudden exposure to osmotic stress will effect an almost immediate reduction in the xylem's hydrostatic pressure. Although xylem hydrostatic pressures

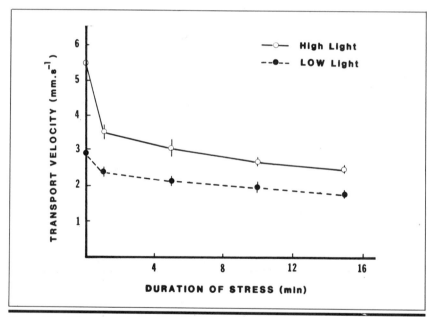

Figure 4.9.   Effect of sudden osmotic stress on heat-pulse-transport velocities. Unstressed plants were maintained for 2 hr or more under the light intensities used in Fig. 4.8, then stressed with solution containing NaCl (-0.8 MPa). Heat-pulse-transport velocities measured between 25 and 45 mm from the seeds were determined before or at the indicated times after stress. Values ($\pm$SD) are the means of five observations. (Data were reproduced with permission from Rayan and Matsuda, 1988; Copyright 1988, The American Society of Plant Physiology.)

could not be determined, it was reasoned that if the hypothesis is valid, sudden osmotic stress should reduce xylem water transport before transpiration. This was verified by data (Fig. 4.9) that show that water-transport velocity is reduced by stress in 1 min, but transpiration of identically grown plants from the same population is not reduced for at least 5 min (Rayan and Matsuda, 1988).

VI. PARTIAL RESTRICTION OF WATER FLOW TO MESOPHYLL CELLS

Sudden stress-caused reduction in xylem water transport without a concomitant change in tissue Ψ would not be likely if the xylem water mixed freely with the water in the symplasm, but some exchange must occur to maintain cell functions. Although it is difficult to rigorously determine the extent of exchange, a semiquantitative estimate was obtained by supplying [^{3}H]water to the lower half of the roots of intact seedlings and then measuring the specific radioactivity of water in various parts of the leaf over time. In order to compare data from several experiments, results were expressed as specific radioactivity ratios (SRR); that is, the ratio of the specific radioactivity of water in a sample to the specific radioactivity of water in the nutrient solution.

Because labeled water entering a given part of a plant is diluted continuously by a large endogenous supply of unlabeled water, the SRR of nonequilibrated tissues can be expected to decrease as distance from the point of labeling increases. This was confirmed by studies (see Fig. 4.10) that show that the SRR of water in the upper roots are higher at all times than the SRR of the nearly adjacent but more distal growing region of the leaves, which reaches 0.7 in 20 hr. The data show also that the SRR of transpired water is higher during the initial labeling period than the SRR of the blade. Such results were used to support the hypothesis that transpiring water in the expanded blade may largely bypass most mesophyll cells in its movement toward the stomata (Matsuda and Riazi, 1981).

Water in barley leaves can be separated into an easily extractable and residual fraction by centrifuging cut segments against an adsorptive surface at $5 \times g$ for 1 min (Rayan and Matsuda, 1988). In the growing region, this easily extractable water is considered apoplastic (xylem + cell wall) water because (1) it is extracted under conditions where it is unlikely that symplastic water can be removed, (2) its volume is about 2% of the total water in the tissue, which is about twice the volume of water calculated to be in the lumen of vessels (Table 4.1), and (3) after 24-hr labeling, its SRR is 1.0 rather than the 0.7 obtained for residual water (Rayan and Matsuda, 1988).

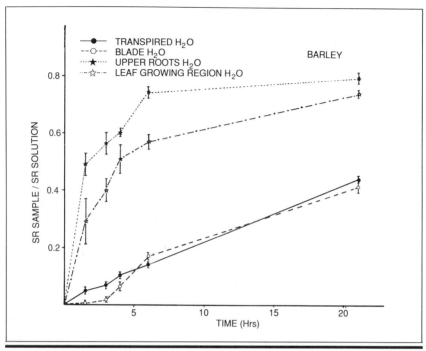

Figure 4.10. Long-term changes in the SRR of water extracted from the upper root and leaf tissues and water collected as transpiration. Lower halves of roots of intact seedlings were immersed in ³H-labeled water and changes in the SRR of the upper roots, the leaf-growing region, the midblade, and transpired water were detected. Values (±SD) are the means of three observations.

The changes in SRR are in the predicted direction if water is brought into tissues via the xylem, and the existence of the distinctly lower SRR values in tissues suggests that apoplastic water does not flow readily across the plasma membrane and into the cells.

The amount of easily extractable water in the growing region can be reduced nearly by half if intact seedlings are stressed for 5 min prior to excision and centrifugation (Fig. 4.11). This finding was combined with labeling experiments with [³H]water in order to infer the source of extractable water obtained before and after the 5-min stress period. These studies showed that when intact seedlings were labeled for 2.5 hr, a 5-min stress increased the SRR from 0.8 to 0.95 while the SRR of the residual water remained unchanged; similar trends were seen in extracts from plants labeled for longer periods (Fig. 4.12). Results of this type can be obtained if

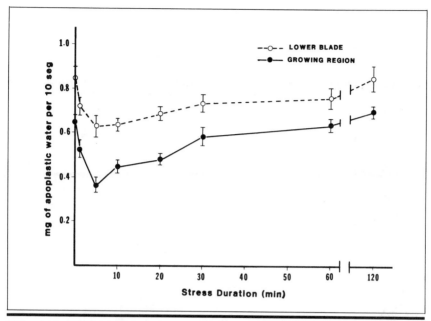

Figure 4.11. Effect of stress on the amount of water extracted by centrifugation of leaf segments. Ten 10-mm-long sections were cut from leaves of unstressed and stressed (-0.8 MPa NaCl) intact seedlings, and water was extracted by centrifugation against an adsorptive surface at $5 \times g$ for 1 min. Water in the lower blade was obtained from sections that were 40–50 mm from the seed. Values ($\pm$ SD) are the means of 10 observations.

the short stress period disrupts the continuity that exists between the cell wall and vessel water and if only the water in vessels is removed by centrifugation.

We suggest the following scenario probably occurs *in vivo*: In unstressed plants, water flows easily from the nutrient solution into the vessels. Because vessel walls have pits there is continuity of water between the vessels and the walls of mesophyll cells, but there is also some resistance to water flow into the wall region and further resistance to water flow across the plasma membrane and into the symplasm. The combined resistances offered by the pits, the cell wall, and the plasma membrane serve to restrict water exchange between the apoplasm and the symplasm. When plants are stressed, there is an immediate reduction in the hydrostatic pressure of the water in the xylem, which is transmitted rapidly to the walls surrounding the mesophyll cells in the growing region. However, this condition also leads to a disruption of the continuity of water that exists between the cell

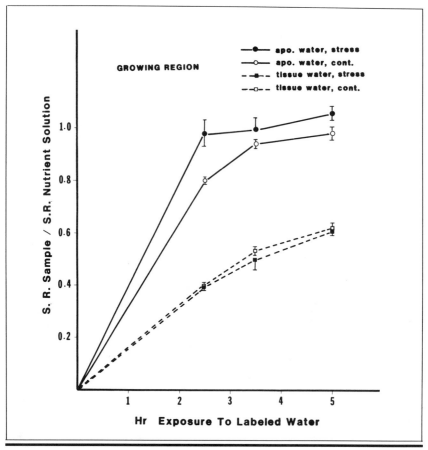

Figure 4.12.   Time-course changes in the SRR of easily extracted and residual tissue water from the growing region of barley leaves. Intact seedlings were exposed to nutrient solutions containing ^{3}H-labeled water for indicated periods and transferred to unlabeled fresh or salinized nutrient solution ($\pm$ 0.8 MPa) for 5 min before excision and measurement of radioactivity. Values ($\pm$SD) are the means of three determinations.

walls and the vessels. The effects of the disruption are maximal after 5 min, but plant cells begin to adjust to restore the water connection that exists between the walls and the vessels and it is conceivable that some rapid physiological responses that occur in stressed plants may be initiated in this period. The date obtained in the studies with labeled water are in general accord with the recent views of stress-caused cavitation (Sperry and Tyree, 1988).

VII. CONCLUDING REMARKS

Water largely passes through plants, and it is a formidable and challenging task to try to explain how shortages of this common but generally inert molecule can regulate plant processes. Water can permeate isolated cells quite readily but its movement in multicellular tissues is often restricted. Furthermore, water deficits are now known to rapidly effect changes in several processes; as a result, it is no longer sufficient to try to explain stress responses simply in terms of their effect on growth.

Although viable mechanisms for explaining how shortages of water regulate growth and other processes are not available, we believe that progress toward achieving a much clearer understanding of plant responses to stress can be made if studies are continued with highly responsive tissues and if acquired concepts are then tested in other plants. Frequently, seedlings are best suited for obtaining basic information; young barley and other grass plants appear ideal for many investigations. In barley, leaf expansion can be measured easily and inexpensively, and growing and other regions can be sampled easily. Also, young barley plants with attached seeds will not only cease leaf elongation almost immediately when stressed but, because of a proximal supply of assimilates, leaf expansion will resume within 2 hr even in plants, that are stressed at -1.0 MPa. In contrast, leaves of older stressed seedlings often do not grow for 24 hr.

Experiments with seedlings have provided additional reasons for believing that the same general mechanism in widely different plants controls several processes. As in field-grown plants, organ expansion rates are related quantitatively to the availability of water to roots. Furthermore, studies with young wheat, barley, squash, and peas have shown that the Ψ, Ψ_π and polyribosome proportions of the growing region are, like growth, directly proportional to the ψ of the nutrient solution (Mason and Matsuda, 1985). Finally, studies using primarily barley leaves have provided these concepts, which should be tested with other plants.

A. Stress Responses Depend on Anatomy and Physiology

Plant tissue responses to water stress depend on the physiological properties of the component cells and on the anatomical features that regulate the transmission of the water-deficit effect to the cells. In young barley seedlings, there are instances when leaf responses to stress depend largely on cell physiological properties, but there are also cases where anatomy is largely responsible for response differences.

In young barley plants stressed for 1 day or more, stress-caused osmotic adjustment is clearly greater in the leaf's 10-mm-long growing region than in the adjacent 10-mm-long nongrowing region. Since the anatomy of the adjacent sections are alike and since similar stress-caused reductions in easily extractable water are obtained throughout the lower regions of the leaf when tested as in Fig. 4.11, we believe that stress signals are passed rapidly to cells in both sections. The higher extent of osmotic adjustment found in the growing zone, therefore, appears to be due to differences in cell physiology. As a starting hypothesis for future studies, we suggest that cells in the growing region normally absorb high amounts of sucrose, which are then converted into wall precursors. We propose also that sucrose import into the growing region continues at high rates when plants are stressed, but since stress reduces elongation, the precursors accumulate and are then metabolized to various hexoses, hexose derivatives, and other osmotically active substances. The adjacent nongrowing region will likely not absorb as much sucrose, and its sugar metabolism will likely differ from the growing area.

In contrast, the low responsiveness of the blade and the high responsiveness of the growing region of the leaf to water stress appear to be due primarily to anatomical differences. The growing region transpires little because stomata are poorly developed and the region itself is protected by a coleoptilar sheath or sheaths of older leaves. Because of the arrangement of mesophyll cells to the vessels that can transport water (Fig. 4.7a), it is reasonable to believe that sudden stress-caused growth cessation will occur because stress will lead to a drop in xylem Ψ, which is then transmitted to the walls of all mesophyll cells. The anatomy of the expanded blade (Fig. 4.7b), however, suggests that transpirational water will bypass most mesophyll cells in its movement from the xylem to the stomata. As a result, most leaf blade cells will not be influenced by changes in the xylem's Ψ. This idea is supported by data that demonstrated that labeled water provided to the roots of intact seedlings will appear as transpirational water before there is significant labeling of the blade (Matsuda and Riazi, 1981).

Although it is difficult to demonstrate unequivocally that response differences to water stress are due to anatomy or to physiology, there are several examples that show that it is critical to be fully aware of the developmental state of any organ being studied. In addition to data obtained with young barley leaves, Barlow (1986) found that the expanded blade of wheat leaves can lose turgor when stressed whereas the turgor of the growing region is maintained. The blade's turgor reduction likely occurs because water loss through the stomata exceeds its ability to replenish the water or to osmotically adjust. In soybean hypocotyls grown in the dark at high humidity, Nonami and Boyer (1989) demonstrated in several ways that the growing region maintains its turgor but expanded regions will lose turgor when the seedlings are stressed. They attributed the expanded regions's turgor loss as due to water removal by the cells in the growing region.

Organs such as leaves vary greatly in their structure and in their response to water or osmotic stress, but the available results indicate that growing regions tend to maintain their turgor when stressed. At present, however, there are no clear data that show how the growing region of light-grown dicot leaves will respond to water deficits. Unlike cereal leaves, light-grown dicot leaves usually have more diffuse growing regions and can lose water through transpiration.

Similarly, expanded regions of leaves can vary greatly in their internal anatomy and while water loss may be controlled by stomatal activity, their response to stress may vary greatly depending on such factors as cell packing, numbers of functional vascular bundles, and stomatal numbers and placement. For example, the Ψ of mesophyll cells of C_4 grasses with many vascular strands, each of which serve two bundle sheath and two mesophyll cells, will likely be closer to that of the xylem Ψ than that of mesophyll cells of dicot leaves that have blades with large intercellular spaces. Although tests may show that the results are not as predicted, we believe that further attempts to link function to anatomy will clarify how plants respond to stress.

B. Cells Respond to the Surrounding Apoplastic Water Potential

Because most short-term and longer-duration responses of plant tissues to water stress cannot be described adequately with existing ideas (e.g., turgor, hormones), it has been necessary to suggest an alternative view. Although our views may change as more information becomes known, it now seems reasonable to believe that plant cells are able to adjust their metabolism to the apoplastic Ψ around them. The supporting evidence derived from both short- and long-duration studies is summarized briefly and a hypothesis is presented below to explain why leaf growth and several other processes occurring in the growing region are related ultimately to the Ψ of the external medium.

1. Short-Term Responses

We now know that growth of corn (Acevedo *et al.*, 1971) and barley (Matsuda and Riazi, 1981) leaves and mung bean hypocotyls (Passos, 1989) will cease almost immediately after the roots of intact seedlings are exposed to osmotic solutions, and similar responses will undoubtedly be seen in many other plants once any technical problems of measuring rapid growth changes are overcome. Such rapid growth responses can occur if cells involved in elongation are able to sense and respond to a sudden reduced apoplastic Ψ, but selective changes in apoplastic Ψ can only occur if the xylem is somewhat isolated from the rest of the cells in the plant. There are now several lines of evidence that suggest that the water flow from the xylem to the mesophyll cells is restricted and that stress will cause xylem Ψ to drop almost immediately.

The hypothesis of stress-caused reduction in xylem Ψ was proposed originally because it seemed to be the only viable idea for explaining why instantaneous growth cessation occurs without a concomitant change in the Ψ or Ψ_π of the growing cells (Matsuda and Riazi, 1981). Subsequent anatomical views and short-term stress experiments have supported and clarified this idea. Anatomically it is clear that there are mature vessels in the growing region (Fig. 4.7) and the total volume of the lumen of vessels is only about 1% of the total volume of the mesophyll (Table 4.1). Thus, a drop in xylem Ψ can occur without a noticeable change in tissue Ψ. Heat-pulse-transport data have confirmed that transpirational water flows exclusively in the vessels' lumen (Rayan and Matsuda, 1988), and that stress will reduce xylem water-transport velocity within 1 min (Fig. 4.9), whereas at least 5 min is required before transpiration rates will drop. Additionally, experiments with [^{3}H]water (e.g., Fig. 4.12) have supported the view that there is resistance of water flow from the vessels to the cell wall and additional resistance across the plasma membrane.

2. Long-Term Control of Plant Processes

In continuously stressed barley seedlings, polyribosome proportions, and Ψ and Ψ_π of the leaf's growing region are directly related to the Ψ of the external medium, whereas proline levels are inversely related. Water status values are not affected to the same extent in the growing regions of squash hypocotyls or pea epicotyls, but in these plants as well, Ψ, Ψ_π and polyribosome proportions in plants stressed for extended periods are also directly related to the Ψ of the external nutrient solution. The existence of such long-term quantitative relationships can be explained if the metabolism of plant cells is being continuously influenced by the Ψ of the apoplasm surrounding them. However, this suggestion also raises a question about how plant cells can respond to changes in the apoplastic Ψ.

Without attempting to suggest a specific mechanism, we hypothesize that the apoplastic Ψ controls the direction of water flow into or out of plant cells, and cells can sense and respond to a change in the direction of water flow. As a starting point for additional experimentation, we propose the following scenario for explaining how growth might originally stop when plants are stressed: Because the growing region maintains positive turgor even during stress, it appears likely that net water influx may be the key to growth. In addition to allowing water refilling of cells, this directional water movement may control the release of critical factors (e.g., wall loosening factors) that permit wall extension. Sudden stress, of course, will reduce the apoplastic Ψ and will therefore cause a reversal in the direction of water flow, which will continue until the cells adjust. When stress is mild, the needed adjustments are minor and may largely involve an oscillation in water flow between the cells and the apoplasm; on the other hand, if stress is severe, adjustment will require sufficient solute accumulation to reduce the cellular Ψ, and this osmotic adjustment may occur as an indirect result of lowered cell expansion.

Of course, these views are speculative, and no attempt was made to explain why changes in cellular processes can be related quantitatively to the degree of stress. Fortunately, because of the availability of highly stress-responsive systems that show quantitative responses to stress intensity, we believe these and other views can be tested.

REFERENCES

Acevedo, E., Hsiao, T. C., and Henderson, D. W. (1971). *Plant Physiol.* **48**, 631–636.

Barlow, E. W. R. (1986). *Aust. J. Plant Physiol.* **13**, 45–58.

Barlow, E. E. R., Munns, R., Scott, N. S., and Reisner, A. H. (1977). *J. Exp. Bot.* **28**, 909–916.

Boyer, J. S. (1985). *Annu. Rev. Plant Physiol.* **36**, 473–516.

Brandle, J. R., Hinckley, T. M., and Brown, G. N. (1977). *Physiol. Plant.* **40**, 1–5.

Cavalieri, A. J., and Boyer, J. S. (1982). *Plant Physiol.* **69**, 492–496.

Cocucci S., Cocucci, M., and Poma-Trecanni, C. (1976). *Physiol. Plant.* **36**, 379–382.

Cosgrove D. J., Van Volkenburg, E., and Cleland, R. E. (1984). *Planta* **162**, 46–54.

Esau, K. (1943). *Hilgardia* **15**, 327–368.

Hanks, R. J., ed. (1982). "Predicting Crop Production as Related to Drought Stress under Irrigation," Utah Agric. Exp. St., Res. Rep. No. 65. Utah State University, Logan.

Hsiao, T. C. (1970). *Plant Physiol.* **46**, 281–285.

Hsiao, T. C. (1973). *Plant Physiol.* **24**, 519–570.

Hsiao, T. C., and Jing, J. (1984). *Plant Physiol.* **75**, J-984.

Hsiao, T. C., Acevedo, E., Fereres, E., and Henderson, D. W. (1976). *Philos. Trans. R. Soc. London, Ser. B* **273**, 479–500.

Mason, H. S., and Matsuda, K. (1985). *Physiol. Plant.* **64**, 95–104.

Matsuda, K., and Riazi, A. (1981). *Plant Physiol.* **68**, 571–576.

Matyssek, R., Maruyama, S., and Boyer, J. S. (1988). *Plant Physiol.* **86**, 1163–1167.

Meyer, R. F., and Boyer, J. S. (1981). *Planta* **151**, 482–489.

Michelena, V. A., and Boyer, J. S. (1982). *Plant Physiol.* **69**, 1145–1149.

Nonami, H., and Boyer, J. S. (1987). *Plant Physiol.* **83**, 596–601.

Nonami, H., and Boyer, J. S. (1989). *Plant Physiol.* **89**, 798–804.

Passos, L. P. (1989). Ph. D. Dissertation, University of Arizona, Tucson.
Rayan, A., and Matsuda, K. (1988). *Plant Physiol.* **87,** 853–858.
Rhodes, P. R., and Matsuda, K. (1976). *Plant Physiol.* **58,** 631–635.
Riazi, A., Matsuda, K., and Arslan, A. (1985). *J. Exp. Bot.* **36,** 1716–1725.
Sands, R., and Correll, R. L. (1976). *Physiol. Plant.* **37,** 293–297.
Sperry, J. S., and Tyree, M. T. (1988). *Plant Physiol.* **88,** 581–587.
Westgate, M. E., and Boyer, J. S. (1984). *Plant Physiol.* **74,** 882–889.

CHAPTER 5

From Metabolism to Organism: An Integrative View of Water Stress Emphasizing Abscisic Acid

Katrina Cornish*
John W. Radin**

*U.S. Department of Agriculture
Agricultural Research Service
Western Regional Research Center
Albany, California

**U.S. Department of Agriculture
Agricultural Research Service
Western Cotton Research Laboratory
Phoenix, Arizona

I. INTRODUCTION

Recurring drought is a pervasive problem throughout most of the world's agriculturally productive regions, and it has been recognized as the single most important limitation to productivity (Boyer, 1982). Plants, especially those species that evolved in arid or semiarid ecosystems, have numerous mechanisms for conserving water or otherwise making the most efficient use of a limited supply. Many of these mechanisms necessarily entail slow growth rates. In the case of perennial plants, natural selection may tend to emphasize survival as the vital characteristic in the face of erratic rainfall because reproduction need not occur every year. Annuals, on the other hand, must reproduce in every life cycle. Similarly, in agricultural production systems, there must be reliable production of harvestable commodities each year (or harvesting interval). This requirement leads us to consider how plants can effectively conserve water without unnecessarily sacrificing their growth potential.

Maximizing production potential means that photosynthesis must respond as rapidly as possible after rainfall. Clearly, if water is conserved by loss of most of the leaves capable of transpiration, or by dormancy during periods of erratic water stress, little flexibility to respond remains when rainfall occurs. The solution to this dilemma is so successful that it is found in all land plants: the occurrence of pores of adjustable size (stomata) in an epidermal cell layer that is otherwise essentially gas tight. The guard cells that control stomatal pore size are complex transducers of environmental and chemical signals, capable of responding to numerous stimuli including light, air temperature and humidity, and soil water availability. The last factor is the one of greatest concern here. How do plants sense water stress, and how do they transmit that information to the stomata where it can elicit a response? Here we describe current concepts of the control of stomata, some metabolic underpinnings of these control mechanisms, how these metabolic patterns may also determine responses of other systems to stress, and finally some quantitative aspects of stomatal performance during drought.

In adopting this approach, we do not mean to imply that all effects of stress are exerted through stomata. They are not. Water stress exerts independent effects on many aspects of plant development and functioning including leaf growth, root growth and turnover, osmotic adjustment (solute accumulation), and so on. Each of these effects is legitimately the subject of its own review. Our approach is dictated by space limitations and the need to describe interrelated characteristics of one stress response in enough detail to convey a feeling for its *biology*.

II. THE ROLE OF ABSCISIC ACID IN STOMATAL RESPONSES TO WATER STRESS

Abscisic acid (ABA) is a sesquiterpenoid first identified on the basis of its leaf abscission-promoting properties (Ohkuma *et al.*, 1965). Reports shortly thereafter showed that, when applied to plants, ABA inhibits transpiration strongly (Mittelheuser and van Steveninck, 1969) and that its concentration increases in leaves during drought (Wright, 1969; Wright and Hiron, 1969). The hypothesis quickly became firmly entrenched that the accumulation of ABA in leaves during water stress is responsible for closing stomata. This hypothesis is bolstered by work with the *flacca* and *sitiens* mutants of tomato. The mutants wilt readily and appear to be unable to accumulate ABA to normal levels (Tal and Imber, 1970; Cornish and Zeevaart, 1988).

In assigning the predominant role in stomatal closure to stress-induced ABA accumulation, this hypothesis does not account for the ubiquitous presence of ABA, sometimes in substantial concentrations, prior to the occurrence of stress. Also, a few early studies indicated that stomata begin to close before there is a detectable increase in ABA (Beardsell and Cohen, 1975; Walton *et al.*, 1977; Henson, 1982). This line of evidence culminated in the belief that increased guard cell sensitivity to ABA, rather than increased tissue concentration of ABA, is the important event controlling stomata (Trewavas, 1981). The original hypothesis was further challenged by reports that stomata respond to extremely small doses of ABA even when the leaf already contains large quantities of the substance (Cummins *et al.*, 1971; Ackerson, 1980).

These apparent inconsistencies are reconciled in a model of ABA-directed stomatal closure, in which stomatal responses to drought are mediated by three sequential steps that respond independently to various environmental stimuli (Radin and Hendrix, 1988). The three steps are (1) ABA release from storage pools in the source (normally considered to be the leaf mesophyll, although increasing evidence also implicates roots as a source of ABA) to the apoplast (extracellular space), where it can be carried by the transpiration stream toward the guard cells; (2) recognition of and response to ABA by the guard cells; and (3) ABA accumulation in plant tissue. Release of ABA during early stages of stress (step 1) appears to trigger initial stomatal closure, but all three steps play an important role (Fig. 5.1). Each step will be considered in turn in the following sections.

A. Partitioning of Abscisic Acid to the Apoplast in Leaves

Abscisic acid has long been known to accumulate in chloroplasts, and first reports indicated that it was synthesized there (Milborrow, 1974). In a series of elegant papers, Hartung and co-workers showed that ABA is synthesized in the cytosol and accumulates in chloroplasts because of intracellular pH gradients (Hartung *et al.*, 1980, 1981; Heilmann *et al.*, 1980; Kaiser and Hartung, 1981). The undissociated carboxylic acid ABAH (pK_a 4.7), being uncharged, can freely permeate most cell membranes and equilibrate among cellular compartments by diffusion. The dissociated anion ABA^- is impermeant. Thus, if ABAH diffuses from a compartment with low pH to

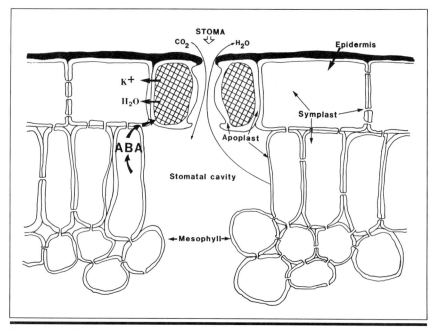

Figure 5.1. Diagram of a leaf showing some of the transport processes occurring during gas exchange. Of particular interest are the three processes concerning ABA: release to the apoplast, response by the guard cells, and accumulation in mesophyll cells.

one with higher pH, some of it will dissociate in that compartment to form ABA^-, which will become trapped. This mechanism of distribution works, even when the pH is far from the pK_a, because of the immobility of the anion. These properties are fairly common for weak acids and are, in fact, the basis for the well-known procedure to determine intracellular pH values using DMO (Walker and Smith, 1975).

Chloroplasts accumulate ABA because photosynthetic metabolism causes stromal alkalinization. When the thylakoid membranes are illuminated, they withdraw H^+ ions from the stroma as an integral part of electron transport and O_2 evolution. This leads to a strong pH rise in the stroma (Heldt *et al.*, 1973). The pH gradient across the thylakoid membranes provides the driving force for ATP synthesis (Forster and Junge, 1985). A detailed discussion of chloroplast reactions is inappropriate here. Nevertheless, it is important to note that, among various other effects, water stress inhibits electron transport and photophosphorylation reactions (Boyer, 1976) and disrupts this H^+ transfer process. Berkowitz and Gibbs (1983a,b) also reported acidification of the stroma of isolated spinach chloroplasts by osmotic stress.

From this discussion, it is apparent that ABA is accumulated in chloro-plasts in the light and that water stress will cause partial or full release of the accumulated ABA. From theoretical considerations, Cowan *et al.* (1982) calculated the redistribution of ABA that should occur upon darkening a leaf and estimated a doubling of the apoplastic pool that is ultimately re-sponsible for stomatal closure. To the extent that water stress and darkness have similar effects on stromal pH, they will cause similar alterations of the apoplastic ABA pool.

As we discussed earlier, water stress initiates stomatal closure before there is significant overall accumulation of ABA in the leaves. The transfer of ABA from the symplast to the apoplast, however, occurs early enough to account for initial stomatal reactions. Hartung *et al.* (1983) provided the first experimental evidence of enhanced ABA release. When leaf slices preloaded with ^{14}C-ABA were osmotically stressed, labeled ABA was trans-ferred to the incubation medium. Cornish and Zeevaart (1985a) pressurized *Xanthium* leaves to dehydrate them and then collected the fluid forced out of the leaf by the excess pressure. This sap was shown to contain apoplastic solutes, based on the quantity and types of sugars present. They found that ABA began to appear in the sap before there was noticeable accumulation in the leaf and in sufficient quantity to initiate stomatal closure. Radin and Hendrix (1988) later obtained results with cotton that agreed closely with Cornish and Zeevaart (1985a).

Hartung *et al.* (1988) collected very small sequential fractions of the pres-sure-induced exudate from cotton leaves. They found strong increases in the pH of the exudate as the leaf became dehydrated. The concentration of ABA increased concomitantly. Treatment of leaves with fusicoccin, a fungal toxin that stimulates H^+-ATPases, decreased apoplastic K^+ and pH and drastically decreased ABA concentration in the exudate. These results indi-cate strongly that ABA is transferred from the symplast to the apoplast dur-ing water stress as a result of altered pH gradients. A plasmalemma-bound ATPase is implicated as a primary site of action of water stress, although the nature of the water stress effect on the ATPase cannot be deduced from available evidence. It is important to note that some ABA was present in the apoplast even in unstressed leaves. The basal concentration appeared to be about 0.1 μM, with stress greatly increasing that concentration (Cornish and Zeevaart, 1985a; Hartung *et al.*, 1988; Hartung and Radin, 1989).

The degree to which ABA is partitioned into the apoplast is influenced by environment and plant development. Radin and Hendrix (1988) fed ^{14}C-ABA to leaves and then isolated the apoplastic sap by application of pres-sure. Growth on N-deficient media increased the release of ABA, as did leaf aging. These experiments did not include measurements of apoplastic pH, so no conclusions can be drawn about mechanisms. Nonetheless, the find-ings are consistent with stomatal behavior in these systems, in that both N deficiency and leaf aging increase stomatal responsiveness to water stress (Radin, 1981).

B. Guard Cell Responses to Abscisic Acid

The three-step model described earlier postulates that ABA released from the leaf mesophyll or from the root is involved in regulating stomatal aperture. Implicit in this model is the failure of guard cells to produce their own ABA, at least at the right time or in sufficient quantities to be physiologically important. Two studies, using completely different methods, support this conclusion. Cornish and Zeevaart (1986) isolated epidermis from leaves and sonicated it to kill and empty all cells but the guard cells. When isolated from water-stressed plants, guard cells contained 8- to 16-fold more ABA than when isolated from unstressed controls. When sonicated epidermis was water-stressed directly, however, the ABA content (combined ABA in the strips and in the incubation medium) increased only about 3-fold. These data imply that most of the ABA in the guard cells originated elsewhere. Behl and Hartung (1986) lysed epidermal cells with an acid treatment instead of sonication. Using the techniques of compartmental efflux analysis on the remaining guard cells, they concluded that guard cell ABA content actually decreases with stress. In line with these findings, Grantz and Schwartz (1988) reported that stomata in isolated epidermis are incapable of "hydroactive" closure (metabolically dependent closure involving ion transport across membranes, stimulated by ABA).

Once ABA is released from mesophyll cells into the apoplast, it is then carried by the transpiration stream flowing through the leaf to its various evaporation sites. These sites are predominantly in the cell walls surrounding the substomatal cavity (Fig. 5.1). Thus, the conclusion seems inescapable that a reasonably high proportion of the released ABA will come into contact with the guard cells. Guard cells are unique in that they have no plasmodesmatal connections to other cells (Sack, 1987). Therefore, symplastic transport of ABA to the guard cells is not possible, and all ABA delivered to the guard cells will come via the apoplast to the outer surface of the plasmalemma. If the leaf is turgid, what little ABA is supplied tends to be accumulated within the guard cells to a high concentration (Behl and Hartung, 1986; Lahr and Raschke, 1988; Baier and Hartung, 1988). This accumulation does not cause stomatal closure, however, because the active site is on the *exterior* of the plasmalemma, inaccessible from the interior (Hartung, 1983). The guard cell itself can perhaps be considered a storage site for ABA: Behl and Hartung (1986) indicated that ABA is released from the guard cells in large quantities during stress, and they speculated that the active sites may be exposed to some of this released ABA.

Little is known about ABA binding sites. A single report has appeared (Hornberg and Weiler, 1984) with evidence for binding sites on the guard cell membrane of very high affinity and specificity for S-ABA, the active enantiomer. Confirmation and characterization of this site have not yet been reported. The apoplast of even unstressed leaves contains about 0.1 μM ABA (Cornish and Zeevaart, 1985a; Hartung *et al.*, 1988; Hartung and Radin, 1989), and so there are limits on the useful properties of any putative binding sites. If the dissociation constant K_d of ABA with its binding site is 10^{-7} or less, then the sites would always be near saturation. The apoplastic concentration of ABA near the guard cells during stress is very difficult to estimate, but it is at least an order of magnitude greater than the basal (unstressed) level (Hartung *et al.*, 1988) and perhaps much higher (Cornish and Zeevaart, 1985a; Hartung and Radin, 1989). Because of this range of possible concentrations near the active site, the appropriate K_d cannot be predicted with certainty except that it must exceed 10^{-7}.

Other hormones, in addition to ABA, have been shown to act on stomata. Cytokinins open stomata in some cases, and both cytokinins and IAA have been reported to antagonize the action of ABA (Cooper *et al.*, 1972; Snaith and Mansfield, 1982). Cytokinins are believed to be produced primarily in the roots. (They are discussed in greater detail in the next section.) Although cytokinins are active on stomata in isolated epidermis (Jewer and Incoll, 1980), and therefore to some extent must act directly on guard cells rather than in the mesophyll, their mode of action is unknown. Cytokinins may have additional effects on mesophyll cells, although there is direct evidence that they do not affect ABA partitioning to the apoplast (Radin and Hendrix, 1988). Similarly, interactions with IAA have been little studied.

C. Accumulation of Abscisic Acid in Leaves

The biochemistry of ABA synthesis in higher plants has always been a difficult problem. The compound is one of many sesquiterpenoids in plants, which as a class are derived from mevalonic acid (MVA) via the isoprenoid pathway. Nonetheless, ^{14}C-MVA incorporation into ABA has usually been very low, less than 0.1% (Milborrow and Robinson, 1973; Loveys *et al.*, 1975). The intractability of this problem stands in contrast to studies of ABA-synthesizing fungi such as *Cercospora rosicola*. In this system MVA is rather easily converted to α-ionylidene acetic acid, 4'-hydroxy-α-ionylidene acetic acid, 1'-deoxy ABA, and finally to ABA (Neill *et al.*, 1982, 1987). In higher plants these intermediates can be converted as far as 1'-deoxy ABA, but seldom beyond that to ABA (reviewed by Walton, 1988). If 1'-deoxy ABA is an immediate precursor of ABA, it must become oxygenated in the 1' ring position. Incorporation of ^{18}O into the ring from $^{18}O_2$ was seen in stressed leaves only after extremely long exposure times and with very low specific activity at these positions (Creelman et al., 1987). The results imply that there are large pools of intermediates between the final product and the point at which ^{18}O is incorporated into the ring.

Studies with $^{18}O_2$ have shown substantial early incorporation of label not into the 1' position, but instead into the carboxyl group of ABA (Creelman and Zeevaart, 1984). These data provide one of several lines of evidence for an "indirect" pathway of ABA biosynthesis involving xanthophylls, a class of carotenoids. Cleavage of xanthophylls, either photolytically or enzymatically, produces sesquiterpenoids of which one, xanthoxin (derived from violaxanthin), can be rapidly converted to ABA even in a cell-free system (Sindhu and Walton, 1987, 1988). The ^{18}O is incorporated into the aldehyde at the 1 position of xanthoxin during oxidative cleavage of violaxanthin.

Other evidence, more physiological than biochemical, also points toward the indirect pathway of ABA biosynthesis. Several mutants of corn are known to be deficient in carotenoid synthesis. Many of these are also viviparous (the seed germinates within the fruit) because vivipary is normally suppressed by ABA and the mutants contain very little ABA (Neill et al., 1986). A similar condition can be induced in wild-type plants by the herbicides fluridone and norflurazon, which act at a specific point in the synthesis of carotenoids (Fong et al., 1983).

This discussion is not meant to imply that all ABA synthesis in higher plants occurs via the indirect pathway. Some sketchy evidence suggests that synthesis in turgid plants may be, in part, via the "direct" (isoprenoid) pathway, whereas synthesis during water stress may be primarily via the indirect pathway (reviewed by Zeevaart and Creelman, 1988). Also, roots may produce a greater fraction of their ABA by the direct pathway than do leaves (Creelman et al., 1987); roots also have a much smaller pool of precursor carotenoids for the indirect pathway. The flacca and sitiens mutants of tomato are unable to convert xanthoxin or ABA aldehyde to ABA whether turgid or stressed (Parry et al., 1988; Sindhu and Walton, 1988), and this is powerful evidence for the indirect pathway in both conditions. Nonetheless, these mutants contained 10–40% as much ABA as the wild type even though their conversion of xanthoxin to ABA in cell-free extracts was virtually completely suppressed. This apparent contradiction has not been explained satisfactorily, although the turnover of ABA may also be affected in these mutants. As both flacca and sitiens contain single-gene mutations (Rick, 1980), any changes at a second locus would have to be indirect.

The catabolism of ABA is equally as important as its synthesis. Pools of ABA are normally turned over rapidly, a process that enables plants to recover from stress quickly once it is relieved. There are apparently two main pathways of ABA catabolism: ABA can be converted to its glucose ester (ABA-GE), or to phaseic acid (PA) and then to dihydrophaseic acid (DPA). This latter pathway predominates in most species, and in most tissues containing ABA one can also find a buildup of one of these catabolites or their glucose conjugates (Zeevaart and Creelman, 1988). Abscisic acid may also be converted to 7'-hydroxy ABA but in most, or perhaps all, cases this reaction is limited to the unnatural enantiomer R-ABA, which is usually fed as part of a racemic mixture in isotope studies (Boyer and Zeevaart, 1986).

Water stress stimulates the rates of both biosynthesis and degradation of ABA. When intact *Xanthium* plants were stressed, PA began to accumulate in the leaves before ABA, the latter accumulating only after its rate of synthesis surpassed that of its catabolism (Cornish and Zeevaart, 1984). The new elevated steady-state level of ABA in water-stressed tissues is determined ultimately by the new balance between the rates of its synthesis and degradation. The capacity for rapid conversion of ABA to PA appears to be substrate-induced in the barley aleurone system (Uknes and Ho, 1984), and preliminary feeding studies with wild-type and wilty tomatoes indicate a similar mechanism in leaves during water stress (K. Cornish and J. A. D. Zeevaart, unpubl. data). In the same experiments, water stress inhibited conversion of PA to DPA in tomato leaves but this was independent of the level of ABA in the tissue. Water stress also appears to inhibit PA catabolism in barley leaves (Cowan and Railton, 1987). The level of ABA-GE also increased during water stress but usually accounted for a smaller portion of ABA catabolites than did compounds of the PA pathway (Cornish and Zeevaart, 1984). This stress-induced biosynthesis of ABA-GE in tomato was independent of the ABA level (K. Cornish and J. A. D. Zeevaart, unpubl. data). The control of ABA metabolism during water stress, as described here, is diagrammed in Fig. 5.2.

Upon relief of stress, the rate of ABA synthesis declines but conversion to PA remains rapid until the ABA level has returned to a value near its prestress level. Phaseic acid is degraded more slowly and, after stress is relieved, concentrations in the tissue remain high longer than ABA concentrations (Cornish and Zeevaart, 1984). Phaseic acid can also be released into the apoplast in even greater amounts than ABA (Radin and Hendrix, 1988). Although the activity of PA on stomata is considerably less than that of ABA (Sharkey and Raschke, 1980), in special circumstances like these it may have a biological function. In contrast, metabolism of ABA-GE is much less dynamic than that of PA; ABA-GE synthesis comes to a halt on relief of stress and, unlike PA, the compound is not then degraded (Cornish and Zeevaart, 1984). Compartmentation studies have shown that ABA-GE accumulates in vacuoles as an inert storage product (Bray and Zeevaart, 1985).

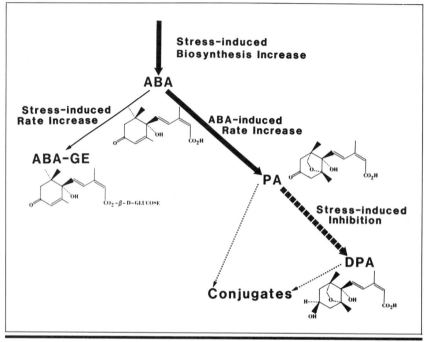

Figure 5.2. A model of the regulation of ABA metabolism during water stress.

Environmental variables other than soil water availability can also influence the balance between synthesis and degradation of ABA. One of the
most important of these variables is temperature. When cotton leaves are
at 35°C or warmer, their stomata tend to open to very high diffusive capacity, and they transpire extremely rapidly (Radin *et al.*, 1987; Radin, 1989).
These conditions are associated with a greatly decreased leaf ABA content
under both well-watered and water-stressed conditions (Radin *et al.*, 1982;
Radin, 1989). Feeding ^{14}C-ABA to leaf discs at 20°C or 35°C demonstrated
that ABA catabolism through PA and DPA was increased at the higher temperature, with an apparent Q_{10} of about 2 (Radin and Hendrix, 1986). Although ABA synthesis was not similarly studied, the low endogenous ABA
levels observed at high temperature led to the postulate that this increased
turnover was not matched by a similar increase in synthesis. This single
environmental variable may account for the great difference in water relations between field-grown and growth chamber-grown plants (e.g., Jordan
and Ritchie, 1971), as the latter typically experience much cooler temperatures. Similar stomatal responses to temperature are seen in some other species, especially those adapted to very warm climates (e.g., alfalfa; Carter

and Sheaffer, 1983). Often this response gives rise to a "crossover temperature"; that is, an air temperature above which the leaf is cooler than the air because of the great evaporative cooling, and below which the leaf is warmer than the air (Raschke, 1975). Mahan and Upchurch (1988) have discussed this behavior in terms of plant adaptation to high temperatures.

The role of the balance between synthesis (including import) and degradation (including export) is seen also in the accumulation of ABA in very young leaves. The capacity to metabolize ABA increases with leaf age, and young leaves contain the greatest amount of ABA under either turgid or stressed conditions (Cornish and Zeevaart, 1984; Cowan and Railton, 1987; Jordan *et al.*, 1975; Raschke and Zeevaart, 1976). Despite this difference, stomata of old leaves are often more responsive to stress than stomata of young leaves because of age-dependent increases in the release of ABA to the apoplast (Radin and Hendrix, 1988).

III.   MESSENGERS FROM THE ROOT SYSTEM TO THE STOMATA

The discussion in Section II considers only the leaf tissues as participants in the stomatal closure process. However, evidence has arisen that stomatal closure can be more closely related to conditions in the soil than to conditions in the leaf, and it has been postulated that roots transmit information to stomata in a kind of "feed-forward" control system (reviewed by Schulze, 1986). Blackman and Davies (1985) grew corn plants with the root system split between two pots. When water was withheld from one pot, stomata closed partially even though the other pot was supplying all the water needed by the plant. Leaf water potential increased slightly as a result of the stomatal closure, clearly showing that water potential was not controlling these stomatal reactions. The bulk concentration of ABA in the leaves did not increase during the experiment. Stomatal closure associated with partial root drying could be prevented by applying cytokinins to the leaf tissue, implicating decreased cytokinin movement from root to leaf as a possible messenger (Blackman and Davies, 1985). Zhang *et al.* (1987) found a similar stomatal response to partial root drying in *Commelina communis*. In this species, though, ABA accumulated in the roots and in epidermal peels removed from the leaves (epidermis could not be stripped from the corn leaves), again without accumulation in the leaf as a whole (Zhang *et al.*, 1987). This evidence suggests that root-produced ABA can also be carried to the leaf epidermis in the transpiration stream as a messenger to stomata. Because ABA and cytokinins are antagonists, an increase in the former and a decrease in the latter would provide two coordinated messages from the root system and would increase the degree of control exerted by roots over the stomata.

Cytokinins have long been recognized as a group of phytohormones that are produced in the roots and carried to the shoot in the transpiration stream (Itai and Vaadia, 1965). Cytokinins can stimulate stomatal opening, but usually in plants that are environmentally stressed in some fashion. Cotton plants grown on a mildly deficient N or P supply show enhanced stomatal responsiveness to water stress (Radin *et al.*, 1982; Radin, 1984). This shift in stomatal response to stress results, in part, from enhanced stomatal sensitivity to ABA. When a cytokinin is applied to leaves of these nutrient-deficient plants, it reduces stomatal responsiveness to ABA to the level in well-nourished plants (Radin and Hendrix, 1988). Cytokinins had no effect on the release of leaf ABA to the apoplast (Radin and Hendrix, 1988). Taken together, these results indicate that roots may influence stomata through two concurrently operating factors; one factor (cytokinin level) "conditions" the stomatal response to the other factor (ABA). The "feed-forward" nature of this control system enhances sensitivity to the environment. This is easily seen in nutrient-deficient cotton plants; the stomata close well *before* turgor in the leaf cells declines to zero (Radin *et al.*, 1982; Radin, 1984).

Unlike the cytokinins, ABA has only recently been recognized as a root-produced hormone. Its accumulation in roots is stimulated by dehydration and it accumulates even in excised roots, indicating its independence from synthesis in the shoot (e.g., Cornish and Zeevaart, 1985b). Roots can accumulate ABA to a staggering degree of dehydration, up to 80 or 90% loss of fresh weight (Fig. 5.3). It must be remembered, though, that roots have a much greater capacitance than do shoots because of greater elasticity and succulence. Therefore, even very large changes in water content of roots may correspond to relatively small changes in root water potential. The amount of ABA produced in roots is small compared to that produced in the shoot, but it represents a much greater relative increase in concentration than in the shoot (Cornish and Zeevaart, 1985b). Abscisic acid functions as a messenger because it is carried directly by the transpiration stream to a site close to the guard cells, apparently without being sequestered or inactivated along the way (Zhang *et al.*, 1987). When ABA is administered to leaves through the transpiration stream, concentrations as low as 0.1 μM can initiate stomatal closure and, if cytokinins are diminished at the same time, 0.01 μM may suffice (Radin and Hendrix, 1988). The concentrations released from roots can exceed these values (Loveys, 1984).

The process governing ABA release from root cells into the xylem has been little studied. It is clear that ABA is partitioned in root cells along pH gradients as in leaves (Behl *et al.*, 1981). However, fusicoccin does not affect

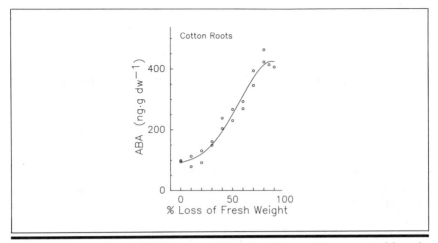

Figure 5.3.   Accumulation of ABA in cotton (*Gossypium hirsutum* L.) roots excised from the plant and air-dried to varying degrees. (From K. Cornish, unpublished.)

ABA release from root cells to the xylem, casting doubt on a role for H^+-ATPases (Hartung and Radin, 1989). In leaves, fusicoccin greatly diminished ABA release (Hartung *et al.*, 1988; see Section IIA). Nevertheless, pH gradients can arise not only by proton transport but also from charge imbalances created by differential transport of cations and anions (Good, 1988). Such a mechanism may operate in roots, creating conditions that promote ABA movement into the xylem. The experimental evidence is poor, though, because data sets seldom contain complete catalogs of all cations and anions present in the xylem, and the required difference between total cations and total anions would be small enough to challenge our ability to detect it.

Roots that are synthesizing ABA release it into the xylem and also readily release large quantities into the surrounding medium. In stressed *Xanthium* roots, the ABA released into the medium increased rapidly, eventually accounting for about 70% of the total ABA produced by the roots (Cornish and Zeevaart, 1985b). Some of this released ABA may be carried passively through the root into the xylem, especially if the transpiration stream itself is apoplastic to some degree. There is no question that roots exposed to exogenous ABA can pass a very large portion of it into the xylem (Markhart, 1982).

In 1988 Munns and King reported that xylem exudate of wheat roots contains a factor that closes stomata. When water was withheld from the soil, the ABA concentration of the exudate increased 50-fold to 0.05 μM, but this level of ABA, when applied to excised leaves, could account for only 1% of the activity. Removal of ABA with an immunoaffinity column did not remove the factor. Even the root exudate from well-watered plants displayed significant activity. Munns and King concluded that root exudates contain a stomatal inhibitor that is not ABA. Other explanations, such as a factor that is converted to ABA or one that enhances release of ABA already present in the leaf (as might happen if the xylem sap has an altered pH), may also be possible. Furthermore, this inhibitor clearly was prevented from acting on the guard cells of the intact turgid wheat plants. This discovery highlights how much more there is to discover about the participation of roots in stomatal responses to environment.

IV.   NONSTOMATAL EFFECTS OF ABSCISIC ACID

In the earlier sections we indicated that initial stomatal closure depends on release of the ABA already present in unstressed tissue. What advantage, then, is there to the massive synthesis of ABA that occurs during stress? There are two possible answers to this question. First, ABA turns over rapidly, and continued synthesis during stress may be necessary to maintain levels that are effective at keeping stomata closed. Second, ABA exerts other effects that may be important in integrated plant responses to water stress, and the level of ABA that triggers these other responses may not correspond to the level required for stomatal closure. For example, applied ABA enhances leaf senescence, especially in the older leaves that are most susceptible to senescence and abscission during drought (Radin, 1981). The level of exogenous ABA promoting senescence appeared to be 10- to 100-fold greater than the level associated with stomatal closure.

Another potentially important effect of ABA centers on the properties of the root system. When ABA is applied to roots, the volume of water flow through the roots is often increased. Such a change is obviously important during water stress because it can improve the plant's water balance. This increased transport has been ascribed to either decreased hydraulic resistance of the roots (Glinka and Reinhold, 1971) or enhanced ion transport that increases the osmotic forces driving water flow through the root (Karmoker and van Steveninck, 1978). Although responses can be highly variable (Pitman and Wellfare, 1978), with increases in water transport seen only at low flow rates (Fiscus, 1981), the typical response may be an increase in hydraulic resistance coupled with an increase in ion transport (Davies et al., 1982). Markhart et al. (1979) presented evidence that ABA is affecting a membrane when it alters hydraulic resistance. The root cells involved, the ABA concentration at the active site(s), the points of origin of the ABA, and the means of controlling its delivery to the active sites, have yet to be determined.

ABA also has well-documented effects on the immature embryo. Among those effects with obvious survival value during drought are induction or prolongation of dormancy (Karssen *et al.*, 1983) and induction of storage protein accumulation (Bray and Beachy, 1985); both of these changes hasten seed maturation. The capacity of the immature embryo to synthesize ABA appears to be low, and ABA is imported into the developing embryos from the rest of the plant (Hendrix and Radin, 1984). Even in unstressed plants, these developmental events are regulated by ABA (e.g., Karssen *et al.*, 1983). Water stress causes a large increase in the ABA level of young fruits (Guinn and Brummett, 1988). It seems highly likely that the hastened seed maturation is an indirect result of the greatly enhanced ABA content of the rest of the plant, some of which is translocated to the seeds through the phloem.

Although each of these responses to ABA has been studied numerous times, typically it is by application of ABA to an isolated system. In comparison, virtually no attention has been paid to the systems controlling these functions in an intact plant—where and when the controlling agent arises, the means of transport, the thresholds of response, and so on. This situation is akin to studying guard cells in depth, but with no knowledge of factors affecting the production and delivery of ABA. Such an approach would not have led to insight into stress-induced stomatal closure! The nonstomatal responses of plants to ABA deserve much more attention, including characterization of the roles of the tissues other than the "target" tissue.

V. INTEGRATION OF STRESS EFFECTS ON GAS EXCHANGE IN LEAVES

The theme of this chapter is the integrated, organismal nature of responses to stress. When considering integrated events, a quantitative approach often best assesses the consequences of any initial event. For example, consider the hypothetical question of whether changes in atmospheric humidity can regulate plant growth by altering uptake of some essential mineral element. We must determine transpiration rate as a function of humidity, uptake of the mineral as a function of transpiration rate, and finally, growth rate as a function of mineral uptake. Models incorporating such functions are seldom simple; often a response surface, rather than a single function, is required because the relationships are altered by other variables. Answers to questions such as the one posed above may not be intuitively obvious.

A.   Relationship between Water Loss and CO_2 Uptake

The relationship between water loss and CO_2 uptake has long been modeled in attempts to predict quantitatively how much a change in the former will affect the latter. It is clear that, although CO_2 and water vapor pass through the same portals in the epidermis, their pathways are rather different within the leaf and their fluxes are therefore subjected to different constraints. It follows that stomatal closure will not affect the two processes to the same degree. Evaporation of water is determined by the relationship:

$$E = (e_i - e_a)/(r_s + r_b) = (e_i - e_a)/r_L \qquad (5.1)$$

in which E is the rate of evaporation; e_i and e_a are absolute humidities in the gas phase inside and outside the leaf, respectively; r_s and r_b are diffusive resistances to water vapor of the stomata and the external boundary layer of the air, respectively; and r_L is a composite leaf resistance. The r_L is calculated by measuring E and e_a, and assuming that the air of the substomatal cavity is saturated with water vapor at the temperature of the leaf (i.e., there is zero resistance to water vapor diffusion internal to the stomata).

At the same time, CO_2 diffusion into the leaf and subsequent fixation have been described as follows:

$$A = (c_a - \Gamma)/(r_m + 1.6r_L) \qquad (5.2)$$

in which A is the CO_2 assimilation rate; c_a is the external CO_2 concentration; Γ is the CO_2 compensation point, or the lowest concentration to which photosynthesis can draw down the CO_2 supply; and r_m is a "mesophyll-resistance" term representing all the processes occurring in the mesophyll during photosynthesis. The factor 1.6 is introduced to correct for the different diffusivities of H_2O and CO_2.

As r_m is often much greater than r_L, Eqs. (5.1) and (5.2) imply that stomatal closure should diminish E more than A and thereby improve water use efficiency. This is often the case with mild water stress (e.g., McCree and Richardson, 1987), but the relationship breaks down during more severe stress because of direct effects on chloroplast metabolism (Boyer, 1976).

At steady state, the flux of CO_2 must be the same through each portion of the CO_2 uptake pathway. Therefore, Eq. (5.2) can be separated into three components:

$$A = (c_a - c_i)/1.6r_L \qquad (5.3)$$

$$A = (c_i - c_c)/r_t \qquad (5.4)$$

$$A = (c_c - \Gamma)/r_c = CE(c_c - \Gamma) \qquad (5.5)$$

in which c_i and c_c are CO_2 concentrations in the intercellular gas phase and the chloroplasts, respectively; r_t is the resistance to transport of CO_2 from the gas phase to the chloroplast; and r_c is resistance within the chloroplast. The carboxylation efficiency CE ($= 1/r_c$) is introduced to specify the origin of the resistance in the chloroplast. In some cases, the transport of CO_2 to the chloroplast can be limiting, especially if the mesophyll cells have inadequate surface area to absorb CO_2 rapidly (Evans, 1983; Evans and Seemann, 1984). In most treatments of photosynthesis, however, the transport resistance is assumed to be unchanging and small. As a result, $c_c = c_i$, and Eq. (5.5) is modified to become

$$A = CE(c_i - \Gamma) \qquad (5.6)$$

It is now routine to measure A, c_a, and r_L for calculation of c_i from Eq. (5.3) and to plot A as a function of c_i as represented by Eq. (5.6) (see Figs. 5.4 and 5.5). These $A : c_i$ curves typically have a region with a high positive slope at low c_i and a region with a slope near zero at high c_i. In line with our derivation of Eq. (5.6), the slope at low c_i is assumed to reflect the rate of the carboxylation reaction catalyzed by ribulose-1,5-bisphosphate (RuBP) carboxylase (in C_3 plants) or phosphoenolpyruvate (PEP) carboxylase (in C_4 plants). It has also been stated, for purposes of modeling the biochemistry of photosynthesis (von Caemmerer and Farquhar, 1981), that

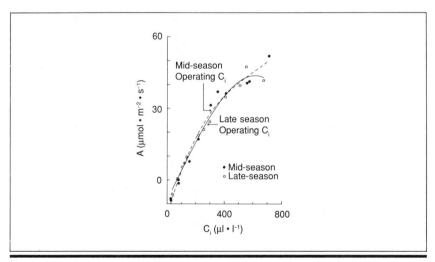

Figure 5.4. $A:c_i$ curves for cotton leaves in the field at two times of the year. Temperatures were much higher in mid-season than in late-season. Hot weather increased the c_i from 269 to 305 $\mu l.l^{-1}$ and increased photosynthetic rate as a result, despite similar photosynthetic capacities of the leaves. Operating c_i is the c_i at ambient CO_2. (From Radin, 1989.)

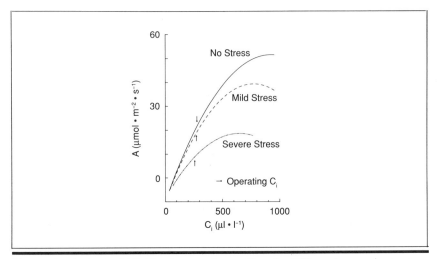

Figure 5.5. $A{:}c_i$ curves for leaves of cotton plants subjected to different degrees of water stress in the field. All measurements were taken during the hot portion of the season. Each curve is a regression based on at least 18 points. The most severe water stress greatly decreased photosynthetic capacity and also decreased c_i from 300 to 260 μl.1^{-1} as the stomata closed. (From Radin, 1989.)

the plateau at high c_i represents a limit to the photosynthetic rate imposed by regeneration of the substrate for carboxylation. This is not strictly the case; in wheat and in *Xanthium*, conditions that limit resynthesis of the CO_2 acceptor also cause deactivation of RuBP carboxylase so that carboxylation and regeneration of RuBP remain balanced and co-limiting (Perchorowicz *et al.*, 1981; Mott *et al.*, 1984).

The degree of stomatal limitation to A is also easily quantified from an $A : c_i$ curve. The actual A at the existing c_i is compared to the A that could be maintained as $c_i \rightarrow c_a$ (i.e., as $r_L \rightarrow 0$). This limitation by stomata is zero when c_i is high and saturating for A, but it increases rapidly as stomatal closure drives c_i below the value for saturation. When plants become water stressed, it follows that stomatal closure should drive down the c_i and greatly increase the degree to which it limits photosynthesis. Usually the stomatal limitation is increased only modestly, because water stress also directly affects photosynthetic capacity at the chloroplast (Farquhar and Sharkey, 1982). Data indicate that photosynthetic capacity can be protected against damage during water stress by osmotic adjustment (Matthews and Boyer, 1984) or by decreased levels of nutrient Mg^{2+} (Rao *et al.*, 1987). Again, effects of treatments on photosynthetic capacity of the mesophyll were accompanied to some degree by effects on stomata. As a result, the c_i

and the stomatal limitation to photosynthesis across treatments were less variable than the stomatal resistance itself.

Cowan (1977) and Cowan and Farquhar (1977) developed a hypothesis that plants "optimize" gas exchange to minimize water loss per unit of carbon gain. The hypothesis was originally formulated to apply to gas exchange over a complete diurnal cycle. However, restating the hypothesis in terms of instantaneous gas exchange patterns is useful. From the shape of an $A : c_i$ curve (Fig. 5.4), one can see that when stomata open to the degree that the c_i nears the plateau region, further stomatal opening produces much smaller photosynthetic gains. On the other hand, if stomata close to the point that c_i is very low, then the photosynthetic rate will be too low to be of much use. Wong *et al.* (1979) first reported evidence for stomatal coordination with mesophyll properties to maintain the c_i slightly below the value for saturation of the $A : c_i$ curve. From the interpretation of $A : c_i$ curves described earlier, this is also the point at which the capacities for carboxylation and regeneration of the CO_2 acceptor are balanced. Thus if the $A : c_i$ curve changes because of a shift in the leaf's environment, then stomata should "tune" the r_L to balance the rate of CO_2 entry with the availability of the CO_2 acceptor in the chloroplast. Such a strategy minimizes loss of photosynthetic potential due to stomata but maintains a reasonable upper limit to E.

B. Abscisic Acid as an Integrator of Gas Exchange

The "tuning" of stomatal behavior to match photosynthesis may be based on messenger(s) from the mesophyll that communicate the leaf's photosynthetic status to the stomata. Cowan *et al.* (1982) suggested that ABA could function as the mesophyll messenger because factors controlling its release from the mesophyll are determined in part by photosynthetic reactions as described earlier. Bradford *et al.* (1983) obtained evidence consistent with this suggestion. They found that the $A : c_i$ curves of wild-type and *flacca* tomatoes were almost identical, but the mutant maintained a much lower r_L and a much higher c_i. Application of ABA caused the *flacca* plants to revert to the normal pattern of gas exchange.

Radin *et al.* (1988) reported that field-grown cotton, like the *flacca* tomato, displays little regulation of c_i. When the plants were grown in a CO_2-enriched atmosphere, both stomatal regulation of c_i and stomatal responsiveness to exogenous ABA were enhanced. In these field-grown plants, high temperature suppressed stomatal coupling to the mesophyll. The $A : c_i$ curves obtained in the field during the hot and cool portions of the season were similar, but high temperature increased ABA turnover and drastically decreased the ABA content of the leaf (Radin, 1989). The r_L was substantially smaller during the hot weather and c_i and A were increased (Fig. 5.4).

These data from both tomato and cotton show that when leaves contain low levels of ABA, or when stomata are relatively insensitive to ABA, the c_i does not behave as if coupled to photosynthetic capacity by a messenger. In cotton, the decreased r_L at high temperature allows greater transpirational cooling of leaves and may be related to the heat resistance of these plants. Even at high temperature, however, water stress decreased photosynthetic capacity, increased leaf ABA content, and restored stomatal regulation of c_i (Radin, 1989; Fig. 5.5). At least during hot weather, "tuning" of gas exchange probably should be considered an acclimation to water stress that limits the rate of water use when the water supply is small.

When intact leaves are treated with ABA, their $A : c_i$ curves are lowered, implying that ABA directly inhibits photosynthesis in the mesophyll (Cornic and Miginiac, 1983; Raschke and Hedrich, 1985). Seemann and Sharkey (1987) and Loske and Raschke (1988) fed ABA to excised leaves, but found no biochemical effects of the ABA at the chloroplast level, except for a decreased PGA/RuBP ratio (indicating a decreased carboxylation rate). As a result, the assumptions underlying $A : c_i$ curves have been reexamined. One important assumption is the uniformity of internal gas concentrations across the leaf. Studies have indicated nonuniform ("patchy") stomatal closure in response to ABA (Downton et al., 1988; Terashima et al., 1988), which in heterobaric leaves violates this important assumption. (Heterobaric leaves have barriers consisting of cells extending from the upper to the lower epidermis at the minor veins. These barriers prevent lateral gas exchange and isolate the areoles, or areas between veinlets.) If the stomata of one areole close completely, its gas exchange with the atmosphere is zero, and A and r_L (both expressed on a leaf area basis) are proportionally decreased and increased, respectively. Because c_i is calculated from the product of A and r_L (see Eq. [5.3]), these errors negate each other and the calculated c_i is unchanged. The result is a decrease in A with no apparent change in c_i. The actual cause of the decreased A is an undetectable decrease in c_i in a portion of the leaf.

Several points must be made. First, not all leaves are heterobaric, and without that characteristic, lateral variability in c_i is unlikely (Terashima et al., 1988). Second, in an amphistomatous leaf (stomata present on both surfaces), patchiness requires that spatial variation in stomatal behavior be the same on both surfaces (Radin et al., 1988). Third, patchiness typically has been studied after applying ABA in high concentrations (10 μM or more) through the transpiration stream, from which it is relatively accessible to guard cells. In an intact plant, the apoplastic concentrations to which stomata respond may be one to two orders of magnitude lower (see discussion in Sections IIA and III). In addition, if the ABA concentration decreases from the apoplast inward (as in a plant fed ABA) instead of from the chloroplast outward (as in an intact plant), the relative concentrations at the guard cells and in the mesophyll chloroplasts may be the reverse of normal. Again, determination of ABA concentrations at the active sites is paramount to an understanding of such effects. Until the apparent effects of ABA directly on the mesophyll are better understood, results must be interpreted cautiously.

Results described here imply that coordination of gas exchange may largely be a reflection of what happens to ABA, not just in water-stressed plants but also in plants subjected to other conditions. We have reached a reasonable understanding of ABA's role in stomatal regulation during water stress. If we want to alter stomatal behavior during water stress by manipulating ABA, it may unfavorably affect nonstomatal systems also regulated by ABA. In addition, it may alter responses to other environmental factors altogether. Knowledge of ABA must now be extended to include its effects on targets other than guard cells and its role when water stress is not a factor. Intelligent choices of strategies depend on research that reveals all such consequences.

VI. MOLECULAR BIOLOGY OF WATER STRESS IN RELATION TO ABSCISIC ACID

The emphasis of this chapter is on organismal responses to water stress. However, recent discoveries of stress effects at the molecular level deserve brief mention. It is becoming clear that water stress alters the expression of a relatively small number of genes. For example, water-stressed *Brassica napus* plants produce a new type of root that is short, very desiccation-tolerant, and that contains 13 unique polypeptides that are not present in roots of normal morphology (Damerval *et al.*, 1988). This number is comparable to the number of polypeptides unique to other root types as well. Guerrero and Mullet (1988) found a set of new mRNAs in wilted pea leaves, some of which appeared before ABA accumulation and some after ABA accumulation. Application of ABA also caused the appearance of a set of new mRNAs that partially matched the set induced by wilting. Bray (1988) went further by studying the spectrum of mRNAs and polypeptides synthesized in water-stressed wild-type and *flacca* tomatoes. Stress caused the appearance or intensification of about 20 individual species of each in the wild-type plants. In the mutant plants, though, some of those 20 species were missing or failed to increase during stress. Application of ABA to the mutant tissue improved the resemblance between its polypeptide and RNA profiles and those of the water-stressed wild-type tissue. These results indicate that at the molecular level, many (but not all) drought-initiated events are responses to ABA. The studies also convey a sense that relatively few genes are influenced by stress or by ABA despite profound changes in tissue metabolism. Therefore, it seems likely that continued work will lead to the identification of the functions of some of those new polypeptides, eventually providing explanations for changes in the properties of intact plants. The economic incentive for such studies is to transfer desirable characteristics by identifying the specific genes from which they arise in donor species and transferring those genes to target species. This goal appears to be close at hand.

REFERENCES

Ackerson, R. S. (1980). *Plant Physiol.* 65, 455–459.
Baier, M., and Hartung, W. (1988). *Bot. Acta* 101, 332–337.
Beardsell, M. F., and Cohen, D. (1975). *Plant Physiol.* 56, 207–212.
Behl, R., and Hartung, W. (1986). *Planta* 168, 360–368.
Behl, R., Jeschke, W. D., and Hartung, W. (1981). *J. Exp. Bot.* 32, 889–897.
Berkowitz, G. A., and Gibbs, M. (1983a). *Plant Physiol.* 71, 905–911.
Berkowitz, G. A., and Gibbs, M. (1983b). *Plant Physiol.* 72, 1100–1109.
Blackman, P. G., and Davies, W. J. (1985). *J. Exp. Bot.* 36, 39–48.
Boyer, G. L., and Zeevaart, J. A. D. (1986). *Phytochemistry* 25, 1103–1105.
Boyer, J. S. (1976). *Philos. Trans. R. Soc. London, Ser. B* 273, 501–512.
Boyer, J. S. (1982). *Science* 218, 443–448.
Bradford, K. J., Sharkey, T. D., and Farquhar, G. D. (1983). *Plant Physiol.* 72, 245–250.
Bray, E. A. (1988). *Plant Physiol.* 88, 1210–1214.
Bray, E. A., and Beachy, R. N. (1985). *Plant Physiol.* 79, 746–750.
Bray, E. A., and Zeevaart, J. A. D. (1985). *Plant Physiol.* 79, 719–722.
Carter, P. R., and Sheaffer, C. C. (1983). *Crop Sci.* 23, 676–680.
Cooper, M. J., Digby, J., and Cooper, P. J. (1972). *Planta* 105, 43–49.
Cornic, G., and Miginiac, E. (1983). *Plant Physiol.* 73, 529–533.
Cornish, K., and Zeevaart, J. A. D. (1984). *Plant Physiol.* 76, 1029–1035.
Cornish, K., and Zeevaart, J. A. D. (1985a). *Plant Physiol.* 78, 623–626.
Cornish, K., and Zeevaart, J. A. D. (1985b). *Plant Physiol.* 79, 653–658.
Cornish, K., and Zeevaart, J. A. D. (1986). *Plant Physiol.* 81, 1017–1021.
Cornish, K., and Zeevaart, J. A. D. (1988). *Plant Physiol.* 87, 190–194.
Cowan, A. K., and Railton, I. D. (1987). *Plant Physiol.* 84, 157–163.
Cowan, I. R. (1977). *Adv. Bot. Res.* 4, 117–228.
Cowan, I. R., and Farquhar, G. D. (1977). *Symp. Soc. Exp. Biol.* 31, 471–505.
Cowan, I. R., Raven, J. A., Hartung, W., and Farquhar, G. D. (1982). *Aust. J. Plant Physiol.* 9, 489–498.
Creelman, R. A., and Zeevaart, J. A. D. (1984). *Plant Physiol.* 75, 166–169.
Creelman, R. A., Gage, D. A., Stults, J. T., and Zeevaart, J. A. D. (1987). *Plant Physiol.* 85, 726–732.
Cummins, W. R., Kende, H., and Raschke, K. (1971). *Planta* 99, 347–351.
Damerval, C., Vartanian, N., and de Vienne, D. (1988). *Plant Physiol.* 86, 1304–1309.
Davies, W. J., Rodriguez, J. L., and Fiscus, E. L. (1982). *Plant, Cell Environ.* 5, 485–493.
Downton, W. J. S., Loveys, B. R., and Grant, W. J. R. (1988). *New Phytol.* 108, 263–266.
Evans, J. R. (1983). *Plant Physiol.* 72, 297–302.

Evans, J. R., and Seemann, J. R. (1984). *Plant Physiol.* **74**, 759–765.
Farquhar, G. D., and Sharkey, T. D. (1982). *Annu. Rev. Plant Physiol.* **33**, 317–345.
Fiscus, E. L. (1981). *Plant Physiol.* **68**, 169–174.
Fong, F., Smith, J. D., and Koehler, D. E. (1983). *Plant Physiol.* **73**, 899–901.
Forster, V., and Junge, W. (1985). *Photochem. Photobiol.* **41**, 183–195.
Glinka, Z., and Reinhold, L. (1971). *Plant Physiol.* **48**, 103–105.
Good, N. E. (1988). *Photosynth. Res.* **19**, 225–236.
Grantz, D. A., and Schwartz, A. (1988). *Planta* **174**, 166–173.
Guerrero, F. D., and Mullet, J. E. (1988). *Plant Physiol.* **88**, 401–408.
Guinn, G., and Brummett, D. L. (1988). *Plant Physiol.* **86**, 28–31.
Hartung, W. (1983). *Plant, Cell Environ.* **6**, 427–428.
Hartung, W., and Radin, J. W. (1989). *Curr. Top. Plant Biochem. Physiol.* **8**, 110–124.
Hartung, W., Gimmler, H., Heilmann, B., and Kaiser, G. (1980). *Plant Sci. Lett.* **18**, 359–364.
Hartung, W., Heilmann, B., and Gimmler, H. (1981). *Plant Sci. Lett.* **22**, 235–242.
Hartung, W., Kaiser, W. M., and Burschka, C. (1983). *Z. Pflanzenphysiol.* **112**, 131–138.
Hartung, W., Radin, J. W., and Hendrix, D. L. (1988). *Plant Physiol.* **86**, 908–913.
Heilmann, B., Hartung, W., and Gimmler, H. (1980). *Z. Pflanzenphysiol.* **97**, 67–78.
Heldt, H. W., Werden, K., Milovancev, H., and Geller, G. (1973). *Biochim. Biophys. Acta* **314**, 224–241.
Hendrix, D. L., and Radin, J. W. (1984). *J. Plant Physiol.* **117**, 211–221.
Henson, I. E. (1982). *Ann. Bot. (London)* [N.S.] **50**, 9–24.
Hornberg, C., and Weiler, E. W. (1984). *Nature (London)* **310**, 321–324.
Itai, C., and Vaadia, Y. (1965). *Physiol. Plant.* **18**, 941–944.
Jewer, P. C., and Incoll, L. D. (1980) *Planta* **150**, 218–221.
Jordan, W. R., and Ritchie, J. T. (1971). *Plant Physiol.* **48**, 783–788.
Jordan, W. R., Brown, K. W., and Thomas, J. C. (1975). *Plant Physiol.* **56**, 218–221.
Kaiser, W. M., and Hartung, W. (1981). *Plant Physiol.* **68**, 202–206.
Karmoker, J. L., and van Steveninck, R. F. M. (1978). *Planta* **141**, 37–43.
Karssen, C. M., Drinkhorst-van der Swan, D. L. C., Breekland, A. E., and Koorneef, M. (1983). *Planta* **157**, 158–165.
Lahr, W., and Raschke, K. (1988). *Planta* **173**, 528–531.
Loske, D., and Raschke, K. (1988). *Planta* **173**, 275–281.
Loveys, B. R. (1984). *New Phytol.* **98**, 563–573.
Loveys, B. R., Brien, C. J., and Kriedemann, P. E. (1975). *Physiol. Plant.* **33**, 166–170.
Mahan, J. R., and Upchurch, D. R. (1988). *Environ. Exp. Bot.* **28**, 351–357.
Markhart, A. H. (1982). *Plant Physiol.* **69**, 1350–1352.
Markhart, A. H., Fiscus, E. L., Naylor, A. W., and Kramer, P. J. (1979). *Plant Physiol.* **64**, 611–614.
Matthews, M. A., and Boyer, J. S. (1984). *Plant Physiol.* **74**, 161–166.
McCree, K. J., and Richardson, S. G. (1987). *Crop Sci.* **27**, 539–543.
Milborrow, B. V. (1974). *Phytochemistry* **13**, 131–136.
Milborrow, B. V., and Robinson, D. R. (1973). *J. Exp. Bot.* **24**, 537–548.
Mittelheuser, C., and van Steveninck, R. F. M. (1969). *Nature (London)* **221**, 281–282.
Mott, K. A., Jensen, R. G., O'Leary, J. W., and Berry, J. A. (1984). *Plant Physiol.* **76**, 968–971.
Munns, R., and King, R. W. (1988). *Plant Physiol.* **88**, 703–708.
Neill, S. J., Horgan, R., and Parry, A. D. (1986). *Planta* **169**, 87–96.
Neill, S. J., Horgan, R., Walton, D. C., and Lee, T. S. (1982). *Phytochemistry* **21**, 61–65.
Neill, S. J., Horgan, R., Walton, D. C., and Mercer, C. A. M. (1987). *Phytochemistry* **26**, 2515–2519.

Ohkuma, K., Addicott, F. T., Smith, O. E., and Thiessen, W. E. (1965). *Tetrahedron Lett.* **29**, 2529–2535.

Parry, A. D., Neill, S. J., and Horgan, R. (1988). *Planta* **173**, 397–404.

Perchorowicz, J. T., Raynes, D. A., and Jensen, R. G. (1981). *Proc. Natl. Acad. Sci. U.S.A.* **78**, 2985–2989.

Pitman, M. G., and Wellfare, D. (1978). *J. Exp. Bot.* **29**, 1125–1138.

Radin, J. W. (1981). *Physiol. Plant.* **51**, 145–149.

Radin, J. W. (1984). *Plant Physiol.* **76**, 392–394.

Radin, J. W. (1989). James M. Brown (ed.) *Proc. Beltwide Cotton Prod. Res. Conf.* National Cotton Council of America, Memphis, Tennessee. pp. 46–49.

Radin, J. W., and Hendrix, D. L. (1986). *Plant Sci.* **45**, 37–42.

Radin, J. W., and Hendrix, D. L. (1988). *Planta* **174**, 180–186.

Radin, J. W., Parker, L. L., and Guinn, G. (1982). *Plant Physiol.* **70**, 1066–1070.

Radin, J. W., Kimball, B. A., Hendrix, D. L., and Mauney, J. R. (1987). *Photosynth. Res.* **12**, 191–203.

Radin, J. W., Hartung, W., Kimball, B. A., and Mauney, J. R. (1988). *Plant Physiol.* **88**, 1058–1062.

Rao, I. M., Sharp, R. E., and Boyer, J. S. (1987). *Plant Physiol.* **84**, 1214–1219.

Raschke, K. (1975). *Annu. Rev. Plant Physiol.* **26**, 309–340.

Raschke, K., and Hedrich, R. (1985). *Planta* **163**, 105–118.

Raschke, K., and Zeevaart, J. A. D. (1976). *Plant Physiol.* **58**, 169–174.

Rick, C. M. (1980). *Rep. Tomato Genet. Coop.* **30**, 2–17.

Sack, F. D. (1987). *In* "Stomatal Function" (E. Zieger, G. D. Farquhar, and I. R. Cowan, eds.), pp. 59–89. Stanford Univ. Press, Stanford, California.

Schulze, E.-D. (1986). *Annu. Rev. Plant Physiol.* **37**, 247–274.

Seemann, J. R., and Sharkey, T. D. (1987). *Plant Physiol.* **84**, 696–700.

Sharkey, T. D., and Raschke, K. (1980). *Plant Physiol.* **65**, 291–297.

Sindhu, R. K., and Walton, D. C. (1987). *Plant Physiol.* **85**, 916–921.

Sindhu, R. K., and Walton, D. C. (1988). *Plant Physiol.* **88**, 178–182.

Snaith, P. J., and Mansfield, T. A. (1982). *J. Exp. Bot.* **33**, 360–365.

Tal, M., and Imber, D. (1970). *Plant Physiol.* **46**, 373–376.

Terashima, I., Wong, S. C., Osmond, C. B., and Farquhar, G. D. (1988). *Plant Cell Physiol.* **29**, 385–394.

Trewavas, A. J. (1981). *Plant, Cell Environ.* **4**, 203–228.

Uknes, S. J., and Ho, T-H. D. (1984). *Plant Physiol.* **75**, 1126–1132.

von Caemmerer, S., and Farquhar, G. D. (1981). *Planta* **153**, 376–387.

Walker, N. A., and Smith, F. A. (1975). *Plant Sci. Lett.* **4**, 125–132.

Walton, D. C. (1988). *Bull. Plant Growth Regul. Soc. Am.* **16**, 7–11.

Walton, D. C., Galson, E., and Harrison, M. A. (1977). *Planta* **133**, 145–148.

Wong, S. C., Cowan, I. R., and Farquhar, G. D. (1979). *Nature (London)* **282**, 424–426.

Wright, S. T. C. (1969). *Planta* **86**, 10–20.

Wright, S. T. C. and Hiron, R. W. P. (1969). *Nature (London)* **224**, 719–720.

Zeevaart, J. A. D., and Creelman, R. A. (1988). *Annu. Rev. Plant Physiol. Plant Mol. Biol.* **39**, 439–473.

Zhang, J., Schurr, U., and Davies, W. J. (1987). *J. Exp. Bot.* **38**, 1174–1181.

CHAPTER 6

Biochemistry of Heat Shock Responses in Plants

Mark R. Brodl

Department of Biology
Knox College
Galesburg, Illinois

I. INTRODUCTION

The environment in which an organism lives is always changing, often becoming inhospitable for normal life processes. Animals are usually motile enough to move into an area where conditions are less harsh. Plants, on the other hand, do not have that option. Instead, they must alter cellular physiologies in an attempt to counter the imbalances created by environmental change. Often these adjustments take place on several levels; many adjustments occur at the level of gene expression. The mechanisms by which organisms change their pattern of gene expression has provided fertile ground for investigative experimentation into cell physiology and gene expression.

Perhaps the most extensively characterized stress response is that produced by heat shock. The response has been demonstrated in organisms from archebacteria throughout the evolutionary hierarchy to humans (Schlesinger *et al.*, 1982). Because of its high degree of evolutionary conservation, aspects of the heat shock response described for one species are often paralleled in other species as well, even in highly divergent species. The heat shock response is induced quite simply by exposing an organism to supraoptimal temperatures (generally 5–10°C above normal growth temperatures for a period of 15 min to a few hours). Classically, heat shock induces the synthesis of an elite set of proteins that were either not present or present only at low levels in unstressed cells, the so-called heat shock proteins (hsps). The response was first documented by Ritossa in 1962 when he observed that within 1 min of temperature elevation, puffs were initiated at new loci on *Drosophila* polytene chromosomes. The puffing of polytene chromosomes has been shown to be indicative of transcriptional activity at that locus (Tissières *et al.*, 1974). In most organisms the synthesis of hsps is favored over the synthesis of many or most of the normal cellular proteins. Ritossa observed that concomitant with the appearance of heat shock puffs there was a regression of previously active puffs (Ritossa, 1962).

As it is now understood, the heat shock response involves the perception of heat via some change in a key molecule or in the internal environment or membranes of the cell. This receptor communicates the stimulus to affect a change in transcriptional activity; genes for hsps are actively transcribed while those for normal cellular proteins become quiescent. The hsp messenger ribonucleic acids (mRNAs) are then vigorously and often preferentially translated. The resulting hsps are thought to play some role in alleviating the deleterious consequences of the heat stress. Just how this is accomplished is uncertain, since few functions have been assigned to hsps. In general, however, cells that have synthesized hsps are able to withstand subsequent, otherwise lethal heat shocks—a state described as thermotolerant.

II.   ELICITATION OF THE HEAT SHOCK RESPONSE

A.   Perception of the Stimulus

The conditions under which hsp expression is elicited provide important clues about perception and regulation of the heat shock response. It is fairly evident that the mechanism for perception is not analogous to hormone receptor mechanisms where there is the presence of an inductive molecule. Because heat shock involves changes in the physical parameter of temperature, it is much more likely that changes in the physiology or biochemistry of heat-shocked cells signal the heat shock event. It appears that induction of the heat shock response may be the result of (1) changes in the oxidative/reductive environment of the cell, (2) changes in ion levels in the cell, (3) the presence of heat-denatured proteins in the cell, or (4) developmental induction (Fig. 6.1).

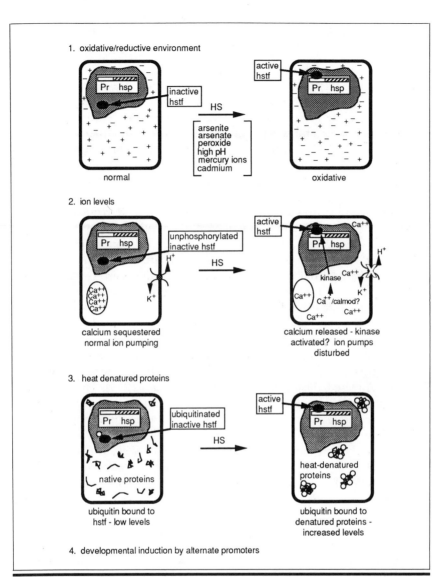

Figure 6.1. Diagrammatic representations of potential mechanisms of the elicitation of the heat shock response. (1) Heat shock shifts the oxidative/reductive balance of the cell toward the oxidative, somehow activating the heat shock transcription factor (hstf) and causing it to bind to the promoters (Pr) of hsp genes (hsp, cross-hatched bar). (2) Heat shock perturbs membrane ion pumps that exchange H^+ and K^+ ions or that sequester Ca^{2+} ions in internal compartments, creating an ion imbalance that activates hstf, perhaps through Ca^{2+}/ calmodulin-activated kinases. (3) Heat shock denatures proteins within the cell that are then tagged by ubiquitin. It is hypothesized that normally hstf is ubiquitinated and thereby inactivated; however, the presence of large amounts of denatured proteins shifts the ubiquitin pool to the cytoplasm, leaving unubiquitinated hstf to activate hsp genes. (4) In many organisms the induction of hsp genes takes place early in development without any increase in temperature. This is apparently due to the activation of alternate promoters.

1. The Oxidative/Reductive Environment of the Cell

Numerous investigators have noted that many other stress conditions elicit the synthesis of at least a subset of the hsps (Sachs and Ho, 1986). Incubation of cells in the presence of arsenite, arsenate, mercury ions, cadmium, hydrogen peroxide, and other sulfhydryl reagents (Ashburner and Bonner, 1979; Lin *et al.*, 1984; Czarnecka *et al.*, 1984; Schlesinger, 1985; Kapoor and Sveenivasan, 1988) mimics, at least in part, the heat shock response, often establishing thermotolerance. These stresses are, in common, oxidative in nature (Fig. 6.1, #1). Furthermore, when mammalian cells are exposed to anaerobic stress, they synthesize a set of glucose-regulated proteins (grps), and when recovering from anoxia, the synthesis of hsps is induced (Hightower and White, 1982; Welch *et al.*, 1983; Sciandra *et al.*, 1984). Whelan and Hightower (1985) noted that exposure of chicken embryo cells to low extracellular pH and 2-mercaptoethanol induces grp synthesis, while high extracellular pH induces hsp synthesis. They proposed that a reductive environment inside the cell induces grp synthesis, while an oxidative environment induces hsp synthesis. In support of this observed change in the oxidative/reductive environment of the cell, Nieto-Sotelo and Ho (1986) noted an increase in glutathione levels during continued heat shock in maize roots. A similar increase in glutathione levels has been reported during heat shock in Chinese hamster V79 cells (Mitchell *et al.*, 1983). This increase in glutathione provides a mechanism for returning the cell's environment to a less oxidized state. Alternatively, the increase in glutathione levels may be necessary to sustain the synthesis of phytochelatins, which are small peptides involved in the binding of heavy metals (Grill *et al.*, 1987). Burke and Orzech (1988) proposed that heat shock makes plant cell membranes leaky to toxic metals that are otherwise excluded from the cell's interior. If metal ion influx is a problem during heat shock, it may serve to rationalize these seemingly disparate inductive stimuli.

2. Changes in Ion Levels in the Cell

Membranes are highly sensitive to alterations in temperature, and it is likely that some of the earliest effects of heat shock impact membrane properties. It has long been noted that exposure to high temperature affects photosynthesis, probably through alteration of the permeability properties of partitioning membranes (Berry and Björkman, 1980). Heat alters thylakoid permeability to protons, thereby disturbing the proton gradient that drives photophosphorylation (Santarius, 1975). Alternatively, alterations in thylakoid permeability for other charge-compensating ions may be involved in reducing photophosphorylation during heat exposure (Emmett and Walker, 1973).

Heat appears to impact the partitioning of ions across membranes in general. Within 5 min of heat shock, ion flux across the plasma membrane of excised maize roots is affected (Cooper, 1985). The influx of K^+ and the efflux of H^+ is arrested, likely resulting in changes in the cell's ionic balance and potentially affecting other biochemical changes. Burke and Orzech (1988) pointed out that heat may alter the binding efficiencies of membrane pumps in a manner analogous to the mode by which heat alters the K_m of enzymes.

Work from studies on animal tissues implicate changes in Ca^{2+} as having an impact on the cell's response to heat shock. Rat hepatocytes have greater thermosensitivity when heat-shocked in a medium containing elevated Ca^{2+} levels (Malhotra et $al.$, 1986). If Ca^{2+} is involved in the induction of the heat shock response, it does not appear that external Ca^{2+} is critical, since lowering the external Ca^{2+} levels by incubation in the presence of EGTA does not abate the response (Malhotra et $al.$, 1986). However, internal Ca^{2+} stores may well be involved in induction. If cells are preincubated in the presence of an EGTA-containing (calcium chelating) medium and exposed to heat shock temperatures that are slightly submaximal with regard to hsp induction, heat shock responses are not observed, and maximal induction requires heat shock treatments that are incrementally more severe (Lamarche et $al.$, 1985). The effects of Ca^{2+} on the induction of the heat shock response make it quite attractive to propose that early perception of the heat shock event is communicated in the cell via some Ca^{2+}–dependent protein kinase (Landry et $al.$, 1988; Burke and Orzech, 1988) (Fig. 6.1, #2). The fact that calmodulin antagonists alter many of the cell's responses to heat shock supports this proposal (Landry et $al.$, 1988).

3. Heat-Denatured Proteins in the Cell

An early report of hsp induction without actual temperature increase documented the synthesis of hsps in chicken embryo fibroblasts in the presence of amino acid analogs (Kelly and Schlesinger, 1978). The presence of "abnormal" proteins appears to induce hsp synthesis in many systems (Ananthan et $al.$, 1986). $Escherichia$ $coli$ possesses a heat-inducible protease, La, a product of the lon gene. When $E.$ $coli$ is transformed with a plasmid containing a $lon/lacZ$ operon fusion, β-galactosidase is expressed when cells are incubated in the presence of amino acid analogs, puromycin, and streptomycin. All of these treatments result in abnormal proteins. Similarly, $Xenopus$ oocytes transformed with a hybrid $Drosophila$ $hsp70/\beta$-gal gene will express β-galactosidase when injected with denatured bovine β-lactoglobulin or bovine serum albumin, though not when injected with native protein (Ananthan et $al.$, 1986). The expression of a nonsense mutation in an actin gene results in hsp synthesis in the indirect flight muscles of *Drosophila* where actin expression is high (Hiromi *et al.*, 1986). Conversely, agents that are putatively capable of protecting proteins from thermal damage, such as glycerol or D_2O, block the induction of hsps by high temperature (White and White, 1987; Edington *et al.*, 1989). Ubiquitin levels in HeLa cell nuclei (mainly associated with histones) drop significantly during heat shock as the cytoplasmic ubiquitin pool increases (Pratt *et al.*, 1989). Such data give support to the hypothesis of Munro and Pelham (1985) that the level of heat-denatured proteins in the cell is responsible for the perception of heat shock. They proposed that heat shock results in a large increase in denatured proteins, which are then tagged by ubiquitin for degradation. This conflicts with another role for ubiquitin, which is to bind to and inactivate a constitutively expressed transcription factor that binds to the 5′ regulatory regions of hsp genes. The swamping-out of the cell's ubiquitin pool by massive increases in denatured proteins results in unubiquitinated transcription factor, which is then free to turn on genes that encode hsps (Fig. 6.1, #3).

Ubiquitin has been demonstrated to be a hsp in chicken embryo fibroblasts and several other animal and plant systems (Bond and Schlesinger, 1985, 1986; Brodl *et al.*, 1987). This could explain the transient nature of the heat shock response, as increased levels of free ubiquitin would permit reestablishment of ubiquitinated transcription factor. The transcription factor for the induction of hsp genes has been isolated and characterized in *Drosophila* (Wu *et al.*, 1987; Weiderrecht *et al.*, 1987), though evidence for ubiquitination has not been reported. Evidence for such a mechanism in plants is somewhat lacking. Incubation in the presence of amino acid analogs does not appear to be an inducer of the heat shock response in plants (Kimple and Key, 1985; M. R. Brodl, unpubl. results).

4. Developmental Induction

Many hsps are expressed during normal development. During sporulation in yeast, hsps of 26 and 84 kDa (hsp 26 and hsp 84) are expressed in the absence of heat shock, while the rest (always induced as a cohort during heat shock) are unexpressed (Kurtz *et al.*, 1986). Oogenesis in *Drosophila* involves the normal expression of hsp 26, hsp 28, and hsp 84 mRNAs in nurse cells; these mRNAs are then transported into the developing oocyte (Zimmerman *et al.*, 1983). It also appears that developmental induction is by a mechanism that is distinct from the induction mechanism for heat shock (Fig. 6.1, #4). The 5′ regulatory region of the gene that encodes hsp 83 (an hsp that is 83 kDa in molecular weight) in *D. melanogaster* contains distinct, independently regulated sequence elements (Xiao and Lis, 1989). This was made evident in studies using a series of deletion lines in which various portions of the 5′ regulatory region were deleted. Certain fragments retained either constitutive (low-level) or developmental inducibility or both, but lost heat inducibility, while in other deletion lines the converse was observed (Xiao and Lis, 1989). The presence of two distinct induction pathways indicates that hsps are not highly specialized proteins that function solely in the heat shock response.

The expression of hsps during normal development occurs in at least one species of higher plants as well. During somatic embryogenesis in carrot, several low-molecular-weight hsps are expressed at normal temperatures during the globular stage of embryo development (Pitto, 1983; Zimmerman *et al.*, 1989). Heat shock does not result in an increase in the mRNA levels for these proteins; however, protein levels are increased during heat shock (Zimmerman *et al.*, 1989). When exposed to heat shock temperatures, the normal development of these embryos is arrested at the globular stage, though the cells remain alive (Zimmerman *et al.*, 1989).

B. Heat Shock Protein Expression

Though the nature of the mechanism for the perception of elevated temperature remains unclear, much work has been done to uncover the molecular mechanism by which hsp gene transcription is initiated. In the 5′ regulatory region of all hsp genes, there is a 10-base sequence (CT-GAA—TTC-AG) of conserved homology (Pelham, 1982). Through deletion–addition studies in *Drosophila*, this sequence (the heat shock element or HSE) has been demonstrated to be crucial for high-temperature induction of hsp gene transcription (Dudler and Travers, 1984; Mestril *et al.*, 1985). Studies involving the placement of this sequence before non-hsp genes results in their thermoinducibility, serving as additional evidence of the role of HSE in hsp gene induction (Bonner *et al.*, 1984). Perhaps most striking is the conservation of this sequence as a regulatory element across widely divergent evolutionary lines. Monkey COS cells and *Xenopus* oocytes transformed with a segment of DNA containing a synthetic HSE spliced before the thymidine kinase gene of herpes simplex results in the induction of thymidine kinase in a heat-regulated fashion (Pelham and Bienz, 1982). The heat shock elements of plants appear to be similar. Genes encoding hsps from maize (Rochester *et al.*, 1986; T.-h. D. Ho, pers. comm.) and soybean (Gurley *et al.*, 1986) contain 5′ elements conforming to the heat shock consensus sequences of animals.

It is postulated that HSEs serve as binding sites for a transcription factor (heat shock transcription factor or hstf). Wu (1984a,b) demonstrated through exonuclease-protection experiments that the HSEs of yeast are protected from nuclease digestion. The hstf has been isolated from yeast (Weiderrecht *et al.*, 1987) and *Drosophila* (Wu *et al.*, 1987; Weiderrecht *et al.*, 1987). The isolated hstf is capable of binding HSEs and generating the same footprint pattern as is seen *in vivo* (Weiderrecht *et al.*, 1987). SDS-poly-acrylamide gel electrophoresis of the products from proteolytic digests of the yeast and *Drosophila* hstfs reveal a high degree of homology between the two factors (Weiderrecht *et al.*, 1987). Yeast hstf has been purified and its gene has been isolated and sequenced (Weiderrecht *et al.*, 1988; Sorger and Pelham, 1988).

It is likely that the hstf from higher plants are also highly similar to that of yeast and *Drosophila*. Cells from tobacco were transformed with a chimeric construct that fused the *hsp 70* promoter of *Drosophila* to the structural gene for neomycin phosphotransferase II (Spena *et al.*, 1985). Transformed callus tissue exhibited expression of the construct in a heat-regulated manner.

C. Expression of Normal Cellular Proteins

Cells subjected to heat shock contain a population of mRNAs encoding hsps and another population of mRNAs encoding normal cellular proteins. The extent of translational activity in these populations is characteristically distinct. Populations of hsp mRNAs are vigorously translated in most organisms (Lindquist and DiDomenico, 1985). The rate of ribosome loading and ribosome transit times are slightly elevated over those seen for normal cellular mRNAs at nonheat shock temperatures, presumably due to increased molecular motion. However, this alone does not account for the observed bias for hsp synthesis found in many organisms (Lindquist, 1980; Ballinger and Pardue, 1983).

The fate of mRNA remaining from normal cellular gene transcription varies with the organism that is heat shocked (Brodl, 1989). In *Drosophila*, normal cellular mRNAs are not translated, yet they remain stable in the cytoplasm and are reactivated during recovery (DiDomenico *et al.*, 1982). This also appears to be true in tomato (Scharf and Nover, 1982) and in soybean (Key *et al.*, 1981). How this translational bias is established is not known. Ribonucleic acid isolated from heat-shocked *Drosophila* and translated in lysates derived from nonheat-shocked cells results in the synthesis of both normal cellular proteins and hsps; the same RNA translated in lysates derived from heat-shocked cells results in the synthesis of only hsps (Storti *et al.*, 1980). However, the addition of a crude ribosome fraction "rescues" the synthesis of normal cellular proteins in heat shock lysates (Scott and Pardue, 1981). These authors stated that this indicates that the mechanism for translational bias for hsps during heat shock resides in the translational machinery and not in the mRNA itself. Other work indicates that the 5′ untranslated regions of hsp mRNA are unusually long, rich in adenine, and have little secondary structure (McGarry and Lindquist, 1985). It is possible that these features might enhance the observed translational bias.

The normal cellular mRNA of many other organisms is translated during heat shock to varying degrees. In yeast, normal cellular mRNAs continue to be translated, yet the synthesis of normal cellular transcripts is arrested (Lindquist, 1981; Lindquist *et al.*, 1982). A gradual decline in normal cellular protein synthesis results as their mRNAs are turned over at their normal rates (Lindquist, 1981). In cultured soybean cells, a similar mechanism appears to operate with regard to the mRNA encoding the small subunit of ribulose-1,5-bisphosphate carboxylase. mRNA levels for this protein decrease during heat shock, presumably at the rate of normal mRNA turnover in the absence of new transcription (Vierling and Key, 1985). At the far end of the spectrum, many normal cellular mRNAs in maize are translated at their normal levels on top of hsp synthesis (Cooper and Ho, 1983).

When isolated aleurone layers from mature imbibed barley seeds are heat shocked, there is a selective cessation of normal cellular protein synthesis; some normal cellular proteins continue to be synthesized while the synthesis of others is arrested (Belanger *et al.*, 1986). The synthesis of the secretory protein α-amylase is arrested during heat shock. The synthesis of this protein represents approximately 40% of the total new protein synthesis in barley aleurone cells (Varner and Ho, 1976). The arrest of α-amylase synthesis during heat shock is correlated with a greater than 85% decrease in the levels of α-amylase mRNA (Belanger *et al.*, 1986). What makes this response remarkable is that in the absence of heat shock α-amylase mRNA is otherwise stable, with a half-life estimated to be greater than 100 hours (Ho *et al.*, 1987). Furthermore, the endoplasmic reticulum (ER) lamellae upon which α-amylase mRNA is translated is disrupted during heat shock (Belanger *et al.*, 1986). Figure 6.2 shows light micrographs of aleurone cells incubated in the absence (A) or presence (B) of GA_3 and in the presence of GA_3 and a 3-hr heat shock (C). The cells were centrifuged at 100,000 xg for 30 min following incubation and then prepared for transmission electron microscopy. Ultracentrifugation causes the organelles to sediment according to their relative densities. The identities of the various bands were established by examination at the ultrastructural level. It can be clearly seen that the ER-containing band present in (B) is no longer present in (C) (following a 3-hr heat shock). From these observations, it has been concluded that the heat shock–catalyzed dissociation of ER lamellae quite possibly prevents the association of α-amylase mRNA into ER-bound ribosomes, leading to a rapid destabilization of this otherwise stable mRNA.

Support for this hypothesis comes from evidence that mRNA encoding a secreted protease of barley aleurone cells is also destabilized during heat shock, yet actin and tubulin mRNA levels (mRNAs that are translated on free polyribosomes) are not affected by heat shock (Brodl and Ho, 1989a). Such a class of mRNAs (those translated on free polyribosomes) may well represent the proteins in barley aleurone cells whose expression is not

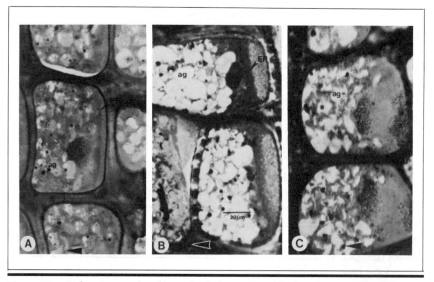

Figure 6.2. Light micrographs of heat-shocked centrifuged barley aleurone cells. Barley aleurone cells were incubated at 25°C in the absence (A) or presence (B) of 1 μM GA₃ for 19 hr or at 25°C in the presence of 1 μM GA₃ for 16 hr then a further 3 hr at 40°C (C). The layers were then centrifuged at 100,000 xg for 30 min prior to specimen preparation. GA₃-stimulated cells (B) proliferate a band of ER lamellae, which is no longer visible after heat shock (C). Arrows indicate centrifugal force vectors. All micrographs are taken at 1000 times magnification. Key: ag, aleurone grain; ER, endoplasmic reticulum; n, nucleus.

affected by heat shock (Brodl and Ho, 1989a). Furthermore, the kinetics of ER dissociation and α-amylase mRNA destabilization are correlated, as are their recovery kinetics (Brodl and Ho, 1989a). Absolute evidence of a direct cause-effect relationship is still lacking, however.

The effect of heat shock on the synthesis of secretory proteins does not appear to be unique to barley aleurone. When parenchyma cells from carrot root tissues are mechanically wounded, the synthesis and secretion of cell wall glycoproteins is stimulated (Showalter and Varner, 1989). When wounded carrot root tissues are heat shocked, there is a destabilization of extensin (a cell wall glycoprotein) mRNA and a concomitant dissociation of ER lamellae (Brodl *et al.*, 1987). In the absence of heat shock, extensin mRNA is at least eight times more stable than during heat shock. Similarly, injections of male *Xenopus* with estrogen induces the synthesis and secretion of vitellogenin from hepatocytes. The mRNA from which this protein is translated has a half-life of 500 hours (Brock and Shapiro, 1983). However, when exposed to heat shock temperatures, there is a cessation of vitellogenin secretion and a rapid destabilization of vitellogenin mRNA; actin mRNA levels remain unaffected during heat shock (Wolffe *et al.*, 1985).

In contrast, when developing cotyledonary tissues from soybeans are exposed to heat shock temperatures, the synthesis of storage proteins is actually slightly enhanced (Mascarenhas and Altschuler, 1985). Similarly, exposure to heat shock results in an enhancement of phytohemagglutinin synthesis in developing cotyledons in common beans (Chrispeels and Greenwood, 1987). In addition, the ER lamellae upon which they are translated are stable during heat shock (Chrispeels and Greenwood, 1987). The function of these cells and the proteins they are producing is distinctly different, however, from the function and the proteins of the cells of barley aleurone, carrot root parenchyma, and *Xenopus* hepatocytes. The bean tissues are producing proteins for internal storage rather than secretion. It may well be that as the cell prioritizes its functions during heat shock, sustaining the export of cellular materials and the expenditure of energy required for their synthesis is incompatible with and secondary to efforts to cope with the immediate dangers of high temperature. Furthermore, agricultural selection has produced bean plants capable of accomplishing pod fill even under adverse environmental conditions.

III. THE INTERCELLULAR ENVIRONMENT OF HEAT-SHOCKED CELLS

Considering the fact that the cell's response to heat shock is so dramatic with regard to the redirection of physiological processes, it is not surprising that heat shock causes rather remarkable changes in cell ultrastructure as well. As has been already noted, heat shock causes the dissociation of ER lamellae in the secretory cells of the barley aleurone layer. The reason for this dissociation is not clear. Potentially, heat shock causes a decrease in the activity or synthesis of enzymes involved in the formation of ER. If aleurone cells are heat shocked in the presence of the translation inhibitor cycloheximide or the transcription inhibitor cordycepin, the ER lamellae are dissociated despite the presence of the inhibitors (Brodl and Ho, 1989b). Furthermore, enzymes involved in the biosynthesis of phosphatidylcholine (the primary phospholipid of the ER) are not appreciably affected by heat shock (Fig. 6.3). Alternatively, heat shock could enhance the activity of phospholipases in the cell. When assays for phospholipase A (which removes fatty acids) and phospholipase D (which removes choline from phosphatidylcholine) are performed, activity appears to increase 1.2- and 1.3-fold, respectively (M. R. Brodl and J. D. Campbell, unpublished data). It is not clear whether such increases in phospholipase activity could account entirely for the dramatic changes in ER ultrastructure observed in heat-shocked barley aleurone cells.

The dissociation of ER may also be the result of problems with mem-

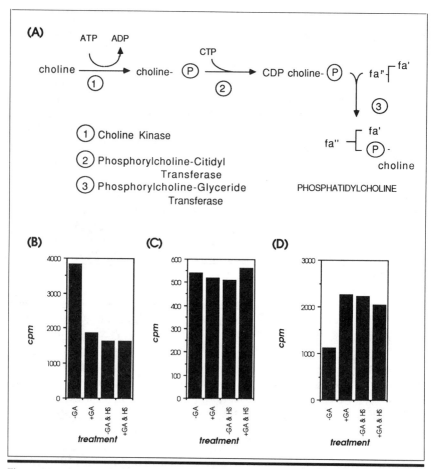

Figure 6.3. Activity of enzymes involved in the biosynthesis of lecithin (phosphatidylcholine) during heat shock. Enzymes were assayed according to the procedures of Johnson and Kende (1971). Heat shock does not appreciably alter the activity of lecithin biosynthetic enzymes, indicating that high temperature does not likely impair the cell's ability to synthesize this main component of the endoplasmic reticulum.

brane fluidity. As has already been discussed, membranes are very sensitive to temperature perturbations. Considering their role in sustaining a rather massive secretory effort, the maintenance of a fairly high degree of fluidity would seem advantageous. It is quite possible that this degree of fluidity is a liability at elevated temperature. In fact, evidence suggests that the fatty acids associated with phosphatidylcholine shift during 3-hr heat shock from longer-chain unsaturated species to shorter-chain saturated species (Brodl *et al.*, 1989). This may well be an adaptive shift, representing an attempt to form membranes with a decreased degree of fluidity that would be capa-

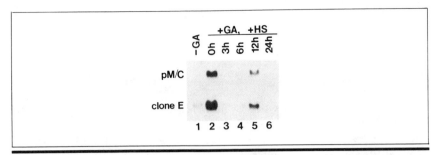

Figure 6.4.   Levels of α-amylase mRNA in the continued presence of heat shock. Aleurone layers were incubated at 25°C in the absence (lane 1) or presence (lane 2) of 1 μM GA₃ for 19 hr or for 16 hr and then for increasingly longer periods at 40°C (lanes 3–6). Ribonucleic acid was isolated from each sample (see Chapter 2), electrophoresed, transferred to "GeneScreen Plus" membranes, prehybridized, and then hybridized with [³²P]-labeled, nick-translated complementary DNA clones for the high (pM/C) and low (clone E) pI isozymes of α-amylase. Extended heat shock results in the partial recovery of α-amylase mRNA levels after 12 hr or heat shock acclimation.

ble of both sustaining secretion and maintaining integrity at high temperature. It is interesting to note that aleurone cells exposed to heat shock temperatures for extended periods are capable of resuming α-amylase synthesis (Belanger *et al.*, 1986), and reforming α-amylase mRNA (Fig. 6.4). It is also interesting to note that following heat shock a new band appears toward the centripetal pole in centrifuged aleurone cells (Fig. 6.2, labeled X). This band has no recognizable structure at the ultrastructural level. It may most logically be composed of lipids from the dissociation of ER lamellae during heat shock, the result of the observed increased phospholipase activities. It is relatively electron dense under the electron microscope, which is characteristic of phospholipids; however, the band has a lighter relative density than the spherosomes, which are triglyceride storage bodies.

In addition to changes in ER ultrastructure, heat shock promotes the formation of cytoplasmic heat shock granules in tomato and corn leaves (Nover *et al.*, 1983) and in chicken embryo fibroblasts (Collier *et al.*, 1988). These heat shock granules are composed of low-molecular-weight hsps, hsp 17 in tomato (Nover *et al.*, 1983), and hsp 24 in chicken (Collier *et al.*, 1988). In tomato, these heat shock granules contain mRNAs that encode normal cellular proteins, and they may be important structures in sequestering normal cellular mRNAs during heat shock (Nover *et al.*, 1989).

Heat shock also causes a pronounced change in the ultrastructural appearance of nucleoli. Specifically, the nucleoli of heat-shocked cells appear to be less granular, and heterochromatin content appears reduced (Welch and Suhan, 1985; Mansfield *et al.*, 1988). This is consistent with observations of heat shock–induced inhibition of RNA processing and ribosome assembly (Neumann *et al.*, 1984).

IV. THE FUNCTION OF HEAT SHOCK PROTEINS

The evolutionary conservation and ubiquity of the heat shock response indicate that hsps play some indispensable role in organisms exposed to high temperature; however, their precise function has yet to be unequivocally described. Data from DNA sequencing of the genes encoding hsps have provided some useful information about some hsp groups, but molecular biological studies have not provided much information about their physiological or biochemical properties. To date, several hsps have been identified as to their general roles in cells. Most appear to be involved with cell protection and repair or both, during or after heat shock.

A. Stabilization of Cellular Constituents

Though far from being decisive, studies examining the intracellular localization of hsps indicate potential roles and help rule out other potential roles. Velazquez *et al.*, (1980), capitalizing on the fact that during shock *Drosophila* exclusively synthesize hsps, used autoradiography to localize the bulk of the hsps in the nucleus, where they are associated primarily with chromatin. They proposed a chromatin-stabilizing role for them there. Other areas of localization appeared in the cytoplasm and cell membrane. Notably, they did not see any hsps localizing in mitochondria, despite strong evidence that mitochondria were very sensitive to heat shock (Sin, 1975; Velazquez *et al.*, 1980). Sinibaldi and Turpen (1985) later published evidence that isolated mitochondria of maize synthesize a 60-kDa hsp. This observation was subsequently shown to be an artifact of bacterial contamination (Nieto-Sotelo and Ho, 1987), confirming the implications of the original localization studies where researchers found no specific associations of hsps with mitochondria.

Studies of the localization of individual hsps have relied on the use of specific polyclonal and monoclonal antibodies as well as on fractionations of heat-shocked cells. As mentioned above, many hsps localize in the nucleus. Cell fractionation studies have demonstrated a strong association of hsp 68, hsp 70 and the low-molecular-weight hsps (hsp 22–28) with nuclear pellets (Levinger and Varshavsky, 1981). These hsps show a high resistance to salt extraction and nuclease treatments, which suggests that they play roles as structural proteins either inside the nucleus as nucleoskeletal elements (Levinger and Varshavsky, 1981) or outside the nucleus as nucleus-associated cytoskeletal elements (Tanguay and Vincent, 1982). Cell fractionation studies of heat-shocked soybean (Lin *et al.*, 1984) and maize

(Cooper and Ho, 1987) cells revealed that hsp 70 and low-molecular-weight hsps also localize with plant nuclei. This nuclear localization is apparently changeable during heat shock and subsequent recovery. Immunofluorescence studies with monoclonal antibodies specific to hsp 70 in *Drosophila* have shown that this protein is primarily localized in the nucleus during heat shock, but that during recovery there is a quantitative relocalization to the cytoplasm. A second heat shock transported hsp 70 rapidly back into the nucleus (Velazquez and Lindquist, 1984). The authors proposed that this behavior of hsp 70 suggests that it has a role in chromatin stabilization during heat shock and that it is recruited into the cytoplasm during recovery to reactivate the normal cellular mRNAs sequestered there, as the resumption of normal cellular protein synthesis is coincident with the relocalization of hsp 70 in the cytoplasm (Velazquez and Lindquist, 1984). Interestingly, hsp 70 induction by arsenite fails to localize this hsp in the nucleus; heat is required to do that (Arragio *et al.*, 1980).

Several heat shock proteins have shown to be localized primarily in the cytoplasm. Heat shock protein 83 of *Drosophila* (and homologous high-molecular-weight hsps of vertebrates) has been demonstrated to be exclusively cytoplasmic (Welch and Feramisco, 1982; Tanguay, 1985). Close inspection of immunocytochemical localizations reveals that in *Drosophila* hsp 83 is concentrated at the cell periphery (Tanguay, 1985). Burdone and Cutmore (1982) presented evidence that hsp 90, which is the human hsp corresponding to hsp 83 of *Drosophila*, appears to play a role in regulating human cell membrane Na^+/K^+–ATPases. As discussed earlier, low-molecular-weight hsps become localized in cytoplasmic heat shock granules (Nover *et al.*, 1983; Collier *et al.*, 1988), which may be involved in protecting normal cellular protein mRNA from turnover during heat shock (Nover, 1988).

Although hsp synthesis has not been demonstrated in plastids or mitochondria, there are reports of the transport of nuclear-encoded hsps into chloroplasts. *In vitro* translations of mRNA isolated from nonheat-shocked and heat-shocked pea plants provides evidence for a heat-inducible protein of 26 kDa. Incubation of the *in vitro*–translated proteins with intact, isolated chloroplasts results in *in vitro* transport and concomitant processing into a 22-kDa mature protein. If the isolated chloroplasts are from heat-shocked plants, the 22-kDa hsp is found in the membrane fraction; if the isolated chloroplasts are from nonheat-shocked plants, the 22-kDa hsp is found in the stroma (Kloppstech *et al.*, 1985). Vierling *et al.* (1986) reported a similar phenomenon in soybean, corn, and pea, both *in vivo* and *in vitro*. The function of this hsp is not yet known, though it might be involved in gene activation, photosynthesis, or other physiological processes of the chloroplast (Vierling *et al.*, 1986). The signal for transport resides in the carboxy terminal portion of this protein and appears to be highly conserved, since antibodies against the carboxy terminal end of pea hsp 21 recognize *Arabadopsis*, petunia, and maize homologs (Vierling *et al.*, 1989). Furthermore, when *in vitro* translation products from mRNA isolated from heat-shocked maize (Vierling *et al.*, 1986) or *Arabadopsis* (Vierling *et al.*, 1989) are incubated with isolated pea chloroplasts, antigenically identified, chloroplast-specific hsps from these species are transported into the pea chloroplasts.

A comparison of the genomic sequences of the small hsps of *Drosophila* (hsp 28, hsp 26, hsp 23, and hsp 22) indicates approximately a 50% homology among them and a comparison with known sequences shows a striking similarity with α-crystallin of the mammalian lens (Ingolia and Craig, 1982). For some time, it had been speculated that this homology with a known structural protein dictated solely a structural role for these proteins during heat shock. However, comparisons of the different classes of lens crystallins with the sequences of other known proteins indicate the crystallin proteins, ancestrally, may have been stable enzymes that were recruited for a structural role because of their inherent stabilities (Wistow and Piatigorsky, 1987). In yeast, enolase has been demonstrated to be a heat-inducible protein (Iida and Yahara, 1985). Enolase may potentially play a role in energy utilization during heat stress and recovery, yet interestingly it is structurally related to the lens crystallin (τ-crystallin) of fish, reptiles, and birds (Wistow and Piatigorsky, 1987).

B. Cell Cleanup or Recovery

There is increasing evidence that many of the hsps may be involved in the process of recovery from heat stress. One of these proteins synthesized in heat-shocked *E. coli* cells is an ATP-dependent protease (Goff *et al.*, 1984). As already mentioned, ubiquitin is a heat-inducible protein (Bond and Schlesinger, 1985; Brodl *et al.*, 1987). Although from different systems, a cell cleanup mechanism that involves the tagging and proteolysis of heat-damaged proteins by ubiquitin and protease, respectively, could well be universal.

In heat-stressed HeLa cells, a 73-kDa heat-inducible protein (antigenically related to hsp 70 in *Drosophila*) has been demonstrated to bind strongly to poly-A sequences (Schönfelder *et al.*, 1985). The authors proposed that the protein may be required for mRNA stabilization and noted that it is required for translation during recovery. In addition, hsp synthesis has been correlated with the resumption of proper RNA splicing after heat shock. In many organisms, the processing of ribosomal RNA precursors

(Ellgaard and Clever, 1971; Rubin and Hogness, 1975) and mRNA precursors (Myrand and Pederson, 1983) is interrupted during heat shock. Preexposure to a mild heat shock, resulting in the synthesis of hsps, will subsequently permit mRNA processing even at otherwise restrictive temperatures (Yost and Lindquist, 1986). Furthermore, if the synthesis is inhibited by cycloheximide, this protection was not observed (Yost and Lindquist, 1986).

In mammalian cells, the 71-kDa hsps are members of a family of structurally related proteins, which includes the 71-kDa clathrin-uncoating ATPase (Chappell *et al.*, 1986). A similar protein in this group of related proteins is a glucose-regulated protein that is localized in the endoplasmic reticulum. It is believed to potentially be involved in dismantling aggregates of IgG that were prematurely formed before secretion (Munro and Pelham, 1986). Furthermore, this family of proteins is capable of binding and, under proper conditions, hydrolyzing ATP (Welch and Feramisco, 1985; Lewis and Pelham, 1985). In yeast, all of the hsp 70 family of genes have been cloned, and the *SSA* subset has been demonstrated to be required for viability. The products of the four genes perform interchangeable functions, and two (*SSA3* and *SSA4*) are transcribed only during heat shock (Werner-Washburne, *et al.*, 1987). When linked to the GAL1 promoter as a construct inserted into *SSA3, SSA4/ssa1, ssa2* lines, the genes can be turned on (present) or off (absent) by regulating the presence of galactose in the growth medium at normal temperatures. In the absence of *SSA* gene products, the protein prepro-α-factor (normally secreted from the cell) and the precursor of the mitochondrial F_1ATPase β-subunit (normally transported into the mitochondria) accumulated in the cytoplasm (Deshaies *et al.*, 1988). These results were simultaneously repeated and demonstrated *in vitro* by Chirico *et al.* (1988). This led to the proposal that proteins of the hsp 70 family are normally required for the unfolding of translocated proteins in an ATPase-dependent manner (Deshaies *et al.*, 1988; Chirico *et al.*, 1988). Pelham (1988) then proposed that when cells are heat shocked, hsp 70 (constitutively expressed at low levels) abandons its role as an unfolding-ATPase to move to the cytoplasm where it functions much like a clathrin-uncoating ATPase to dismantle protein aggregates in heat-shocked cells. There is no concrete evidence for such a role for hsp 70 in higher plants; however, close examination of the effects of heat shock on ER ultrastructure in barley aleurone cells, revealed that within 15 min of heat shock (prior to ER dissociation) there was a conspicuous constriction of the lumen of the ER (Fig. 6.5, compare A and B). The reduction of the ER lumen may represent the arrest of protein translocation into the ER lumen as hsp 70 is recruited elsewhere to perform its disaggregation role.

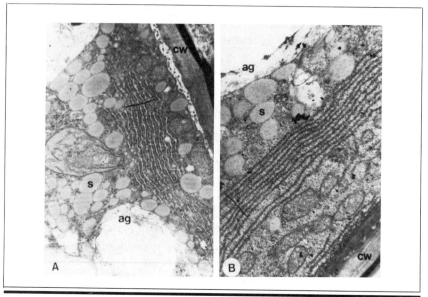

Figure 6.5. Effect of heat shock on the lumen of barley aleurone ER lamellae. The lumen of the ER lamellae in cells treated with 1 μM GA₃ for 16 hr (A) is distended, indicative of active secretory protein synthesis. After 15 min of heat shock (B), the space in the lumen is noticeably reduced, indicating that protein synthesis or translocation has been interrupted. Both micrographs are taken at 12,000 times magnification. Key: ag, aleurone grain; brace, endoplasmic reticulum; cw, cell wall; s, spherosome.

C. Thermotolerance

The expression of hsps has been correlated with the acquisition of thermotolerance in a variety of organisms (Schlesinger *et al.*, 1982). Indeed, Minton *et al.* (1982) reported that the mere presence of quantities of heat-stable proteins can stabilize heat-labile proteins during heat shock, and they suggested that this may be one aspect of thermotolerance imparted by hsps. Incubation of yeast (McAlister and Finkelstein, 1980), soybean (Lin *et al.*, 1984), *Drosophila* (Ashburner and Bonner, 1979), and mammalian cells (Landry *et al.*, 1982; Li and Werb, 1982) at temperatures that induce hsp synthesis will permit these organisms to survive later heat shocks that would otherwise be lethal. Landry *et al.* (1982) demonstrated that the establishment of thermotolerance is correlated with the timing of hsp synthesis. Heat shock proteins can be induced by arsenite, and such induction results in thermotolerance in soybean (Lin *et al.*, 1984) and Chinese hamster fibroblasts (Li, 1983).

The role of hsps in the establishment of thermotolerance is also implied by examples where hsps are not expressed and thermotolerance is not established. In maize, pollen is extremely heat labile and, coincidently, lacks a classic heat shock response (Cooper *et al.*, 1984). A mutation in *Dictyostelium* results in the virtual absence of hsp synthesis during heat shock; the mutation was isolated by screening for *Dictyostelium* deficient in thermal protection (Loomis and Wheeler, 1982).

Despite the strong implications from these observations, there are also observations, though fewer, to the contrary. Pollen from *Tradescantia* is capable of developing thermotolerance without the synthesis of hsps (Altschuler and Mascarenhas, 1982). Hall (1983) demonstrated that yeast cells became thermotolerant even when hsp synthesis was inhibited by cycloheximide, directly contradicting the observations of McAlister and Finkelstein (1980). A similar experiment with Morris hepatoma cells resulted in the same conclusion (Landry and Chrétien, 1983). That same study also demonstrated that arsenite-induced synthesis of hsps failed to establish a thermotolerant state. Barley aleurone cells synthesize hsps vigorously, yet despite their presence, cells are not able to withstand subsequent exposure to lethal temperatures (Brodl and Lanciloti, 1989). These findings do not, per se, discount a role for hsp in the establishment of thermotolerance. Rather, they indicate that the establishment of thermotolerance is more complex than a simple cause-effect relationship. For example, studies in soybean seedlings indicate that arsenite-induced hsps do not properly localize in the cell and that heat itself is, in fact, required for their proper localization (Lin *et al.*, 1984).

V. CONCLUDING REMARKS

The heat shock response is a universal response to elevated temperature that apparently enables cells to survive and recover from an environmental insult that may otherwise prove to be irreparably damaging. Although the means by which temperature increase is perceived is not yet well understood, it is likely that the sensing mechanism involves rapid changes in the physiology or biochemistry of the individual cells, as each cell responds to heat shock individually. Because each cell is both the sensor of and responder to high temperature, the heat shock response has been a valuable tool for studying molecular regulation of gene expression; the complexity of the response (compared to flowering, gravitropism, etc.) is considerably reduced. The recent isolation of the heat shock transcription factor in several organisms should facilitate a "back door" approach for retracing the events that lead to the perception of the stimulus.

Recently, progress has been made toward understanding the function of hsps (Pelham, 1988). The role of ubiquitin in cells not exposed to heat shock has been fairly well characterized (Ciechanover *et al.*, 1984), but its precise role in heat-shocked cells needs further characterization. This is also true for the major heat shock protein, hsp 70 (Chirico *et al.*, 1988; Deshaies *et al.*, 1988). However, little concrete information regarding the function of the highly diverse low-molecular-weight hsps has been obtained (Vierling *et al.*, 1989). A complete understanding of the role of hsps will undoubtedly provide useful information for extending the growth range of major crop plants and for understanding the limits of crop plant adaptation to heat.

REFERENCES

Altschuler, M., and Mascarenhas, J. P. (1982). *In* "Heat Shock: From Bacteria to Man" (M. J. Schlesinger, M. Ashburner, and A. Tissières, eds.), pp. 321–327. Cold Spring Harbor Lab., Cold Spring Harbor, New York.
Ananthan, J., Goldberg, A. L., and Voellmy, R. (1986). *Science* 232, 522–524.
Arragio, A. P., Fakan, S., and Tissières, A (1980). *Dev. Biol.* 78, 86–103.
Ashburner, M., and Bonner, J. J. (1979). *Cell (Cambridge, Mass.)* 17, 241–254.
Ballinger, D. B., and Pardue, M. L. (1983). *Cell (Cambridge, Mass.)* 33, 103–114.
Belanger, F. C., Brodl, M. R., and Ho, T.-h. D. (1986). *Proc. Natl. Acad. Sci. U.S.A.* 83, 1354–1358.
Berry, J., and Björkman, O. (1980). *Annu. Rev. Plant Physiol.* 31, 491–453.
Bond, U., and Schlesinger, M. J. (1985). *Mol. Cell. Biol.* 5, 949–956.
Bond, U., and Schlesinger, M. J. (1986). *Mol. Cell. Biol.* 6, 4602–4610.
Bonner, J. J., Parks, C., Parker-Thornburg, J., Mortin, M. A., and Pelham, H. R. B. (1984). *Cell (Cambridge, Mass.)* 37, 979–991.
Brock, M. L., and Shapiro, D. J. (1983). *Cell (Cambridge, Mass.)* 34, 207–214.
Brodl, M. R. (1989). *Physiol. Plant.* 75, 439–443.
Brodl, M. R., and Ho, T.-h. D. (1989a). (submitted).
Brodl, M. R., and Ho, T.-h. D. (1989b). (submitted).
Brodl, M. R., and Lanciloti, D. F. (1989). (in preparation).
Brodl, M. R., Tierney, M., Ho, T.-h. D., and Varner, J. E. (1987). *Plant Physiol.* 83, 78a.
Brodl, M. R., Grindstaff, K. K., and Campbell, J. D. (1989). (in preparation).
Burdone, R. H., and Cutmore, C. M. M. (1982). *FEBS Lett.* 140, 45–48.
Burke, J. J., and Orzech, K. A. (1988). *Plant, Cell Environ.* 11, 441–444.
Chappell, T. G., Welch, W. J., Schlossman, D. M., Palter, K. B., Schlesinger, M. J., and Rothman, J. E. (1986). *Cell (Cambridge, Mass.)* 45, 3–13.
Chirico, W. J., Waters, M. G., and Blobel, G. (1988). *Nature (London)* 332, 805–810.
Chrispeels, M. J., and Greenwood, J. S. (1987). *Plant Physiol.* 83, 778–784.
Ciechanover, A., Finley, D., and Varshavsky, A. (1984). *J. Cell. Biochem.* 24, 27–53.
Collier, N. C., Heuser, J., Levy, M. A., and Schlesinger, M. J. (1988). *J. Cell Biol.* 106, 1131–1139.
Cooper, P. S. (1985). Ph. D. Thesis, University of Illinois, Urbana.
Cooper, P. S., and Ho, T.-h. D. (1983). *Plant Physiol.* 71, 215–222.
Cooper, P. S., and Ho, T.-h. D. (1987). *Plant Physiol.* 84, 1197–1203.

Cooper, P. S., Ho, T.-h. D., and Hauptmann, R. H. (1984). *Plant Physiol.* 75, 431–441.
Czarnecka, E., Edelman, L., Schöffl, F., and Key, J. (1984) *Plant Molec. Biol.* 3, 45–58.
Deshaies, R. J., Koch, B. D., Werner-Washburne, M., Craig, E. A., and Schekman, R. (1988). *Nature (London)* 332, 800–805.
DiDomenico, B. J., Bugaisky, G. E., and Lindquist, S. (1982). *Proc. Natl. Acad. Sci. U.S.A.* 70, 6181–6185.
Dudler, R., and Travers, A. A. (1984). *Cell (Cambridge, Mass.)* 38, 391–398.
Edington, B. V., Whelan, S. A., and Hightower, L. E. (1989). *J. Cell Physiol.* 139, 219–228.
Ellgaard, E. G., and Clever, U. (1971). *Chromosoma* 36, 60–78.
Emmett, J. M., and Walker, D. A. (1973). *Arch. Biochem. Biophys.* 157, 106–113.
Goff, S. A., Casson, L. P., and Goldberg, A. L. (1984). *Proc. Natl. Acad. Sci. U.S.A.* 81, 6647–6651.
Grill, E., Winnacker, E. L., and Zenk, M. H. (1987). *Proc. Natl. Acad. Sci. U.S.A.* 84, 439–443.
Gurley, W. B., Czarnecka, E., Nagao, R. T., and Key, J. L. (1986). *Mol. Cell. Biol.* 6, 559–565.
Hall, B. G. (1983). *J. Bacteriol.* 156, 1363–1365.
Hightower, L. E., and White, F. P. (1982). In "Heat Shock: From Bacteria to Man" (M. J. Schlesinger, M. Ashburner, and A. Tissières, eds.), pp. 369–377. Cold Spring Harbor Lab., Cold Spring Harbor, New York.
Hiromi, Y., Okamoto, H., Gehring, W. J., and Hotta, Y. (1986). *Cell (Cambridge, Mass.)* 44, 293–301.
Ho, T.-h. D., Nolan, R. C., Lin, L.-s., Brodl, M. R., and Brown, P. H. (1987). In "Molecular Biology of Plant Growth Control" (J. E. Fox and M. Jacobs, eds.), pp. 34–49. Liss, New York.
Iida, H., and Yahara, I. (1985). *Nature (London)* 315, 688–690.
Ingolia, T. D., and Craig, E. A. (1982). *Proc. Natl Acad. Sci. U.S.A.* 79, 2360–2364.
Johnson, K. D., and Kende, H. (1971). *Proc. Natl. Acad. Sci. U.S.A.* 68, 2674–2677.
Kapoor, M., and Sveenivasan, G. M. (1988). *Biochem. Biophys. Res. Commun.* 3, 1097–1102.
Kelley, P. M., and Schlesinger, M. J. (1978). *Cell (Cambrige, Mass.)* 15, 1277–1286.
Key, J. L., Lin, C. Y., and Chen, Y. M. (1981). *Proc. Natl. Acad. Sci. U.S.A.* 78, 3526–3530.
Kimple, J. A., and Key, J. L. (1985). *Trends Biol. Sci.* 10, 353–356.
Kloppstech, K., Meyer, G., Schuster, G., and Ohad, I. (1985). *EMBO J.* 4, 1901–1909.
Kurtz, S., Rossi, J., Petko, L., and Lindquist, S. (1986). *Science* 231, 1154–1157.
Lamarche, S., Chrétien, P., and Landry, J. (1985). *Biochem. Biophys. Res. Commun.* 131, 868–876.
Landry, J., and Chrétien, P. (1983). *Can. J. Biochem. Cell Biol.* 61, 428–437.
Landry, J., Berner, D., Chrétien, P., Nicole, L. M., Tanguay, R. M., and Marceau, N. (1982). *Cancer Res.* 42, 2457–2461.
Landry, J., Crête, P., Lamarche, S., and Chrétien, P. (1988). *Radiat. Res.* 113, 426–436.
Levinger, L., and Varshavsky, A. (1981). *J. Cell Biol.* 90, 793–796.
Lewis, M. J., and Pelham, H. R. B. (1985). *EMBO J.* 4, 3137–3143.
Li, G. C. (1983). *J. Cell. Physiol.* 115, 16–22.
Li, G., and Werb, Z. (1982). *Proc. Natl. Acad. Sci. U.S.A.* 79, 3218–3222.
Lin, C. Y., Roberts, J. K., and Key, J. L. (1984). *Plant Physiol.* 74, 152–160.
Lindquist, S. (1980). *J. Mol. Biol.* 137, 151–158.
Lindquist, S. (1981). *Nature (London)* 294, 311–314.
Lindquist, S., and DiDomenico, B. J. (1985). In "Changes in Eukaryotic Gene Expression in Response to Environmental Stress" (B. G. Atkinson and D. B. Walden, eds.), pp. 71–90. Academic Press, New York.

Lindquist, S., DiDomenico, B., Bugaisky, G., Kurtz, S., Petko, L., and Sonoda, S. (1982). *In* "Heat Shock: From Bacteria to Man" (M. J. Schlesinger, M. Ashburner, and A. Tissières, eds.), pp. 167–175. Cold Spring Harbor Lab., Cold Spring Harbor, New York.

Loomis, W. F., and Wheeler, S. (1982). *Dev. Biol.* **90**, 412–418.

Malhotra, A., Kruuv, J., and Lepock, J. R. (1986). *J. Cell Physiol.* **128**, 279–284.

Mansfield, M. A., Lingle, W. L., and Key, J. L. (1988). *J. Ultrastruct. Mol. Struct. Res.* **99**, 96–105.

Mascarenhas, J. P., and Altschuler, M. (1985) *In* "Changes in Eukaryotic Gene Expression in Response to Environmental Stress" (B. G. Atkinson, and D. B. Walden, eds.), pp. 315–326. Academic Press, Orlando, Florida.

McAlister, L., and Finkelstein, D. B. (1980). *Biochem. Biophys. Res. Commun.* **93**, 819–824.

McGarry, T. J., and Lindquist, S. (1985). *Cell (Cambridge, Mass.)* **42**, 903–911.

Mestril, R., Rungger, D., Schiller, P., and Voellmy, R. (1985). *EMBO J.* **4**, 2971–2976.

Minton, K. W., Karmin, P., Hahn, G. M., and Minton, A. P. (1982). *Proc. Natl. Acad. Sci. U.S.A.* **79**, 7107–7111.

Mitchell, J. B., Russo, A., Kinsella, T. J., and Glatstein, E. (1983). *Cancer Res.* **43**, 987–991.

Munro, S., and Pelham, H. R. B. (1985). *Nature (London)* **317**, 477–478.

Munro, S., and Pelham, H. R. B. (1986). *Cell (Cambridge, Mass.)* **46**, 291–300.

Myrand, S., and Pederson, T. (1983). *Mol. Cell. Biol.* **3**, 161–171.

Neumann, D., Scharf, K. D., and Nover, L. (1984). *Eur. J. Cell Biol.* **34**, 254–264.

Nieto-Sotelo, J. and Ho, T.-h. D. (1986). *Plant Physiol.* **82**, 1031–1035.

Nieto-Sotelo, J. and Ho, T.-h. D. (1987). *J. Biol. Chem.* **262**, 12288–12292.

Nover, L., Scharf, K. D., and Neumann, D. (1983). *Mol. Cell. Biol.* **3**, 1648–1655.

Nover, L., Scharf, K. D., and Neumann, D. (1989). *Mol. Cell. Biol.* **9**, 1298–1308.

Pelham, H. R. B. (1982). *Cell (Cambridge, Mass.)* **30**, 517–528.

Pelham, H. R. B., (1988). *Nature (London)* **332**, 776–777.

Pelham, H. R. B., and Bienz, M. (1982). *EMBO J.* **1**, 1473–1477.

Pitto, L., Lo Schiavo, F., Giuliano, G., and Terzi, M. (1983). *Plant Molec. Biol.* **2**, 231–237.

Pratt, G., Deveraux, Q., and Rechsteiner, M. (1989). *In* "Stress Induced Proteins" (M. L. Pardue, J. R. Feramisco, and S. Lindquist, eds.), pp. 149–159. Liss, New York.

Ritossa, F. (1962). *Experientia* **18**, 571–573.

Rochester, D. E., Winer, J. A., and Shah, D. M. (1986). *EMBO J.* **5**, 451–458.

Rubin, G. M., and Hogness, D. S. (1975). *Cell (Cambridge, Mass.)* **6**, 207–213.

Sachs, M. M., and Ho, T.-h. D. (1986). *Annu. Rev. Plant Physiol.* **36**, 363–376.

Santarius, K. A. (1975). *J. Therm. Biol.* **1**, 101–107.

Scharf, K. D., and Nover, L. (1982). *Cell (Cambridge, Mass.)* **30**, 427–437.

Schlesinger, M. J. (1985). *In* "Changes in Eukaryotic Gene Expression in Response to Environmental Stress" (B. G. Atkinson and D. B. Walden, eds.), pp. 183–195. Academic Press, New York.

Schlesinger, M. J., Ashburner, M., and Tissières, A., eds. (1982). "Heat Shock: From Bacteria to Man." Cold Spring Harbor Lab., Cold Spring Harbor, New York.

Schönfelder, M., Horsch, A., and Schmid, H.-P. (1985). *Proc. Natl. Acad. Sci. U.S.A.* **82**, 6884–6888.

Sciandra, J. J., Subjeck, J. R., and Hughes, C. S. (1984). *Proc. Natl. Acad. Sci. U.S.A.* **81**, 8343–4847.

Scott, M. P., and Pardue, M. L. (1981). *Proc. Natl. Acad. Sci. U.S.A.* **78**, 3353–3357.

Showalter, A. M., and Varner, J. E. (1989). *In* "The Biochemistry of Plants: A Comprehensive Treatise" (A. Marcus, ed.), vol. 15. Academic Press, San Diego, California, pp. 485–520.

Sin, Y. T. (1975). *Nature (London)* **258**, 159–160.

Sinibaldi, R. M., and Turpen, T. (1985). *J. Biol. Chem.* **260**, 15382–15385.

Sorger, P. K., and Pelham, H. R. B. (1988). *Cell (Cambridge, Mass.)* **54**, 855–864.

Spena, A., Hain, R., Ziervogel, U., Saedler, H., and Schell, J. (1985). *EMBO J.* **4,** 2739–2743.

Storti, R. V., Scott, M. P., Rich, A., and Pardue, M. L. (1980). *Cell (Cambridge, Mass.)* **22,** 825–834.

Tanguay, R. M. (1985). *In* "Changes in Eukaryotic Gene Expression in Response to Environmental Stress" (B. G. Atkinson and D. B. Walden, eds.), pp. 91–113. Academic Press, New York.

Tanguay, R. M., and Vincent, M. (1982). *Can. J. Biochem.* **60,** 306–315.

Tissières, A., Mitchell, H. K., and Tracy, U. M. (1974). *J. Mol. Biol.* **84,** 389–398.

Varner, J. E., and Ho, T.-h. D. (1976). *Symp. Soc. Dev. Biol.* **34,** 173–194.

Velazquez, J. M., and Lindquist, S. (1984). *Cell (Cambridge, Mass.)* **36,** 655–662.

Velazquez, J. M., DiDomenico, B. J., and Lindquist, S. (1980). *Cell (Cambridge, Mass.)* **20,** 679–689.

Vierling, E., and Key, J. L. (1985). *Plant Physiol.* **78,** 155–162.

Vierling, E., Mishkind, M. L., Schmidt, G. W., and Key, J. L. (1986). *Proc. Natl. Acad. Sci. U.S.A.* **83,** 361–365.

Vierling, E., Harris, L. M., and Chen, Q. (1989). *Mol. Cell. Biol.* **9,** 461–468.

Weiderrecht, G., Shuey, D. J., Kibbe, W. A., and Parker, C. S. (1987). *Cell (Cambridge, Mass.)* **48,** 507–515.

Weiderrecht, G., Seto, D., and Parker, C. S. (1988). *Cell (Cambridge, Mass.)* **54,** 841–853.

Welch, W. J., and Feramisco, J. R. (1982). *J. Biol. Chem.* **257,** 14949–14959.

Welch, W. J., and Feramisco, J. R. (1985). *Mol. Cell. Biol.* **5,** 1229–1237.

Welch, W. J., and Suhan, J. P. (1985). *J. Cell Biol.* **101,** 1198–1211.

Welch, W. J., Garrels, J. I., Thomas, G. P., Lin, J. J.-C., and Feramisco, J. R. (1983). *J. Biol. Chem.* **258,** 7102–7111.

Werner-Washburne, M., Stone, D. E., and Craig, E. A. (1987). *Mol. Cell. Biol.* **7,** 2568–2577.

Whelan, S. A., and Hightower, L. E. (1985). *J. Cell. Physiol.* **125,** 251–258.

White, F. P., and White, S. R. (1987). *J. Neurochem.* **48,** 1560–1565.

Wistow, G., and Piatigorsky, J. (1987). *Science* **236,** 1554–1556.

Wolffe, A. P., Perlman, A. J., and Tata, J. R. (1984). *EMBO J* **3,** 2763–2770.

Wu, C. (1984a). *Nature (London)* **309,** 229–234.

Wu, C. (1984b). *Nature (London)* **311,** 81–84.

Wu, C., Wilson, S., Walker, B., Dawid, I., Paisley, T., Zimarino, V., and Ueda, H. (1987). *Science* **238,** 1247–1253.

Xiao, H., and Lis, J. T. (1989). *Mol. Cell. Biol.* **9,** 1746–1753.

Yost, J. H., and Lindquist, S. (1986). *Cell (Cambridge, Mass.)* **45,** 185–193.

Zimmerman, J. L., Petri, W., and Meselson, M. (1983). *Cell (Cambridge, Mass.)* **32,** 1161–1170.

Zimmerman, J. L., Apuya, N., Darwish, K., and O'Carroll, C. (1989). *Plant Cell* (in press).

CHAPTER 7

Reduced Cell Expansion and Changes in Cell Walls of Plant Cells Adapted to NaCl

Ray A. Bressan*
Donald E. Nelson*
Naim M. Iraki** +
P. Christopher LaRosa*
Narendra K. Singh†
Paul M. Hasegawa*
Nicholas C. Carpita**

+Present address:
Department of Botany
Hebrew University
Jerusalem, Israel

†Department of Botany and Microbiology
Auburn University
Auburn, Alabama

*Center for Plant Environmental Stress
Physiology
Department of Horticulture
Purdue University
West Lafayette, Indiana

**Department of Botany and Plant Pathology
Purdue University
West Lafayette, Indiana

I. INTRODUCTION

A. Osmotic Stress and Growth Inhibition

When plants are exposed to a dry or saline environment, their growth is inhibited. This is an obvious phenomenon that has been observed throughout history and has probably been responsible for the demise of numerous civilizations. Armed with an understanding of the thermodynamic basis of water utilization by plants, scientists have long attributed this growth reduction to an imbalance in plant water relations that results in turgor loss and subsequent growth reduction. It is obvious that wilted plants have sufficient turgor loss to inhibit growth and that desiccation and saline stress can produce wilted plants.

However, is the reduced growth caused by these stresses always the result of reduced turgor? This has been a hard question to answer because we may not be able to predict reliably that reduced turgor will result in reduced growth. In other words, if the plant is not wilted, some turgor pressure remains. Do we therefore expect the remaining turgor to be sufficient to drive growth at unreduced rates? Plant physiologists approximate the relationship between turgor and growth with the following equation (Cosgrove, 1986):

$$G_r = \phi(P - Y) \qquad (7.1)$$

where G_r is the growth rate, ϕ is the extensibility coefficient, P is the turgor pressure, and Y is the yield threshold pressure. We can see from this relationship that constant growth could be maintained at reduced turgor if ϕ increased or perhaps if Y is decreased. Therefore, knowing only that osmotic stress had reduced turgor would not allow us to conclude that growth rate would be reduced. Nevertheless, careful consideration of these various parameters has provided convincing evidence that under certain conditions, especially immediately after stress is imposed, turgor loss can cause reduced rates of cell expansion (see Hsiao, 1987). However, the long-term effects of osmotic stress on growth have taken on renewed importance in recent years as more and more reports have firmly documented the ability of plants to adjust osmotically after exposure to osmotic stresses. These reports have clearly shown that osmotic adjustment can lead to total restoration of turgor pressure in the presence of osmotic stress. But growth rates are still inhibited.

In 1982 we reported that cultured plant cells adapted to desiccation stress exhibited a slower rate of cell enlargement even though osmotic adjustment had produced turgor pressures higher than prestress levels (Handa *et al.*, 1982; Bressan *et al.*, 1982). We suggested that this reduced growth rate was due to altered properties of the cell wall or wall-loosening mechanism (Bressan *et al.*, 1982). Meyer and Boyer (1981), Matsuda and Riazi (1981) and Cavalieri and Boyer (1982) reported similar reductions in growth without loss of turgor in the growing tissues of whole plants subjected to desiccation stress, and Meyer and Boyer suggested that a reduced growth rate may actually be part of the osmotic adjustment process. Boyer and co-workers (1985) vigorously promoted the hypothesis that the effect of desiccation on water potential gradients between the water source and enlarging cells reduces growth without decreasing turgor. Since these earlier reports, it gradually has become widely acknowledged that growth reduction that is induced by dessication stress is not a simple consequence of turgor loss. We have indicated that this growth reduction certainly involves another mechanism besides turgor loss, which is most likely actively regulated by controlling cell wall extensibility through some signal or induction process involving the perception of a desiccating environment (Bressan *et al.*, 1982; Handa *et al.*, 1983; Hasegawa *et al.*, 1984). We have pointed out that such an active growth-reduction response to desiccation could logically contribute to a plant's having a water-conservation survival strategy (Binzel *et al.*, 1985, 1987, 1988; Hasegawa *et al.*, 1984). Restoration of normal growth rates after osmotic adjustment under desiccating conditions would jeopardize the plant's survival by accelerating the depletion of a limited amount of soil moisture. This would be especially true considering that an uninhibited growth rate would greatly increase the transpiring leaf surface. Thus there seems to be truth to both sides of an old argument dating back to the early data of Briggs and Shantz (1914) on the "water requirement" of plants: Plants under desiccation stress are small because they lack water; this is because the lack of water induces a reduced growth response that results in small plants that will use less water and thereby will survive better.

We reported in 1985 (Binzel *et al.*, 1985) that cultured cells adapted to NaCl also exhibited a reduced growth rate after osmotic adjustment had resulted in turgor pressures above prestress levels. This was a somewhat surprising result since a saline environment should not impose a constraint on water usage as long as adjustment of cell water relations allows for sufficient water uptake. In other words, in contrast to the desiccation environment where a limited water supply is essentially "mined" by the plant, the saline environment offers an unlimited supply of "hard-to-extract" water. Therefore, if osmotic adjustment could keep pace with cell expansion, there should be no need to initiate a reduced growth rate to conserve water.

It is clear that the ability of plants to cope with excessive ions involves both ion transport and growth characteristics of plant cells. We should keep in mind that an obvious balance between the osmotic potential produced by solute (ion) transport and cell expansion is maintained during plant cell expansion (Zimmerman, 1978). When glycophyte cells are exposed to high levels of salt, this balance is apparently disrupted in the sense that cell growth or expansion rates are decreased substantially even though ion transport rates are able to maintain adequate turgor levels.

We have discussed before (Singh *et al.*, 1989a) how decreased growth rates after exposure to NaCl may be detrimental by decreasing the potential size of the pool (vacuole) available for removal of ions from the cytoplasm. However, since the actual ion transport rates and how they are affected by cell expansion are unknown, we cannot exclude the possibility that restricted cell expansion is actually a necessary adjustment to the high salt environment (see Binzel *et al.*, 1989). In other words, if the cell expansion rate of adapted cells was to increase, could the ion transport rate increase accordingly? Rapid transport of cytoplasmic Na^+ into the vacuole probably alone cannot keep the Na^+ concentration in the cytoplasm below toxic levels. This transport is working against a high Na^+ gradient. Therefore, transport of Na^+ out through the plasmalemma is most likely an important additional process needed to prevent ion toxicity (Reinhold *et al.*, 1989). This presents a dilemma for the cell, however, in that Na^+ and Cl^- counterions are needed for osmotic adjustment and growth. Therefore, the expulsion mechanism for Na^+ ions at the plasmalemma works against the accumulation of ions for growth.

To sustain a high growth rate and the accompanying osmotic adjustment, ions must be fluxed through a small cytoplasmic ion pool into a large vacuolar pool. The larger the difference between these two pool sizes (the larger the degree of osmotic adjustment required, or the greater the stress imposed), the greater the flux must be to drive vacuolar osmotic adjustment. The requirements of maintaining a small cytoplasmic ion pool (to prevent toxicity), a large vacuolar ion pool (to provide osmotic adjustment to the high NaCl outside the cell), and a rapid expansion rate of the vacuole volume may demand an ion flux rate through the cytoplasm that is unattainable.

In addition, compatible organic solutes must be synthesized and used in the cytoplasm to balance vacuolar osmotic adjustment. A high rate of cell expansion would also require an increased rate of organic solute accumulation in the cytoplasm. A reduced rate of expansion may turn out to be the mechanism by which a sustainable ion flux rate through the cytoplasm or a sufficient organic solute synthesis rate is achieved during growth.

Altering the growth reduction response in glycophytic plants by genetic manipulation would be greatly facilitated by a more complete understanding of the molecular basis of this response. We first suggested (Binzel *et al.*, 1985) that the growth reduction response to both desiccation and high salt likely involves altered extensibility of the cell wall since adapted cells have an altered relationship between turgor and cell expansion (Binzel *et al.*, 1985; Bressan *et al.*, 1982). Therefore, a better understanding of the involvement of cell wall biochemistry in this growth inhibition is needed.

B. The Involvement of Cell Wall Biochemistry in Growth

The cell walls of plants are rigid networks of polysaccharide, protein, and phenolic compounds that regulate cell volume, size, and shape. To grow, plant cells must expand this network and deposit new material upon and within the loosened matrix. Plant growth and development relies on biochemical factors that discretely alter cell wall architecture. Recent advances in the ability to probe gene expression have prodded investigations of the role of plant hormones in controlling this development. The induction and repression by auxins (2,4-D and IAA) of translation and even transcription of specific messenger RNAs (mRNAs) encoding several polypeptides possibly associated with controlling the rate and pattern of cell expansion have been documented (Hagen *et al.*, 1984; Theologis and Ray, 1982; Walker and Key, 1982; Zurfluh and Guilfoyle, 1982). Although studies of gene expression and identification of polypeptides whose synthesis is altered by plant growth regulators give important insight into the regulation and control of cell development, the function of these polypeptides and whether they represent "growth-essential" or "growth-specific" proteins have not been determined. Answers to these questions must await information on the biochemistry of the cell expansion process. Very simply, if proteins that participate directly in the expansion of the cell wall could be identified, then the molecular bases for their expression could be probed directly. Studies of the molecular basis of adaptation to salinity stress with respect to the mechanisms of altered cell expansion and further differentiation are virtually nonexistent (Sachs and Ho, 1986).

C. The Cell Wall Polymers and Cell Expansion

An understanding of the complexities of the primary cell walls of higher plants is just emerging. The wall is a highly organized matrix of extremely large insoluble polymers. Knowledge of the primary wall consists primarily of a catalog of the protein and carbohydrate polymers selectively extracted by solvents or partial enzymic digestions (Bacic *et al.*, 1988). There are continuing efforts to obtain structural information on novel wall polymers, and eventually we hope to construct functionally accurate three-dimensional models of the walls of higher plants. These studies have been aided extensively by advances in chemical, spectroscopic, and enzymic analysis of the linkage structure of the polymers that comprise the wall, and these analyses have enlightened the view of the wall as a dynamic structure. Synthesis of matrix polymers is stage-specific; certain polymers are made only during cell division, others are made only during cell expansion, and still others are made when cell expansion ceases and the wall is locked mechanically into form during further differentiation.

D. The Chemical Structure of the Cell Wall

The primary wall of *dicotyledonous* species, such as tobacco, is modeled in Fig. 7.1. Basically, cellulosic microfibrils coated with hemicellulosic xyloglucan and some arabinoxylan, which extend farther through the matrix cross-linking other microfibrils, are embedded in a gel matrix of pectic substances (Bacic *et al.*, 1988). The pectins consist of a mixture of polygalacturonic acid and rhamnogalacturonan, the latter polymer containing side groups of arabinans, galactans, and type I and II arabinogalactans (Bacic *et al.*, 1988; Jarvis, 1984). The polygalacturonic acid can form gels by cross bridging with Ca^{2+} ions cross-linked further through hydroxycinnamic acids (Fry, 1983; Rees, 1977). The degree of esterification can influence the extent of Ca^{2+} gel structure, and the size and extent of side-group substitution can affect gel flexibility, porosity, interaction with hemicelluloses, and even the mobility of proteins and enzymes within the gel matrix (Iraki *et al.*, 1989b; Jarvis, 1984).

The principal hemicellulose of dicots is xyloglucan, a linear $(1 \rightarrow 4)\beta$-D-glucosyl chain substituted at regular intervals on the O-6 with xylosyl units, some of which often are further substituted with β-D-galactosyl-, α-L-fucosyl-$(1 \rightarrow 2)\beta$-D-galactosyl units and other small side groups attached to the O-2 of the xylosyl units (Fig. 7.2). Xyloglucan comprises about 80% of the total hemicellulose, but an acidic arabinoxylan, a $(1 \rightarrow 4)\beta$-D-xylosyl–unit backbone substituted mostly at the O-2 with arabinofuranosyl linkages, also constitutes much of the remainder (Darvill *et al.*, 1980). Xyloglucans hydrogen-bond tightly to the cellulose microfibrils and are probably

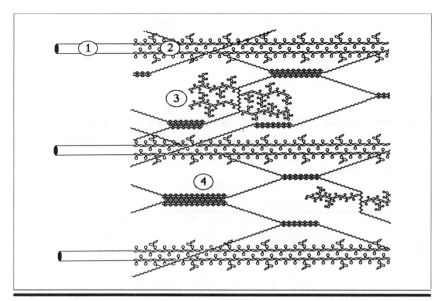

Figure 7.1. Structural model of the primary cell wall of dicots. Cellulose microfibrils (1) are coated with a monolayer of xyloglucan (2). Additional xyloglucan and arabinoxylan molecules may span the microfibrils (not shown). The cellulose-xyloglucan framework is embedded in a gel of polygalacturonic acids, cross-linked in part by Ca^{2+} ions (4). Additional polymers containing mostly neutral sugars constitute major side groups and are attached to the rhamnosyl units of rhamnogalacturonan (3). Not shown is the hydroxyproline-rich glycoprotein, extensin, which can cross-stitch the cellulose fibrillar network. (From Carpita, 1987.)

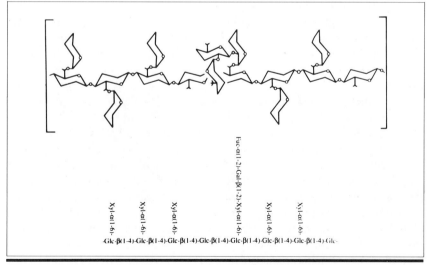

Figure 7.2. Molecular model of the hepta- and nonasaccharide-repeating unit structure of the dicot xyloglucan portion of the cell wall matrix. The model is based on both partial enzymic digestion of the macromolecule and methylation analysis of oligomers.

organized onto the microfibrils shortly after cellulose synthesis (Hayashi *et al.*, 1987). Amounts of xyloglucan are nearly equal to amounts of cellulose (Bacic *et al.*, 1988; Hayashi *et al.*, 1987), and considering that the cellulose microfibril consists of a crystal of at least 36 glucan chains, xyloglucan not only must comprise a "monolayer" on these crystals but may also span to neighboring microfibrils to constitute a true rigid matrix. Alteration of the matrix could then be controlled through hydrolysis of the xyloglucan by endo-β-D-glucanases.

The extracellular matrix also contains many proteins, such as the hydroxyproline-rich glycoproteins (HRGPs), extensin and arabinogalatan-proteins (AGPs). Extensin, a structural HRGP, comprises a significant amount to the primary cell wall of dicots (Cooper *et al.*, 1987). The "polyproline II–like" tight helix is reinforced by intramolecular isodityrosine residues (Fry, 1982), presumably formed through an extracellular peroxidase and producing a rod-like protein long enough to span the wall anticlinally (Stafström and Staehelin, 1986). Several soluble extensin proteins have been identified in the walls of cells and tissues (Cooper and Varner, 1984), and isodityrosine formation coincides with insolubilization of these precursors (Cooper *et al.*, 1987). Polymerization in solution is possible using peroxidase and H_2O_2, but the biphenyl dityrosine, not isodityrosine, is formed (Cooper and Varner, 1984). Either specific peroxidases are required or some steric constraints provided by the acidic pectins may direct this linkage *in vivo* (Cooper and Varner, 1983). Intermolecular isodityrosine linkages or covalent linkages of protein–protein or protein–polysaccharide are suspected of producing a rigid inextensible wall. Although extensin accumulation more likely signals irreversible cessation of growth (Sadava and Chrispeels, 1973), (Fry (1980) added that suppression of peroxidase secretion by growth regulators may delay extensin cross-linking during cell expansion.

While the protein structure of extensin is well documented and some of the extensin genes have been cloned, an appreciation for other wall proteins fulfilling structural roles is just now emerging. Condit and Meagher (1986) inadvertently isolated a clone of a putative cell-wall glycine-rich protein from genomic libraries of *Petunia*. From a protein sequence deduced from the clone and hydropathy plotting, they proposed a flattened β-pleated sheet structure with bulky phenylalanine and histidine groups projecting out from one side of the plane. The actual structure, the actual function, its abundance, and its location are unknown because the data are deduced from a genomic clone. However, walls enriched in glycine have certainly been identified (Rackis *et al.*, 1961; Melin *et al.*, 1979; Dreher *et al.*, 1980). Further, Bozarth *et al.* (1987) found a 28-kDa glycine-rich polypeptide induced on water stress in soybean hypocotyls and a novel proline-rich protein was induced by wounding carrot tissue (Tierney *et al.*, 1988), indicating that regulation of the synthesis of wall proteins may constitute a response to mechanical, environmental, and even biotic stress (Esquerre-Tugaye *et al.*, 1979). Pathogen invasion induces the synthesis of extensin (Showalter *et al.*, 1985) but its role in host-pathogen interactions is obscure.

Thus the biochemistry of the cell wall is complex and its role in cell expansion, although undisputed, is not well understood. Since osmotically adapted plant cells exhibit dramatically reduced growth or expansion rates even though turgor is more than maintained by osmotic adjustment, we have investigated the changes in cell wall biochemistry that occur as a result of adaptation in these cells.

II. MATERIALS AND METHODS

A. Plant Cell Culture

Nicotiana tabacum cv. W-38 cells were maintained as previously described (Binzel *et al.*, 1985). The cells that adapted to medium containing various concentrations of NaCl in grams per liter of medium are designated S-O, S-25, etc. These cells were grown at least 100 cell generations in their respective salt concentrations before use in the experiments.

B. Measurements of Cell Wall Tensile Strength

General theoretical and practical aspects of these determinations were reported earlier (11). These experiments compared unadapted and NaCl-adapted cells. Cells in the logarithmic phase of the culture cycle were transferred to fresh medium, and 20 ml of the cell suspension were placed in a small beaker; a nitrogen gas decompression bomb (Parr Instruments) was gently stirred in. Nitrogen gas was introduced slowly (to avoid heating to the desired pressure. After equilibration for 15 min, the suspension was jettisoned into a large cylinder. An aliquot was mixed with one-half volume of 20% chromic acid in a 3-ml Reacti-vial (Pierce) and stirred gently for 1 hr. The cells that survived decompression were visualized by bright-field microscopy and counted with a hemacytometer. Samples of the same fresh cell suspension were also used to determine incipient plasmolysis in a graded series of NaCl solutions, and turgor pressure was calculated as the difference between the water potential of the growth medium (determined by vapor pressure psychrometry) and the osmotic potential causing incipient plasmolysis of 50% of the cell population. We found this method to be the most reliable for determining osmotic potential. Measurements were made in less than 30 min. The amount of NaCl uptake during this time was negligible and at least 1 hr further incubation was required before noticeable deplasmolysis resulted from the uptake of the exogenous NaCl. Cells at this phase of culture growth were spherical to ellipsoidal; the cell radii were estimated from cell dimensions (Binzel et al., 1987, 1988). Wall thickness was about 0.1 μm for both unadapted and adapted cells, based on calculations from electron micrographs.

C. Isolation of Cell Wall Material for Carbohydrate Analysis

Cells in the stationary growth phase were filtered in coarse sintered-glass funnels, rinsed briefly in iso-osmolar mannitol, and frozen in liquid N_2. Acetone powders of the total material were prepared to precipitate soluble polysaccharides as well as insoluble components of the wall. This treatment, of course, resulted in precipitation of substantial amounts of soluble protein, but its presence did not interfere with carbohydrate analyses. Samples of dry powders (1.5 g) were extracted sequentially as follows: once with 125 ml of ice-cold 5-mM EDTA for 12 hr with constant stirring; twice with 125 ml of 0.5% ammonium oxalate, pH 6.5, at 100°C for 1 hr each; once with 50 ml of 0.1-M KOH for 4 hr; and twice with 4-M KOH overnight each. The EDTA and ammonium oxalate solutions each extracted pectic substances by chelation of Ca^{2+} cross-linking the uronosyl units, and the 0.1-M KOH removed additional pectin, perhaps by hydrolysis of ester linkages or other weak alkali-labile bonds (Fry, 1986; Jarvis, 1984). The 4-M KOH extracted hemicelluloses by disrupting hydrogen bonding or breaking unidentified covalent linkages (Bacic et al., 1988). The KOH solutions contained $NaBH_4$ (3 mg/ml) to prevent end peeling (Aspinall et al., 1961), and the extraction was carried out under N_2 with continuous stirring. The KOH extracts were neutralized with glacial acetic acid. All fractions were filtered through Whatman GF/F glass-fiber filters, dialyzed against running deionized water overnight at room temperature, and lyophilized. The remaining material containing hemicellulose and cellulose was washed with deionized water and lyophilized. The cellulose content in the samples of the material was deterimined according to Updegraff (1969). All results expressed represent the average values from usually three, but sometimes two, different experimental samples. Variance was always less than ±5%.

D. Preparation of Cell Walls for Amino Acid Analysis

Suspension-cultured cells (Binzel et al., 1985) were collected by filtration through a nylon mesh filter and homogenized with a Tekmar Tissumizer in 6 vol of either 50-mM KPi, pH 7.0, or water. The debris was collected again by filtration and washed sequentially with the following: chloroform :-methanol (1 : 1, v/v) at 80°C for 3 hr, water, 200-mM CaCl$_2$ (100 ml/100 g original fresh weight with stirring for 3 hr), 3.0-M LiCl (100 ml/100 g original fresh weight with stirring for 3 hr), 1% sodium dodecyl sulfate (SDS) at 100°C for 3 hr, water, methanol, and acetone. The acetone-washed wall material was then dried and stored at room temperature.

E. CaCl$_2$ and LiCl Extracts of Cell Wall Material

The CaCl$_2$ and LiCl washes from the wall material were concentrated by partial lyophilization and then dialyzed against water with several changes for 24 hr. This usually caused a precipitate to form. The dialysate was separated from the precipitate by centrifugation, and both were lyophilized and stored at $-20°C$ before the protein was extracted with the SDS extraction buffer (Singh *et al.*, 1985).

F. Protein Separation and Amino Acid Analysis

Proteins from various wall fractions were extracted with a sample extraction buffer and separated by sodium dodecyl sulfate polyacrylamide gel electrophoresis (SDS-PAGE) as described by Singh *et al.* (1985). Amino acid composition of the cell wall and other material was performed after hydrolysis with 6-N HCl at 130°C for 24 hr in sealed reaction vials. Charred debris from the hydrolysate was removed by filtration over glass fiber. The hydrolysates were partially purified by Dowex-50 cation exchange chromatography before derivatization and quantitation by gas chromatography as described by Rhodes *et al.* (1986).

G. Isolation of Proteins from the Culture Medium

Stationary phase or near-stationary phase cells were filtered through nylon mesh, and the cell-free medium was collected. The medium was lyophilized and redissolved in a small volume of water before dialysis against water for 24 hr with several changes of the water. The precipitate that formed during dialysis was separated from the soluble dialysate, and both were lyophilized and kept at room temperature. Proteins were analyzed by SDS-PAGE as described by Singh *et al.* (1985).

H. Preparation of Antibodies Against Wall-Bound and Medium Proteins and Western Blotting

Proteins from medium and CaCl$_2$ extractions of cell wall fragments were separated by preparative SDS-PAGE using 20-cm × 20-cm × 1.5-mm gels. Bands containing the 29-kDa protein from the medium and the 40-kDa protein from the CaCl$_2$ extracts were cut from the gels, and the proteins were electroeluted using a BioRad Mini Protean-II Electro-Eluter. Electroeluted protein was rerun on SDS-PAGE to verify the homogeneity of the protein sample. Approximately 30 μg of each protein were injected subcutaneously into young Bantam chickens. Two booster injections were given at 2-week intervals. Eggs were collected and IgY was purified (Song et al., 1985) from the eggs after the third injection of antigen. The specificity of IgY from the immunized chickens was determined by dot blotting the purified antigen onto nitrocellulose and reacting with the purified IgY from the eggs from preimmunized and immunized chickens. Antigen–antibody complex was detected with alkaline phosphatase-conjugated rabbit anti-chicken antibody (Jackson Biochemicals) using Nitro blue tetrazolium and 5-bromo-4-chloro-3-indolyl phosphate as substrate according to the BioRad kit instructions (No. 170-6509). Western blots were made essentially as described in the BioRad Western blot instructions after transferring the proteins from the acrylamide gels to the nitrocellulose by electroblotting for 1 hr with a semidry electroblotter. Staining the electroblotted gel with Coomassie blue confirmed the transfer of proteins to the nitrocellulose. Molecular weights of the blotted proteins were estimated using BioRad prestained molecular weight standards.

I. Determination of Cell Volumes during Growth of Suspension-Cultured Cells

Cell volumes were estimated by determining the fresh weight minus the dry weight of the culture samples, after collecting the cells by vacuum filtration and dividing by the total number of cells in the samples of equivalent weight. Cell number was determined using a hemacytometer to count the cells after cell clumps were separated by treatment with 15% chromic acid at 65°C for 30 min. This method overestimates cell volume because there is no correction for extracellular volume including cell-free space. However, with proper filtration of the cells, we found that the extracellular volumes were quite constant from sample to sample. Some differences in extracellular volume between salt-adapted and unadapted cells would be expected because of differences in cell size (Binzel et al., 1985), however, this effect should also be relatively constant since cell samples are taken over time.

J. Leaf Area of Regenerated Plants

Crosses between W-38 tobacco plants and plants regenerated from adapted cells (S-25 plants) and all subsequent crosses were made in the greenhouse and plants were transplanted to the field in a random block design. After 4 months of growth, leaves were harvested from sample plants in the field and the leaf area of each leaf measured with a LiCor model LI-3000 leaf area meter. After determining the developmental stage (distance from the apex) at which leaves of maximum expansion could be sampled, all field-grown plants were evaluated by measuring the leaf area of the tenth leaf from the apex.

III.   RESULTS

A.   Cell Enlargement of Cultured Cells Adapted to High Levels of NaCl

Tobacco cells adapted to 428-mM NaCl have a greatly reduced ability to gain fresh weight, although their dry-weight growth rate is similar to un-adapted cells (Binzel *et al.*, 1985). This is due to a reduced rate of cell en-largement (Fig. 7.3). It is clear from these data that both the rate of cell expansion and the final volume of adapted cells are reduced compared to

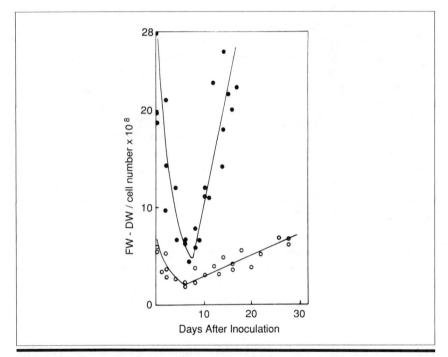

Figure 7.3.   Changes in cell volume during the growth of unadapted cells (●) and cells adapted to 428-mM NaCl (○). Cell volume was estimated by determining the fresh weight (g) minus the dry weight (g) of an aliquot of the cell culture and calculating the number of cells in that aliquot. Enough cells were counted in each aliquot to provide ± 2% error. (After Iraki *et al.*, 1989a.)

unadapted cells. In fact, the maximum cell volume decreases as the cells are adapted to increasing NaCl concentrations (Hasegawa *et al.,* 1986). We have pointed out before that adapted cells appear to lack a cell-elongation phase during development (Binzel *et al.,* 1985, 1988). As seen in Table 7.1, the length-to-width ratio of adapted as compared to unadapted cells clearly indicates failure of adapted cells to elongate.

B.   Physical Alterations in the Cell Wall (Tensile Strength)

The pressure required to break 50% of the cell population (breaking pressure) was measured empirically (Table 7.2). Tensile strength is defined mechanically as the tensile force per cross-sectional area required for breakage (Preston, 1974). For steel rods, plant fibers, and cylinders cut from wood, $N{\cdot}m^2$ are typical units of tensile strength and reflect the external force exerted in a single dimension required to break the material. Our measurements are based on the internal pressures required to burst the cell wall. Values obtained from stretching large cells or pieces of tissue are difficult to compare with those estimated for living cells, especially because the cell wall is not a homogeneous material of defined thickness but a polylamellate matrix of cellulose microfibrils cross-linked with noncellulose polysaccharides and protein. We defined tensile strength as the ability to withstand the tangential force per unit of wall thickness resulting from the cell's internal pressure. This force can only be estimated based on breaking pressure, cell diameter, and wall thickness (Carpita, 1985). Turgor pressures exerted by the cells correspond to enormous tensions because tangential stresses imposed by the pressure of large cells are borne by extremely thin cell walls. Breaking pressures were 20 bar for walls of unadapted cells but only 3 bar for walls of NaCl-adapted cells (Table 7.2). Accounting for the contribution of turgor pressure, the total pressure required to break the walls of unadapted cells was 24 bar and 17 bar for NaCl-adapted cells (Table 7.2).

Table 7.1.   Degree of Expansion of Adapted[a] and Unadapted (S-0) Tobacco Cells at Stationary Phase as Indicated by the Length-to-Width Ratio[b]

Cell type	Length-to-width ratio						
	1.0–1.5	1.5–2.0	2.0–2.5	2.5–3.0	3.0–3.5	3.5–4.0	>4.0
	(Percentage of total cells)						
S-0	10	21	12	22	8	8	18
S-25	96	4	0	0	0	0	0

[a]Adapted to 428 = mM NaCl, S-25
[b]n = 39 and 71 for S-0 and S-25 cells, respectively.

Table 7.2. Tensile Strength of Unadapted (S-0) and NaCl-Adapted (S-25) Tobacco Cells

| Cell line | Turgor[a] pressure (bar) | Breaking[b] pressure (bar) | Cell size[c] | | Wall[d] thickness (μm) | Tensile[e] strength (bar) |
			Length (μm)	Width (μm)		
S-0	4	20	179	71	0.1	4,260–10,740
S-25	14	3	54	46	0.1	1,960–2,300

[a]Turgor pressure was calculated from incipient plasmolysis of 50% of the cells in a graded series of NaCl and the water potential of the medium.
[b]Breaking pressure was that pressure in excess of turgor pressure required to burst 50% of the cell population by nitrogen gas decompression. Values are the average of three independent experiments.
[c]The length and width of ellipsoid and spherical cells were measured empirically in a population of cells at the midlogarithmic stage of growth. Values are the mean of at least 36 samples.
[d]Cell wall thickness was estimated from electron micrographs.
[e]Tensile strength was estimated using an equation that reflects the difference in the planar areas of the cell and the cell wall on which the force applied by P, the sum of the turgor and the breaking pressure; the radius is estimated from the cell's length and width and wall thickness. The range reflects the difference in length and width.

Wall thickness was about 0.1 μm, but even though total pressures contributing to cell breakage were only about 40% higher in adapted cells, the estimated tensile strengths were twofold to fivefold lower because of the smaller size of the cells (Table 7.2). For comparison, mechanical properties of differentiated cells in tissues must comprise a larger range of determinants than strictly those of the cell walls of isolated spherical cells. For example, the tensile properties are related also to the orientation and organization of the cells in the tissue, the cementing of adjacent cells by the middle lamellae, and the orientation of the cellulose microfibrils in the individual cells (Preston, 1974). For tissues, tensile strengths range from 800 to 5000 bar for sisal fibers, depending on relative humidity (Spark *et al.*, 1958), and from 1300 to 3300 bar for *Pinus radiata* late wood, depending on the orientation of the fiber cells in the wood to the stress axis (Wardrop, 1951). Despite such marked differences in the determinants of strengths, these values were surprisingly comparable to those we estimated for the primary walls of the tobacco cells (Table 7.2).

Tensile strength was lower in adapted cells, but elevated turgor pressures exhibited by these cells rose to values that only approach these breaking pressures (Table 7.2). The NaCl-adapted cells had greatly reduced ability to expand and higher turgor pressures even though the tensile strength was substantially lower (Tables 7.1 and 7.2).

The apparent contradiction that adapted cells have much higher turgor pressures and decreased extensibility yet weaker cell walls is explained readily because the mechanical weakness of the wall has little or nothing to do with extensibility. These data illustrate well that the chemical determinants of wall expansion and wall tensile strength are different.

C.   Cell Wall Polysaccharide Structure

The sequential extraction of pectic substances from the acetone powders yielded material corresponding to 70 and 42% in the unadapted and NaCl-adapted cells, respectively (Fig. 7.4). The low recovery by the adapted cells was a result of the loss of low-molecular-weight material, principally reducing sugars and amino acids precipitated by the acetone, during dialysis of the first extract. The lower proportion of the total amounts of wall recovered from NaCl-adapted cells in the acetone powders reflected this enhanced accumulation of solutes at the expense of wall synthesis (Fig. 7.4).

The NaCl-adapted cells had only about half the amount of cell wall polysaccharides per gram of dry weight as the unadapted cells (Fig. 7.4) Not only was the total amount of cell wall greatly reduced on adaptation to saline stress, but there were also differences between adapted and unadapted cells in the distribution of the various wall fractions. Cell walls of adapted cells had much lower proportions of cellulose and, consequently, higher proportions of hemicellulose than those of unadapted cells (Fig. 7.4). The proportions of total pectin in the walls of adapted and unadapted cells

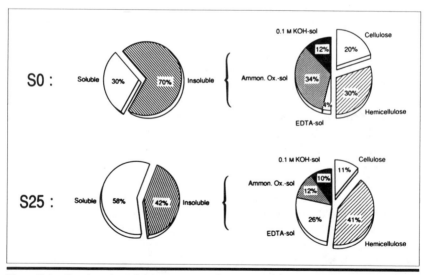

Figure 7.4.   Fractionation of the acetone powders prepared from unadapted cells (S-0) and cells adapted to 428-mM NaCl (S-25). The proportions of insoluble material remaining after dialysis of the acetone powders are represented in the diagrams to the left. These insoluble materials were then fractionated further to give representative pectic substances, hemicellulose, and cellulose, shown on the right. (After Iraki et al., 1989a.)

were about the same, but the organization and composition of the material was markedly different. Cold EDTA extracted only a small portion of pectin from unadapted cells; most of the pectin was extracted sequentially with either hot ammonium oxalate or dilute KOH (Fig. 7.4). In contrast, a larger proportion of the pectin of the adapted cells was extracted by cold EDTA. These results demonstrate that changes in the organization of the pectins were induced by saline stress, whereby a fraction of the Ca^{2+}-insolubilized pectin was bound more loosely and, hence, was extracted more easily from walls of adapted cells.

Although the mass of the wall was lowered, we found that considerable amounts of material accumulated in the extracellular medium during normal growth and that the composition of this material provided additional insight into activities occurring in the cell wall (Iraki *et al.*, 1989c). When soluble polymers from the incubation medium of NaCl-adapted and unadapted cells were separated on Sepharose 4B-200, four major fractions were resolved: (1) a high-molecular-weight polysaccharide (HM_rP)enriched in uronic acid, (2) a homogeneous arabinogalactan (protein) fraction of about 40,000 mol wt, (3) a small group of oligosaccharide "fragments" of xyloglucan, and (4) a heterogeneous group of proteins (Fig. 7.5a). Although secretion of arabinogalactan (protein) was unaffected on adaptation to stress, secretion (or release) of the HM_rP was blocked (Fig. 7.5b). The oligosaccharide xyloglucan fragments, which likely indicate the activity of a cell wall endo-$(1 \rightarrow 4)\beta$-D-glucanohydrolase, were notably absent (Fig. 7.5b) from the medium of adapted cells. The medium was enriched in protein; however, the amounts at the stationary phase of culture reached about 10 times that of an equal number of unadapted cells (Iraki *et al.*, 1989a; Singh *et al.*, 1989a)!

D. Insoluble Protein Content and Composition of Cell Walls

The insoluble protein content of the cell wall increased as the cells adapted to higher levels of NaCl. Cells adapted to 690-mM had about threefold more protein per milligram of cell wall than unadapted cells (Fig. 7.6). This protein did not change in overall amino acid composition much as cells adapt to salt, except for the hydroxyproline content (Table 7.3). As a percentage of the total amino acids, hydroxyproline decreased almost 10-fold at the highest level of adaptation (Fig. 7.7). The hydroxyproline content of the walls of unadapted cells was highest at the end of their growth cycle after cell elongation was completed (Fig. 7.8). Since the cells grew very slowly at this stage, there appeared to be a negative correlation between growth rate and the hydroxyproline content of the walls (Fig. 7.9). How-

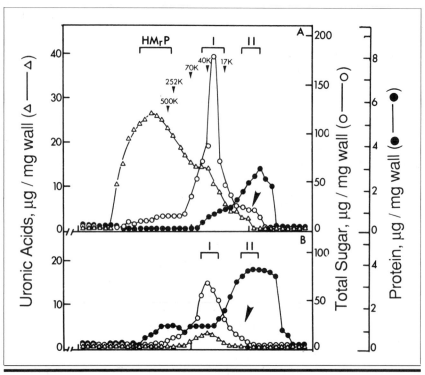

Figure 7.5. Gel chromatography on Sepharose 4B-200 of extracellular polymers from (A) unadapted cells and (B) cells adapted to 428-mM NaCl. Symbols are (○) total sugar, (●) protein, and (△) uronic acid. Arrows indicate the position of xyloglucan oligomers. (After Iraki *et al.*, 1989c.)

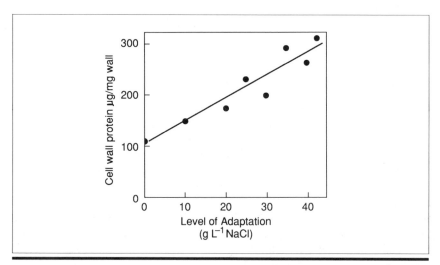

Figure 7.6. Amount of insoluble protein in cell walls of cells adapted to increasing levels of NaCl. Cells were adapted as described by Binzel *et al.* (1985). Insoluble protein was measured from the amounts of total amino acids released from 6-M HCl hydrolysis of purified cell walls.

Table 7.3. Amino Acid Composition[a] of Insoluble Protein from Washed Cell Walls of Cells Adapted to Increasing Concentrations of NaCl[b]

Amino acid	Cell type (mol %)							
	S-0	S-10	S-20	S-25	S-30	S-35	S-40	S-42.5
Alanine	8.0	9.7	9.5	9.7	9.8	10.4	9.5	9.8
Glycine	7.9	8.8	9.8	9.8	9.8	9.7	9.8	7.2
Valine	7.5	8.4	8.2	8.2	7.4	5.6	8.0	5.8
Threonine	5.8	6.0	6.3	6.0	5.8	7.1	6.1	6.1
Serine	8.5	7.0	7.7	7.5	7.3	7.4	7.1	7.6
Leucine	9.5	10.4	10.5	10.4	10.8	11.0	10.2	10.6
Isoleucine	5.5	6.0	6.2	6.2	5.3	6.3	6.1	6.2
Proline	5.1	5.1	5.3	5.2	5.4	4.2	5.2	5.8
Hydroxyproline	6.9	3.1	1.3	1.8	0.9	1.2	0.6	0.8
Methionine	0.7	0.7	0.9	0.9	0.2	0.6	0.8	0.5
Asp + Asn	9.5	9.9	10.0	9.7	10.0	10.8	9.9	10.3
Phenylalanine	4.5	4.4	4.5	4.6	4.9	4.9	4.5	4.6
Gln + Glu	9.5	9.9	9.5	9.7	10.2	10.8	9.8	10.7
Lysine	7.6	5.0	5.7	7.6	6.2	3.7	4.8	6.2
Tyrosine	2.0	1.8	1.1	1.5	0.9	2.8	2.7	2.3
Arginine	3.3	2.6	1.9	1.4	2.8	2.1	2.5	3.8
Histidine	0.4	0.2	0.9	0.6	1.3	1.1	0.7	0.6
Cystine	0.2	0.1	0.4	0.1	0.4	0.2	0.2	0.2

[a]As a mole percentage of total amino acids.
[b]from 0 to 42.5 g/L.

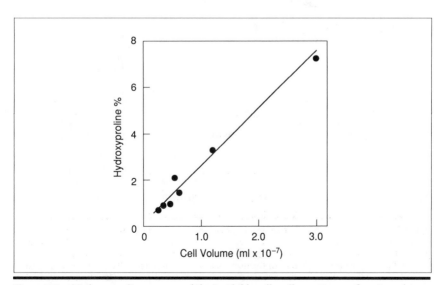

Figure 7.7. Hydroxyproline content of the insoluble cell wall protein as a function of maximum cell volume. Cell volumes are average volumes of cells adapted to various levels of NaCl (as in Fig. 7.2). Hydroxyproline was measured after hydrolysis (as in Fig. 7.2).

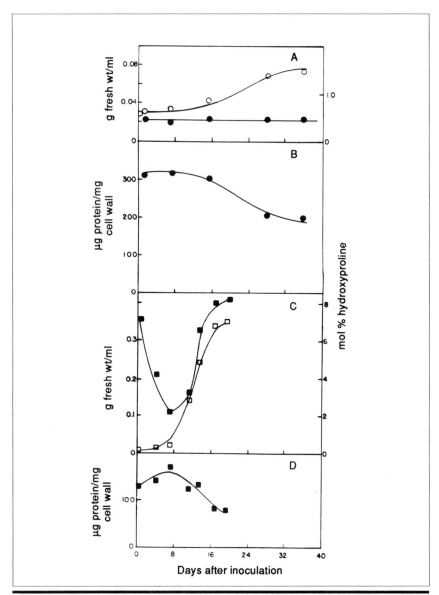

Figure 7.8.   Total insoluble protein and hydroxyproline content of washed cell walls of unadapted and adapted (428-m*M* NaCl) cells over the entire cell culture cycle. (A) Growth in grams of fresh weight (○) and the hydroxyproline content of wall protein (●) of adapted cells. (B) Total insoluble protein content of the insoluble walls of adapted cells. (C) Growth in grams of fresh weight (□) and the hydroxyproline content of the wall protein (■) of unadapted cells. (D) Total insoluble protein content of the cell walls of unadapted cells.

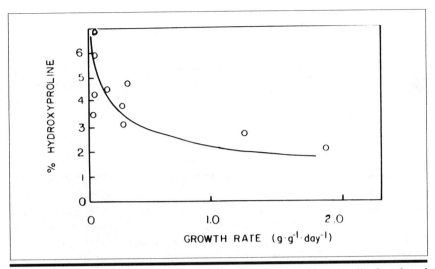

Figure 7.9. Hydroxyproline content of insoluble protein from washed cell walls of unadapted cells as a function of relative fresh weight growth rates of cells at different stages of the cell culture cycle. Relative growth rates were calculated from the following:

$$\frac{(W_2 - W_1)/W_1}{T_2 - T_1}$$

where W_1 is the fresh weight of the culture at any given day (T_1), and W_2 is the weight at the next day of harvest (T_2).

ever, adapted cells grew more slowly than unadapted cells and had wall hydroxyproline levels that were always lower (Fig. 7.8). There must be another mechanism that slows their growth. Increased deposition of HRGP is apparently associated with cessation of a cell-elongation phase of development, which is absent in adapted cells.

E. Proteins Released into the Medium

The unadapted cultured cells released a number of proteins into the medium during growth. However, adapted cells released about seven times as much protein into the medium (Table 7.4). Osmotin (Singh *et al.*, 1987a,b, 1989b) was one major protein found in the medium of adapted cells (data not shown). The released protein from unadapted cells was rich in hydroxyproline. The released protein from adapted cells remained rich in hydroxyproline and similar in overall amino acid composition to that released by unadapted cells (Table 7.5). The high hydroxyproline content of this protein indicates a high arabinogalactan protein (AGP) or extensin content or both (Fincher *et al.*, 1983).

Table 7.4. Protein Content of Cell Walls and Protein Released into Medium by Stationary Phase–Adapted (S-25) and Unadapted (S-0) Cells

Cell type	μg insoluble protein (mg cell wall)$^{-1}$	μg Ca^{2+} + Li$^+$ extracted protein (mg cell wall)$^{-1}$	μg medium protein (g fresh wt)$^{-1}$
S-0	108	12.3	314
S-25	226	9.0	2,048

Some of the released proteins were unique to the adapted cells (Fig. 7.10) including a 29-kDa protein, which is an abundant protein found only in the medium of adapted cells (Iraki et al., 1989c). Antibodies were raised against the 29-kDa protein in chickens. Immunoblotting of ionic-bound proteins from the walls and proteins from the medium and total cell and leaf extracts could not detect this protein from any source except the medium of adapted cells (Fig. 7.11). Its abundance in medium from adapted cells and scarcity elsewhere suggests that it is actively released into the medium specifically by adapted cells. However, if this protein is only a fragment of cell or wall protein, antibodies against the released fragment may not recognize the mature form of the protein.

Table 7.5. Amino Acid Composition[a] of Protein Released into the Medium and of CaCl$_2$- + LiCl-Extracted Cell Wall Proteins from Adapted and Unadapted Cells

Amino acid	Medium		CaCl$_2$ + LiCl extracted	
	S-25 (mol %)	S-0 (mol %)	S-25 (mol %)	S-0 (mol %)
Alanine	14	11	11	4
Glycine	9	11	9	7
Valine	7	5	7	6
Threonine	3	5	6	6
Serine	3	7	3	8
Leucine	5	6	2	5
Isoleucine	3	2	9	2
Proline	8	9	3	9
Hydroxyproline	26	17	3	23
Methionine	ND[b]	ND	ND	ND
Asp + Asn	5	5	6	2
Phenylalanine	3	3	2	2
Glu + Gln	5	5	16	7
Lysine	9	9	12	12
Tyrosine	ND	ND	ND	2
Arginine	2	2	3	5
Histidine	ND	ND	5	1
Cystine	ND	ND	ND	ND

[a]As mole percentage of total amino acids.
[b]ND—not detectable.

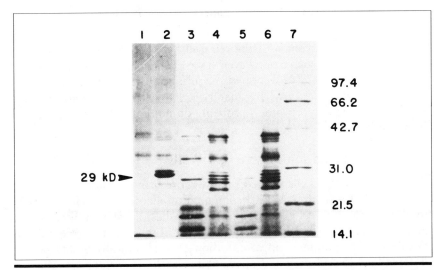

Figure 7.10. SDS-PAGE (10% acrylamide) separation of proteins released into the medium by unadapted (lanes 1, 3, 5) and adapted (lanes 2, 4, 6) cells. Equal amounts of protein were added in lanes 1 and 2 and in lanes 3 and 4. Protein from an equal fresh weight of cells was added in lanes 5 and 6. Lanes 3–6 represent total SDS soluble protein from dialyzed and lyophilized medium. Lanes 1 and 2 are the proteins that remain soluble during dialysis. Molecular weight markers are shown in lane 7.

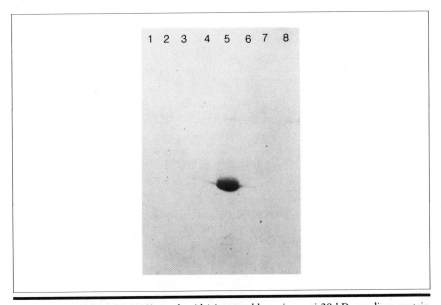

Figure 7.11. SDS-PAGE (10% acrylamide) immunoblot using anti-29-kDa medium protein of SDS soluble proteins from leaves of plants regenerated from adapted (S-25, lane 1) and unadapted (lane 2) cells; from total SDS soluble proteins from adapted (lane 3) and unadapted (lane 4) cultured cells; SDS soluble proteins from the medium of adapted (lane 5) and unadapted (lane 6) cells; and SDS soluble proteins extracted (0.2-M CaCl$_2$) from the washed walls of adapted (lane 7) and unadapted (lane 8) cells.

It is possible that much of the released hydroxyproline-rich protein is soluble extensin precursor. However, none of the major proteins separated by SDS-PAGE (Fig. 7.10) were hydroxyproline-rich, including the 29-kDa protein specifically released by adapted cells (data not shown). The hydroxyproline-rich glycoproteins probably do not enter the gel during electrophoresis.

F.   Protein Bound Ionically to Cell Walls

Walls of unadapted cells yielded slightly more protein when washed with either 0.2-M CaCl$_2$ or 3-M LiCl compared to adapted cells. However, the total protein released by CaCl$_2$ and LiCl washes of walls of adapted and unadapted cells was only a minor component of the protein of the cell wall (Table 7.4). The amino acid composition of total ionically bound protein differed between adapted and unadapted cells. Most striking was the dramatic reduction of hydroxyproline from the walls of adapted cells (Table 7.5). The lower hydroxyproline content of the ionically bound protein of adapted cells may reflect the loss of extensin precursor to the medium prior to deposition into the wall (Iraki *et al.*, 1989c). The high alanine content of this protein compared to unadapted cells suggests that the HRGP remaining

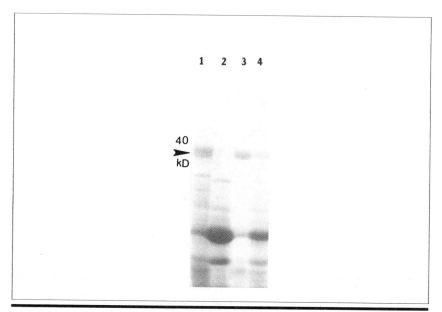

Figure 7.12.   SDS-PAGE (11% acrylamide) separation of proteins extracted with 0.2-M CaCl$_2$ (lanes 1 and 2) and subsequently with 3.0-M LiCl (lanes 3 and 4) from washed cell walls of unadapted (lanes 1 and 3) and adapted (S-25, lanes 2 and 4) cells.

on the walls of adapted cells is AGP (Iraki 1989c), typical of more dicots (Fincher *et al.*, 1983).

A 40-kDa protein was found as a major ionically bound protein on the walls of unadapted cells. This protein was almost undetectable on the walls of adapted cells (Fig 7.12). Antibodies raised against this protein in chickens were used to detect (Western blots) its presence in cell extracts and in the medium of adapted and unadapted cells. An immunoreactive protein of 40 kDa could be detected in cell extracts and the medium from both unadapted and adapted cells (Fig. 7.13). The absence of this protein from the walls of adapted cells could have been caused by its elution from the wall by the high salt medium. However, it appeared in the medium of unadapted cells although at lower levels than in the medium of adapted cells (Fig. 7.13). It may fail to bind to walls of adapted cells because of an alteration in the walls of adapted cells or in the protein itself.

Isoforms of the 40-kDa protein and the 29-kDa protein from the medium were detected after two-dimensional gel electrophoresis (Fig. 7.14). The 40-kDa protein is very basic with an isoelectric point (pI) near 10. Three isoforms were detected, one of which was present only in the medium from unadapted cells. Two very acidic isoforms of the 29-kDa protein were detected; one form appeared only in the medium of adapted cells.

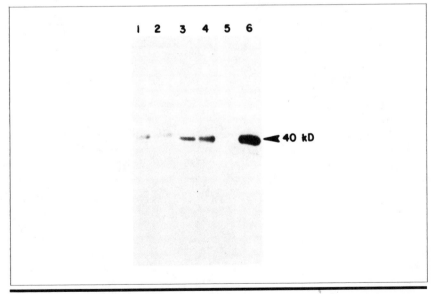

Figure 7.13. SDS-PAGE (10% acrylamide) immunoblot using anti-40-kDa ionically bound wall protein of total SDS soluble protein from adapted (S-25, lane 1) and unadapted (lane 2) cells; from the medium of unadapted (lane 3) and adapted (lane 4) cells; and from 0.2-M CaCl₂-extracted cell wall proteins from washed cell walls of adapted (lane 5) and unadapted (lane 6) cells. Equal amounts of protein were added to each lane.

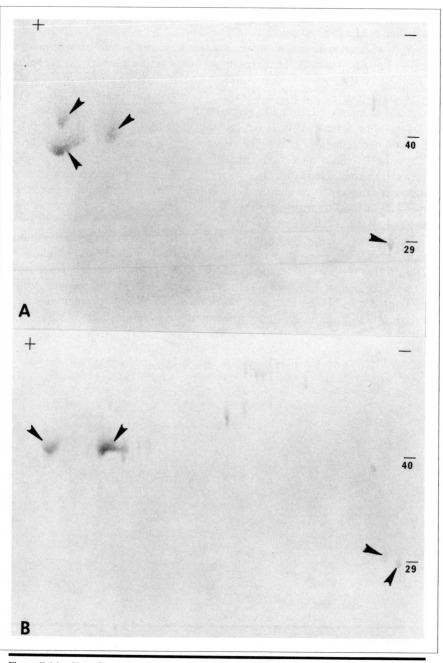

Figure 7.14. Two-dimensional separation of medium proteins of (A) unadapted and (B) adapted (S-25) cells on PAGE, followed by immunoblotting with first anti-40-kDa and then with anti-29-kDa IgY.

G. Enlargement of Cells in Leaves of Plants Regenerated from Salt-Adapted Cells

We reported earlier that plants regenerated from adapted cells have smaller leaves and are able to survive salt better than wild-type plants (Bressan *et al.*, 1987; Hasegawa *et al.*, 1986). We have provided evidence that indicates that the smaller leaves are a result of decreased cell division and a reduced rate of cell expansion (Hasegawa *et al.*, 1986). Thus, the reduced ability of adapted cells to expand is a stable characteristic that is transmitted to regenerated plants (Bressan *et al.*, 1987; Hasegawa *et al.*, 1986). Preliminary evidence that the small leaf phenotype is inherited by sexual progeny and may have a partial maternal basis of inheritance has also been reported (Bressan *et al.*, 1987). We now have further evidence of the heritability of the reduced leaf-area phenotype through three backcrosses to the wild-type parent (Fig. 7.15) and the continued occurrence of a maternal effect on its inheritance (Table 7.6). As seen in Fig. 7.15, the backcross-3 population continued to contain individuals with leaves that had smaller areas than those of any plants observed in the wild-type population.

IV. DISCUSSION

A. Reduced Cell Expansion

The failure of cultures of salt-adapted cells to gain fresh weight as rapidly as cultures of unadapted cells suggests that these cells expand more slowly than unadapted cells. However, a reduced rate of fresh weight growth could

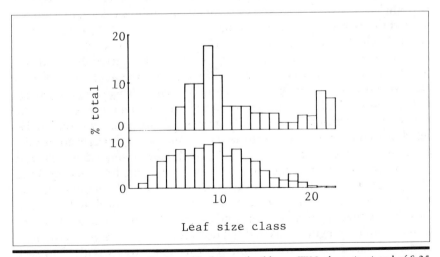

Figure 7.15. Population distribution of leaf sizes of wild-type W38 plants (top) and of S-25 × W38 BC₃ plants (bottom). Area categories (1–22) are in 50-cm² intervals and represent square centimeters per leaf.

Table 7.6.   Leaf Areas[a] of Fully Expanded Leaves of Wild-type W-38 Tobacco Plants and Third Backcross Plants of S-25 × W38

Genotype	Mean leaf area	Genotype	Mean leaf area[c]
W38	616	S-25 × W38	478
W38 × S-25 + S-25 × W38[b]	491	W38 × S-25	523

[a]Leaf area is cm^2/leaf.
[b]Reciprocal crosses were made in the third backcross. The first two backcrosses were with S-25 as the female parent.
[c]Each vertical pair of means is significantly different based on ANOVA results at the 0.05 level.

be caused by decreased cell division, which appears to occur in adapted cells and in leaves of plants regenerated from adapted cells (Bressan *et al.*, 1987). However, it is clear that reduced cell division does not account for the reduction in rate of fresh weight gain, since the average size of adapted cells increases much less rapidly than that of unadapted cells (Singh *et al.*, 1987a; Fig. 7.3). Since the slowly expanding salt-adapted cells always have as much or more turgor than unadapted cells, they appear to have substantially reduced cell wall extensibility. Extensibility is a poorly understood property of plant cell walls, and the biochemical basis of this change in extensibility is unknown.

B.   Alterations in the Physical Properties of the Cell Wall

In 1983 we observed that cells adapted to the osmotic stress provided by polyethylene glycol (PEG) partitioned more carbon toward osmotic adjustment at the expense of synthesis of the cell wall (Handa *et al.*, 1983); NaCl-adapted cells behaved similarly (Binzel *et al.*, 1987; Iraki *et al.*, 1989a). Decreased extensibility, or "stiffening," of the wall occurred despite diversion of carbon away from polymer synthesis. This apparent contradiction was explained by measurements of tensile strength. Using our nitrogen gas decompression method, we determined the pressure required to break the cell walls and calculated the tensile strength based on estimations of the cell dimensions (Iraki *et al.*, 1989a). Two important features were readily apparent from these data: (1) the tensile strength of the unadapted cells was two to five times that of the adapted cells, and (2) the breaking pressures of unadapted cell walls were always higher than the turgor pressures exerted by the cells (Iraki *et al.*, 1989a). Hence, the physical and chemical factors that govern tensile strength and extensibility may be quite different. Loss of tensile strength of adapted cell walls correlated with the loss of mass of the cellulose-extensin framework. This structure was sacrificed to save carbon (Iraki *et al.*, 1989a). The adapted cells remained locked into form, even though more than adequate turgor was maintained and wall strength was weakened. While it is clear to us that the extensin network plays little role in wall dynamics during expansion, the physical and biochemical bases for microfibril separation resulting in wall expansion are open to many other suggestions.

C. Biochemical Alterations in Cell Walls That May Affect Cell Expansion

There is considerable data supporting the hypothesis that xyloglucans constitute the principal load-bearing bonds and that enzymic cleavage of those bonds permits wall expansion (Cleland, 1981; Taiz, 1984). Additional data indicate that changes in the entire wall matrix, including alterations in the pectin gel conformation and interactions with proteins, also accompany hydrolysis of xyloglucans (Tepfer and Taylor, 1981; Terry and Bonner, 1980; Terry et al., 1981).

Although turnover of xyloglucans seems to be under auxin control, how that control is exerted is unknown. Studies with auxin-depleted sections have demonstrated that there are at least two separate responses to auxin: in one response the activities of nascent enzymes can be modified directly through physiological control, and a second response appearing much later that may be responsible for continued development in which gene expression is modified (Vanderhoef and Dute, 1981).

Notable is our observation that xyloglucan fragments expected from endoglucanohydrolase activity were missing in the medium of the adapted cells (Fig. 7.5 A and B). From our initial studies of wall "autolysis," the amounts of xyloglucan fragments released from adapted cell walls was much lower than those from unadapted cells, indicating that absolute activity was decreased (Iraki et al., 1989c). Because several enzymes might participate in turnover of the numerous wall polymers, further analyses of the digestion products have exploited both chemical analyses (such as methylation analysis) to determine the linkage structure of the released polymer and enzymic analyses, using "restriction endoglycosidases" (Carpita, 1987) to yield unique oligomers diagnostic of the larger polymer fragments released. For xyloglucan, methylation analyses yield the amounts of t-xyl, 4- and 4,6-glucose diagnostic of xyloglucan. Trichoderma cellulase (Kato and Matsuda, 1980b) or Driselase (Fry, 1986) yield the disaccharide, isoprimerevose, or α-D-xylosyl-$(1 \rightarrow 6)$-D-glucose, equally diagnostic of xyloglucan. Our "autolysis" products revealed that the release of oligomers containing xylose and glucose was substantially reduced in adapted cells. Methylation analyses of these oligomers confirmed that t-xyl, 4-glc, and 4,6-glc linkages predominated.

Even though hydrolysis of xyloglucans can be quantified specifically in autolysis experiments, the influence of NaCl on the expression and activities of specific enzymes can be masked by loss of unknown physiological controls on homogenization or by lack of sufficient native substrate. For this reason, the true activities of the endo-glucanases can be assayed only after extraction of the protein from the wall matrix and assay against purified substrate *in vitro*. In preliminary experiments, we have examined hydrolase activities of $CaCl_2$–extractable enzymes toward total tobacco cell hemicellulose (0.1–4.0M KOH extracts), a mixture of xyloglucan, arabinogalactan, and glucuronoarabinoxylan. Several hydrolytic activities were found as detected by an increase in reducing equivalents, but exo- versus endo-glycosidase activities, substrate specificities, and other properties have not been characterized.

D.   Changes in Insoluble Hydroxyproline-Rich Glycoprotein

Insolubilized hydroxyproline-rich glycoproteins (HRGPs) are thought to play an important role in the strengthening of plant cell walls (Wilson and Fry, 1986) and have been suggested to be important in regulating cell expansion rate, susceptibility to disease, and other processes (Wilson and Fry, 1986).

We observed earlier that the hydroxyproline content of the soluble protein from adapted cells was reduced almost 20-fold and indicated that this might result from either reduced amounts of extensin precursor or arabinogalactan protein (Singh *et al.*, 1985). Since walls of unadapted cells contain more HRGP and expand more, it appears that an increased HGRP content of the wall does not decrease its extensibility. However, we found that the tensile strength of cell walls of the adapted cells is decreased, which could be due to the decreased HRGP content of these walls (Iraki *et al.*, 1989a). Therefore, we have suggested that the major role of insolubilized HRGP in the cell wall may be to provide mechanical strength (Iraki *et al.*, 1989a; Singh *et al.*, 1989a). From these data, HGRP appears to have little involvement in the regulation of extensibility and the growth process except that expansion growth would be unlikely after extensive deposition of HRGP. However, a subpopulation of HRGP proteins may exist in the wall and be involved in decreasing cell wall extensibility. In addition, the degree or nature (inter- versus intra-molecular) of isodityrosine cross-linking may be altered. Finally, we cannot exclude the possibility that other proteins with other unknown cross-linking may be altered during adaptation and limit cell expansion.

Adapted cells fail to elongate (Table 7.1); this may represent an altered developmental pattern of adapted cells. Part of this developmental pattern may include the failure to deposit HRGP into the cell wall by the insolubilization process. Thus the genetic basis for both decreased cell expansion and decreased HRGP deposition may be found in an altered developmental program that manifests itself in a syndrome of physiological and biochemical changes.

E. Ionically Bound Cell Wall Proteins

Several researchers have investigated the possibility that ionically bound cell wall proteins may have some role in extensibility of the wall (Huber and Nevins, 1981; Melan and Cosgrove, 1988; Nevins *et al.*, 1987). Because of conflicting results using antibodies raised against ionically bound wall proteins to inhibit cell expansion, a role for these proteins in cell wall extension has not been firmly established. We proposed that decreased extensibility in adapted cells is directly related to impaired xyloglucan metabolism (Iraki *et al.*, 1989c), but the enzymic basis for this impairment has not been established.

We found some distinct differences in the occurrence of ionically bound proteins on the walls of adapted and unadapted cells (see Fig. 7.12; Singh *et al.*, 1989a). Even though it is possible that the dramatic absence of the 40-kDa protein from the walls of adapted cells may be the result of elution by the high-salt medium, this elution may still result in an inhibited extension process if this protein has an important function in controlling extension of the wall. A low-abundance enzyme, such as an endo-glucanase, could be displaced from the wall similarly. Judging from the Western blot of the 40-kDa protein (Fig. 7.13), where equal amounts of protein from the medium of adapted and unadapted cells resulted in the detection of nearly equal amounts of the 40-kDa protein, more of this protein is apparently released by the adapted cells since they release into the medium about seven times as much total protein per gram fresh weight of cells (Table 7.4). The antibody raised against the 40-kDa wall protein cross-reacts with a few smaller proteins that are also present in the medium and on the walls of adapted and unadapted cells. It is not known whether these proteins are derived from the 40-kDa protein or are only antigenically related.

One of the abundant proteins found ionically bound to the cell wall is osmotin (Singh *et al.*, 1989b). Since this protein is also released into the medium (Singh *et al.*, 1989b) and one isoform is very basic, it may equilibrate with binding sites on the wall after release from the cells.

F. Protein Released into the Medium

Adapted cells release considerably more protein into the medium than unadapted cells (Table 7.4) and some of these proteins are detectable at very low levels in the medium of unadapted cells. This includes a 29-kDa polypeptide, which is preferentially soluble during dialysis of medium proteins (Fig 7.10). Failure to detect the 29-kDa protein from the cells or walls of either adapted or unadapted cells with anti-29-kDa antibody (Fig. 7.11) may reflect a very low concentration of this protein in the cells. Possibly this protein is transported by vesicles to the plasma membrane and released without any intracellular accumulation. It may be incorporated into the insoluble portion of the wall of unadapted cells but not adapted cells and thus appear in the medium of adapted cells.

G.   Osmotic Stress-Induced Growth Reduction of Adapted Cells and Reduced Leaf Expansion of Regenerated Plants

It is clear that stress-induced reduction of turgor is insufficient to explain reduced growth rates caused by desiccation or salt stress.

We have presented evidence that plants regenerated from salt-adapted cells have leaf cells with constitutively lower rates of expansion (Bressan *et al.*, 1987). The leaves of these plants are smaller than wild-type leaves and the characteristic is inherited through several sexual generations (Bressan *et al.*, 1987, Fig. 7.15). Since the final size of the leaf cells is unchanged, we have concluded that there are fewer cell divisions and slower cell expansion, which results in smaller leaves with normal-sized cells. It seems that the smaller cell phenotype of adapted cells is a reversible change and the slower cell expansion characteristic becomes permanent in both adapted cells and plants regenerated from them (Bressan *et al.*, 1985, 1987). It is interesting to note that Van Volkenburg (1987) found that the leaf cells of plants treated with salt expanded more slowly but reached normal size. Van Volkenburg suggested that cell expansion and division are most likely coregulated. These observations could be explained if cells did not divide again until they regained a certain volume. Slower expanding cells would thus divide more slowly. If cell divisions occurred only over a fixed developmental period, the result of slower cell expansion would be fewer total cells that could expand to normal individual cell volumes but less total volume of the cell population. Salt-adapted cells do exactly this, as do the leaf cells of plants regenerated from them.

Whether reduced cell expansion rate is the result of a stress-induced change in cell wall metabolism remains to be proven. We have presented some evidence that cell wall and medium protein alterations are correlated with stress-induced growth reduction. Whether any of these protein changes persist in adapted cells removed from the stress or in plants regenerated from these cells remains to be seen. Since both adapted cells removed from stress and the regenerated plants retain reduced growth characteristics (Bressan *et al.*, 1987), the expression of these proteins in these cells and plants should prove interesting.

ACKNOWLEDGMENTS

We wish to thank the following persons for their excellent assistance: Jean Clithero, David Rhodes, Glenda McClatchey, Robert Rietveld, Marla Binzel, and Moshe Reuveni.

REFERENCES

Aspinall, G. O., Greenwood, C. T., and Sturgeon, R. J. (1961). *J. Chem. Soc.,* pp. 3667–3677.

Bacic, A., Harris, P. J., and Stone, B. A. (1988). *In* "The Biochemistry of Plants" (J. Preiss, ed.), Vol. 14, p. 297. Academic Press, San Diego, California.

Binzel, M. L., Hasegawa, P. M., Handa, A. K., and Bressan, R. A. (1985). *Plant Physiol.* 84, 118–125.

Binzel, M. L., Hasegawa, P. M., Rhodes, D., Handa, S., Handa, A. K., and Bressan, R. A. (1987). *Plant Physiol.* 84, 1408–1415.

Binzel, M. L., Hess, F. D., Bressan, R. A., and Hasegawa, P. M. (1988). *Plant Physiol.* 86, 607–614.

Binzel, M. L., Hess, F. D., Bressan, R. A., and Hasegawa, P. M. (1989). *In* "Biochemical and Physiological Mechanisms Associated with Environmental Stress Tolerance" (pp. 139–157). (J. H. Cherry, ed.). Springer-Verlag, Berlin (in press).

Boyer, J. S., Cavalieri, A. J., and Schulze, E.-D. (1985). *Planta* 163, 527–543.

Bozarth, C. S., Mullet, J. E., and Boyer, J. S. (1987). *Plant Physiol.* 85, 261–267.

Bressan, R. A., Handa, A. K., Handa, S., and Hasegawa, P. M. (1982). *Plant Physiol.* 70, 1303–1309.

Bressan, R. A., Singh, N. K., Handa, A. K., Kononowicz, A., and Hasegawa, P. M. (1985). *In* "Plant Genetics" (M. Freeling, ed.), pp. 755–769. Liss, New York.

Bressan, R. A., Singh, N. K., Handa, A. K., Mount, R., Clithero, J., and Hasegawa, P. M. (1987). *In* "Drought Resistance in Plants: Physiological and Genetic Aspects" (L. Monti and E. Porceddu, eds.), pp. 41–58. Comm. Eur. Commun., Luxembourg.

Briggs, L. J., and Shantz, H. L. (1914). *J. Agric. Res.* 3(1), 1–64.

Carpita, N. C. (1985). *Plant Physiol.* 79, 485–488.

Carpita, N. C. (1987). *In* "Physiology of Cell Expansion during Plant Growth" (D. J. Cosgrove and D. P. Knievel, eds.), pp. 28–45. Am. Soc. Plant Physiol., Rockville, Maryland.

Cavalieri, A. J., and Boyer, J. S. (1982). *Plant Physiol.* 69, 492–496.

Cleland, R. E. (1981). *Encycl. Plant Physiol., New Ser.* 13B, 255–273.

Condit, C. M., and Meagher, R. B. (1986). *Nature (London)* 323, 178–181.

Cooper, J. B., and Varner, J. E. (1983). *Biochem. Biophys. Res. Commun.* 112, 161–167.

Cooper, J. B., and Varner, J. E. (1984). *Plant Physiol.* 76, 414–417.

Cooper, J. B., Chen, J. A., VanHolst, G.-J., and Varner, J. E. (1987). *Trends Biochem. Sci.* 12, 24–27.

Cosgrove, D. J. (1986). *Annu. Rev. Plant Physiol.* 37, 377–405.

Darvill, J. E., McNeil, M., Darvill, A. G., and Albersheim, P. (1980). *Plant Physiol,* 66, 1135–1139.

Dreher, M. L., Weber, C. W., Bemis, W. P., and Berry, J. W. (1980). *J. Agric. Food Chem.* 28, 364–366.

Esquerre-Tugaye, M. T., Lafitte, C., Mazau, D., Toppan, A., and Touzé A. (1979). *Plant Physiol.* 64, 320–326.

Fincher, G. B., Stone, B. A., and Clarke, A. E. (1983). *Annu. Rev. Plant Physiol.* 34, 47–70.

Fry, S. C. (1980). *Phytochemistry* 19, 735–740.

Fry, S. C. (1982). *Biochem. J.* 204 449–455.
Fry, S. C. (1983). *Planta* 157, 111–123.
Fry, S. C. (1986). *Annu. Rev. Plant Physiol.* 37, 165–186.
Hagen, G., Kleinschmidt, A., and Guilfoyle, T. (1984). *Planta* 162, 147–153.
Handa, A. K., Bressan, R. A., Handa, S., and Hasegawa, P. M. (1982). *Plant Physiol.* 69, 514–521.
Handa, S., Bressan, R. A., Handa, A. K., Carpita, N. C., and Hasegawa, P. M. (1983). *Plant Physiol.* 73, 834–843.
Hasegawa, P. M., Bressan, R. A., Handa, S., and Handa, A. K. (1984). *HortScience* 19(3), 371–376.
Hasegawa, P. M., Bressan, R. A., and Handa, A. K. (1986). *HortScience* 21(6), 1317–1324.
Hayashi, T., Marsden, M. P. F., and Delmer, D. P. (1987). *Plant Physiol.* 83, 384–389.
Hsiao, T. C., and Jing, J. (1987), *In* "Physiology of Cell Expansion during Plant Growth" (D. J. Cosgrove and D. P. Knievel, eds.), pp. 180–192. Am. Soc. Plant Physiol., Rockville, Maryland.
Huber, D.J., and Nevins, D. J. (1981). *Physiol. Plant.* 53, 533–539.
Iraki, N. M., Bressan, R. A., Hasegawa, P. M., and Carpita, N. C. (1989a). *Plant Physiol.* 91 39–47.
Iraki, N. M., Singh, N. K., Bressan, R. A., and Carpita, N. C. (1989b). *Plant Physiol.* 91, 48–53.
Iraki, N. M., Bressan, R. A., and Carpita, N. C. (1989c) *Plant Physiol.* 91, 54–61.
Jarvis, M. C. (1984). *Plant, Cell Environ.* 7, 153–164.
Kato, Y., and Matsuda, K. (1980a) *Agric. Biol. Chem.* 44, 1751–1758.
Kato, Y. and Matsuda, K. (1980b). *Agric. Biol. Chem.* 44, 1759–1766.
Matsuda, K., and Riazi, A. (1981). *Plant Physiol.* 68; 571–576.
Melan, M. A., and Cosgrove, D. J. (1988). *Plant Physiol.* 86, 469–474.
Meyer, R. F., and Boyer, J. S. (1981). *Planta* 151, 482–489.
Nevins, D. J., Hatfield, R., Hoson, T., and Inouhe, M. (1987). *In* D. J. Cosgrove and D. P. Knievel, eds.), "Physiology of Cell Expansion during Plant Growth" pp. 122–132. Am. Soc. Plant Physiol., Rockville, Maryland.
Preston, R. D. (1974). "Physical Biology of Plant Cell Walls," pp. 327–382. Chapman & Hall, London.
Rackis, J. J., Anderson, R. L., Sasame, H. A., Smith, A. K., and Van Etten, C. H. (1961). *J. Agric. Food Chem.* 9, 409–412.
Rees, D. A (1977). "Polysaccharide Shapes. Outline Series in Botany Series." Chapman & Hall, London.
Reinhold, L., Braun, Y., Hassidiam, M., and Lerner, H. R. (1989). "Biochemical and Physiological Mechanisms Associated with Environmental Stress Tolerance" (J. H. Cherry, ed.). (pp. 121–130). Springer-Verlag, Berlin.
Rhodes, D., Handa, S., and Bressan, R. A. (1986). *Plant Physiol.* 82, 890–903.
Sachs, M. M., and Ho, T.-H. D. (1986). *Annu. Rev. Plant Physiol.* 37, 363–376.
Sadava, D., and Chrispeels, M. J. (1973). *Dev. Biol.* 30, 49–55.
Showalter, A. M. *et al.* (1985). *Proc. Natl. Acad. Sci. U.S.A.* 82, 6551.
Singh, N. K., Handa, A. K., Hasegawa, P. M., and Bressan, R. A. (1985) *Plant Physiol.* 79, 126–137.
Singh, N. K., Bracker, C. E., Hasegawa, P. M., Handa, A. K., Buckel, S., Hermodsen, M. A., Pfankoch, E., Regnier, F. E., and Bressan, R. A.(1987a). *Plant Physiol.* 85, 529–536.
Singh, N. K., LaRosa, P. C., Handa, A. K., Hasegawa, P. M., and Bressan, R. A. (1987b) *Proc. Natl. Acad. Sci. U.S.A.* 84, 739–743.
Singh, N. K., LaRosa, P. C., Nelson, D., Iraki, N., Carpita, N. C., Hasegawa, P. M., and Bressan, R. A. (1989a). *In* "Biochemical and Physiological Mechanisms Associated with Environmental Stress Tolerance" (J. H. Cherry, ed.). (pp. 173–194). Springer-Verlag, Berlin

Singh, N. K., Nelson, D. E., LaRosa, P. C., Bracker, C. E., Handa, A. K., Hasegawa, P. M., and Bressan, R. A. (1989b). In "Biochemical and Physiological Mechanisms Associated with Environmental Stress Tolerance" (J. H. Cherry, ed.). (pp. 67–87). Springer-Verlag, Berlin

Song, C.-S., Yu, J.-H., Bai, D. H., Hester, P. Y., and Kim, K.-H. (1985). *J. Immunol.* **135**, 3354–3359.

Spark, L. C., Darnborough, G., and Preston, R. D. (1958). *J. Text. Inst.* **49**, T309–T316.

Strafström, J. P., and Staehelin, L. A. (1986). *Plant Physiol.* **81**, 234–241.

Taiz, L. (1984). *Annu. Rev. Plant Physiol.* **35**, 585–657.

Tepfer, M., and Taylor, I. E. P. (1981). *Can. J. Bot.* **59**, 1522–1525.

Terry, M. E., and Bonner, B. A. (1980). *Plant Physiol.* **66**, 321–325.

Terry, M. E., Jones, R. L., and Bonner, B. (1981). *Plant Physiol.* **68**, 531–537.

Theologis, A., and Ray, P. M. (1982). *Proc. Natl. Acad. Sci. U.S.A.* **79**, 418–428.

Tierney, M. L., Wiechert, J., and Pluymers, D. (1988). *Mol. Genet.* **211**, 393–399.

Updegraff, D. M. (1969). *Anal. Biochem.* **32**, 420–424.

Vanderhoef, L. N., and Dute, R. R. (1981). *Plant Physiol.* **67**, 146–149.

Van Volkenburgh, E. (1987). In "Physiology of Cell Expansion during Plant Growth" (D. J. Cosgrove and D. P. Knievel, eds.), pp. 193–201. Am. Soc. Plant Physiol., Rockville, Maryland.

Walker, J. C., and Key, J. L. (1982). *Proc. Natl. Acad. Sci. U.S.A.* **79**, 7185–7189.

Wardrop, A. B. (1951). *Aust. J. Sci. Res., Ser. B* **4**, 391–416.

Wilson, L. G., and Fry, J. C. (1986). *Plant, Cell Environ.* **9**, 239–260.

Zimmerman, U. (1978). *Annu. Rev. Plant Physiol.* **29**, 121–148.

Zurfluh, L. L., and Guilfoyle, T. J. (1982). *Plant Physiol.* **69**, 332–337.

CHAPTER 8

Gene Expression during Adaptation to Salt Stress

John C. Cushman*
E. Jay DeRocher[+]
Hans J. Bohnert*[+]

[+]Department of Molecular and Cellular
Biology
University of Arizona
Tucson, Arizona

*Department of Biochemistry
University of Arizona
Tucson, Arizona*

I. INTRODUCTION

Plants experience many kinds of biological and physical stress. The cumulative effects of such stress determine species distribution, reproductive success in a given environment, and, in the case of crop plants, productivity. The physical environment imposes many different types of stress including high irradiance, heat shock, cold and freezing, flooding and anaerobiosis, heavy metals, and water stress in the form of drought and high or fluctuating ionic strength. In agriculture the cumulative effects are important from an economic standpoint, since the synergistic contributions of multiple stresses ultimately determine the productivity of major crop species.

Experimentally, as long as no integrative approach exists for attempting to understand these stresses, each factor needs to be targeted individually. Our knowledge of stress phenomena and, to a more limited extent, our understanding of molecular processes that occur in plants under stress has advanced dramatically over the last two decades. Most progress has been made in the molecular analysis of the response to light, heat stress, and anaerobiosis. Biochemical and genetic studies and especially molecular studies of water in the form of drought or salt stress have, however, received inadequate attention in relation to the important effect that they have on plant performance. We do not yet know what genetic makeup distinguishes some plants or varieties from others with respect to stress tolerance or stress resistance, or why some plants thrive in situations where others are eliminated.

In this short review we focus on the biochemistry of salt stress and on recent advances in using molecular techniques that describe how plants respond to drought and salt stress. Both of these stress factors evoke similar types of plant responses, because both share the element of osmotic stress, yet there are likely to be distinct differences in the response. The available literature in this area is slim, in fact, only over the last 5 years have reports of investigations using molecular biology techniques to study salt stress become available. Reviews on drought or salt stress with a physiological, biochemical, and metabolic focus have been published (Flowers *et al.*, 1977; Rains, 1979; Jefferies, 1981; Hanson and Hitz, 1982; Morgan, 1984; Greenway and Munns, 1980; Rhodes, 1987; Epstein and Rains, 1987; Cheeseman, 1988; Csonka, 1989; Cherry, 1989). The literature cited in these articles has been essential in writing this chapter, and we are indebted to these authors for their rationalization of phenomena. As with many areas in plant research, targeting responses to salt stress is complex because many different plants are investigated. While this cannot be avoided, we think that exemplary studies of complex mechanisms could enhance research if plant models were utilized more. We therefore include a short summary of work on specific salt-stress responses in the ice plant (*Mesembryanthemum crystallinum*). This plant, which has been studied extensively by several groups, can be considered a model for salt-stress specific responses.

II. DEFINITION OF THE PROBLEM

In this review we concentrate on approaches to studying and understanding responses on the level of gene expression to unphysiologically high or fluctuating concentrations of sodium (chloride) in plants. The term *unphysiological* refers to a value that is specific for a species or a cultivar but has no absolute value. Surely an outrageously high amount of NaCl will eliminate any plant; a variety of terms have been introduced to try to distinguish degrees of salt tolerance. Glycophytes perform best at very low levels of salt. If increases in salt over a narrow range are tolerated without an immediate loss in productivity, the plants are commonly termed *salt tolerant*. In contrast, halophytes, which are often subdivided into several more categories, perform better when some salt (mio-halophytes) or high amounts of salt (eu-halophytes) are present.

The mechanisms by which plants achieve salt tolerance are exceedingly complex and defy facile definition. Attempts to define salt or drought tolerance at the genetic level have been difficult because of the multigenic character of the phenotypes (Epstein *et al.*, 1980; Shannon, 1985). Breeding barley, wheat, and several other crop plants for increased salt tolerance has demonstrated the multigenicity of the trait (Norlyn, 1980; Dvorak *et al.*, 1985; Ramage, 1988). Varieties have been bred that tolerate salt levels approaching seawater strength; however, these are agronomically irrelevant because of a severe reduction in yield. There must be a fundamental difference between halophytes and glycophytes that still eludes genetic analysis. An alternative to breeding with extant crop species would be either to search for alternative crops and to develop their potential or to search for halophytic ancestors of crop plants and to combine the species characters (Dvorak *et al.*, 1985).

Attempts to define salt tolerance at the physiological level of the whole plant consider metabolism under salt stress as a problem of meganutrient physiology (Cheeseman, 1988). Cheeseman pointed out the difficulties that exist in defining *salt tolerance* as a term that can be helpful in studying mechanisms. Consequently, such an approach is restricted to monitoring (differences in) salt acquisition, transport, and distribution within the plant and partitioning within the cells and to relating these parameters to readjustments of other aspects of metabolism. Exposure to salt has been shown to lead to sodium uptake via the root cell plasmalemma to a variable degree (salt excluders versus includers or transporters). Several membrane transport systems—for example, ATP-dependent Na^+/H^+ transport systems—appear to be involved (Reinhold *et al.*, 1989) and an increasing number of data suggest that a large proportion of the salt is stored in the vacuoles of glycophytes and salt-tolerant plants (Binzel *et al.*, 1988). Growth characteristics of salt-adapted glycophytes are being studied (Singh *et al.*, 1989) in order to find factors that distinguish tolerant and sensitive plants. Various glands in many plant genera, including the glycophyte wheat and halophytic plants in the genus *Poaceae,* are involved in the excretion of salt (Oross and Thomson, 1982). Other plants, such as the obligate halophyte *Salicornia,* appear to maintain a stream of water through their vascular system by means of salt transport, extreme accumulation in the leaves, and excretion. While all plants show similar and specific reactions to desiccation, four metabolic responses that require more physiological study using the tools of molecular genetics stand out: (1) changes in carbon metabolism, (2) accumulation of compatible osmolytes, (3) the ability to partition sodium (or the rapid buildup of this ability), and (4) changes in energy metabolism and growth under conditions of salt stress and drought. It is likely that the efficacy of the response by any of these pathways determines the tolerant phenotype, however, it also appears to be equally likely that these metabolic responses could be consequences of an unknown molecular mechanism and not the cause for tolerance.

Studies of salt and drought tolerance at the level of responses of individual genes are complicated by the complexity of plant genetic responses that are regulated independently of the stress in a diurnal, developmental, or organ- or tissue-specific manner. The stress adds but one more parameter. The use of cells in culture has been an attempt to simplify analysis, although the measurement of cellular responses is potentially flawed due to genetic instability or the inability to correlate cellular responses with the whole plant aspect. Molecular biology provides a valuable set of tools for dissecting the tolerance phenotype in a reductionist manner. It is hoped that, once all the pieces are understood, the whole process leading to tolerance may be reconstructed and understood at a mechanistic level. To embark on such a task requires decisions, largely educated guesses, about which aspects of the phenomenon to focus on. Several groups have come up with different approaches. One avenue often taken is to compare the molecular responses that distinguish adapted and nonadapted glycophytes, either in whole plants or cells in culture. Metabolic pathways to be used as yardsticks for the isolation of genes and the regulation of their expression have been identified. A "blind approach" that has been used frequently is the search for, and isolation of, genes that are expressed only in the stressed state.

Another approach is to focus on the stress responses of halophytes, especially on the molecular responses of facultative halophytes; that is, plants that perform equally well under conditions of either low or high salt in the environment. Adaptation in such species is conceivably not a problem that is associated with genetic changes during adaptation, but rather an established or preformed option that has evolved in the species. The expression of genes for metabolic and developmental pathways that are known or surmised to be important in establishing a tolerance phenotype can be studied in the unstressed halophyte. Application of salt, while not strictly a stress for a halophytic species, should reveal changes at the gene expression level through which metabolic pathways, growth characteristics, and development are adjusted.

In this review we concentrate primarily on the metabolic responses commonly associated with salt stress, as outlined above, and on changes in the expression of genes modulated during stress. We discuss approaches to detect genes whose expression might be crucial for plant survival under salt stress and then outline one approach that emphasizes the use of a model system. In addition, we attempt to point out future areas that should be studied at the biochemical and molecular levels.

III. METABOLIC RESPONSES

A. Carbon Metabolism

Most plants have evolved some mechanisms to cope with a decrease in the availability of water, such as avoidance by rapid completion of ontogeny, exclusion or excretion of salt, intracellular partition or storage, tolerance, and even dependence. Increases in carbohydrate accumulation and in the composition of carbohydrates associated with salinity have been known for many years (Cheeseman, 1988). The production and accumulation of polyols, such as glycerol or sucrose, provides nonstructural carbon in the cytoplasm acting in osmotic adjustment and turgor maintenance. The provision of carbon to act as an osmoticum will be a cost for the organism, however, since this carbon is not available for anabolic reactions. In fact, the exclusive use of polyols for such purpose is restricted to a few organisms (e.g., *Dunaliella* sp.).

One adaptation to salt stress or drought stress is the use of alternate metabolic pathways that allow the plant to gain some additional advantage in survival either through more efficient carbon gain processes such as C_4 photosynthesis, which allows greater CO_2 fixation under conditions of high light and temperature (Nelson and Langsdale, 1989), or crassulacean acid metabolism (CAM), which allows plants to fix CO_2 at night when evaporative water loss is minimal. Both C_4 and CAM plants switch the primary mode of CO_2 fixation to the enzyme phosphoenolpyruvate carboxylase, PEPCase, which ultimately results in the concentration of carbon dioxide. Crassulacean acid metabolism may hold important implications for the analysis of osmotic stress tolerance in that it is a well-characterized physiological adaptation to water stress found in plants of tropical origin that typically grow in warm, dry climates (Kluge and Ting, 1978). The diagnostic characteristic of this pathway is diurnal fluctuation of organic acids (mainly malate) and reciprocal diurnal fluctuation of storage carbohydrates. CAM plants open their stomata at night when the majority of CO_2

uptake occurs and close them during the day to avoid excess evaporative water loss (Osmond and Holtum, 1981; Ting, 1985). A variety of plant species switch from a C₃ mode of photosynthetic carbon metabolism to CAM in response to various environmental stimuli (Ting and Rayder, 1982). In some plants this switch can be induced by salt or water stress (Winter and von Willert, 1972; von Willert et al., 1976a,b), by changes in photoperiod (Brulfert et al., 1982a), or as part of their normal developmental program (Jones, 1975; Brulfert et al., 1982b; Sipes and Ting, 1985). In *M. crystallinum* this metabolic transition is accompanied by substantial increases in the activity of a set of carbon metabolism enzymes of the glycolytic and CAM pathways (Holtum and Winter, 1982).

B. Compatible Osmolytes

Most organisms counter increases in extracellular salt concentration by intracellular increases in "compatible" osmolytes, which often accumulate to high concentrations in the cytoplasm and serve to adjust the osmotic potential of cells (Hanson and Hitz, 1982; Yancey et al., 1982). These osmolytes are low-molecular-weight compounds that are normal constituents of cell metabolism and are usually found in small amounts. Such accumulated compounds are amino acids such as proline and glutamine, modified amino acids such as glycine-betaine, and more complex biogenic amines (Hanson and Hitz, 1982; Csonka, 1989) and polyols, such as trehalose, glycerol, or inositol derivatives. In bacteria, such as *Salmonella* and *E. coli,* and yeast cells, the role of these osmolytes has been studied in detail (Csonka, 1989; Higgins et al., 1987; 1988; Rhodes and Handa, 1989). Accumulation of one or more of these osmolytes due to salt stress has been found in all kingdoms (Yancey et al., 1982; Flores and Galston, 1982; Csonka, 1989). The ability to accumulate these compatible solutes was once thought to be an indicator of salt tolerance (Flowers et al., 1977). While this may be the case, research conducted over the last 10 years suggests that the relationship between stress and the accumulation of any compound is complex. It may be that the accumulating compound is the end product of a pathway fulfilling a metabolic need, such as maintaining energy charge, rather than the necessity of accumulating the particular compound. The role of proline accumulation in adaptation to salt or water stress, for example, is the subject of some dispute. Proline has been strongly implicated as a compatible osmotic solute in bacteria on the basis of genetic evidence (Csonka, 1989). In plants direct genetic evidence for an osmotic role for proline is lacking. While some researchers have found that increased proline accumulation occurred in more salt-tolerant cell lines than in more sensitive lines (Handa et al., 1986), others have concluded that there is no correlation between proline accumulation and salt tolerance (Hanson and Hitz, 1982; Moftah and Michel, 1987; Chandler and Thorpe, 1987).

Comparison of proline biosynthesis in cell suspension cultures of halophytic and nonhalophytic origin has been performed for several halophytic and glycophytic species (Watad *et al.*, 1983; Handa *et al.*, 1986; Rhodes *et al.*, 1986; Binzel *et al.*, 1985). It appears that this biochemical response is a cellular trait that can be elicited in culture and that behaves similarly in whole plants (Rodriguez and Heyser, 1988; Treichel, 1986).

C. Compartmentation of Salt

New techniques such as X-ray microanalysis and compartmental efflux methods have added new insight into how cells organize their ionic compositions and deal with osmotic burdens. In the plants studied, sodium is excluded from the cytoplasm and cell organelles and is primarily sequestered in the vacuole. Other forms of compartmentation, such as export into the cell wall, appear to be of limited significance. In this way a gradient in osmotic pressure may be maintained that allows net water influx into cells. Tobacco cells in culture that have been adapted to growth in high NaCl for many generations accumulate NaCl to a concentration of approximately 800 mM in the vacuole, while cytoplasmic concentrations are maintained in the range of 100 mM (Binzel *et al.*, 1988). The cytoplasmic concentration of unadapted cells amounted, in contrast, to approximately 25 mM. This ability to partition NaCl appears to be a general ability of plants, whether they are halophytes or glycophytes. In the case of extreme halophytes, such as *Salicornia* or *Atriplex* (Braun *et al.*, 1986), which maintain high NaCl for growth, a different situation might be found. One view of the distinction between the two types of plants is the assumption that halophytes will have a constitutive genetic and molecular system in place to deal with high (or fluctuating) salt, while glycophytes would need selection or adaptation to deal with the environmental stress.

From this short discussion, it can be concluded that the study of membrane transport processes by which sodium is effectively sequestered is of utmost importance for the molecular description of salt tolerance. Sodium first has to pass through the plasma membrane. Maintaining high concentrations of sodium in the vacuole necessitates an additional transport system located in the tonoplast membrane. Transport systems through both membranes are ATP-requiring Na^+/H^+ transporters or antiporters whose biochemistry is being studied by several groups (Hassidim *et al.*, 1986; Blumwald *et al.*, 1987; Struve and Lüttge, 1987, Lai *et al.*, 1988; Randall and Sze, 1987; Bremberger *et al.*, 1988; Garbarino and DuPont, 1988; Wang *et al.*, 1989; Randall and Sze, 1989). Analysis is also being done at the nucleic acid level. Genes for subunits of both the plasma-membrane and the tonoplast ATPases are being characterized (Bowman *et al.*, 1988; Harper *et al.*, 1989; Pardo and Serrano, 1989). Bremberger *et al.* (1988) have shown that activity increases of the tonoplast ATPase in *M. crystallinum* are due to *de novo* synthesis of proteins of the enzyme complex and structural changes of the ATPase molecule. Further analysis of this enzyme, which is essential for salt partitioning and investigations of possible alterations in the composition of the tonoplast membrane, is urgently required. It will be of great interest to determine whether essential differences exist between ATPases of glycophytes and halophytes. Information about the primary structure of these important protein complexes and ion transporters may yield insight into possible differences in function. Channels and active systems mediating transport of ions across membranes are important, however, not only with respect to tolerance reactions of plants. Their involvement in overall growth and development is vital to understanding the whole plant (Schroeder and Hedrich, 1989).

Details of the physiological mechanisms, such as osmotic gradients between consecutive cell layers in tissues, are being studied using patch clamp techniques (Hedrich *et al.*, 1988). More biochemical investigations are needed to understand the structure of the tonoplast and plasma membranes, especially with respect to differences between membranes that distinguish plant species with different tolerances to water stress.

D. Hormonal Control

In many cases a correlation between water deficit in the form of drought, desiccation, or salt stress response and hormone metabolism has been observed. The plant hormone abscisic acid (ABA) plays a major role in modulating plant responses to water stress (Davies and Mansfield, 1983). The relationship between salt, water stress, or desiccation in the case of maturing embryos and levels of endogenous ABA has been known for some time (Quatrano, 1986). Physiological studies have shown that endogenous ABA levels increase in plant tissue subjected to water stress by high osmoticum, NaCl, or drying (Jones *et al.*, 1987). Under these stress conditions, specific genes are induced that may play roles in controlling intracellular osmolarity or other protective functions (Finkelstein and Crouch, 1986; Ramagopal, 1987a,b).

Recent experiments have sought to clarify the role(s) of ABA in relation to water stress. Abscisic acid has been found to accelerate the rate of adaptation to NaCl in tobacco cell suspensions (LaRosa *et al.*, 1985, 1987). It has also been observed that on exposure to ABA, cultured tobacco cells synthesize a 26-kDa protein (Singh *et al.*, 1987a). The expression pattern of this ABA-induced 26-kDa protein is transient unless cells are also exposed to NaCl stress. The ABA-induced protein is immunologically cross-reactive to 26-kDa proteins of several plants. It is assumed that ABA is involved in the normal induction of the synthesis of the 26-kDa protein and the presence of NaCl is necessary for the protein to accumulate.

The 26-kDa protein has been termed *osmotin* because it is synthesized and accumulated in cells undergoing gradual osmotic adjustment to salt or desiccation stress (Bressan *et al.*, 1985). Osmotin is preferentially localized to the vacuole within vacuolar inclusion bodies (Singh *et al.*, 1987b, 1989). An osmotin gene has also been isolated from tomato, which encodes a protein having a predicted product of 24 kDa (King *et al.*, 1988). The messenger RNA (mRNA) for this clone is induced by salt in tomato suspension cells during the late-log growth phase and is considerably more abundant in the roots of salt-stressed plants. The osmotin gene products from tobacco and tomato share 92% amino acid identity. Like osmotin from tobacco, the tomato gene product is localized to the cytoplasmic (vacuolar) fraction. It exhibits, as does the tobacco protein, an amino terminal signal sequence indicating that the protein is transported through a membrane to be targeted into the vacuole.

Several other genes that respond specifically to drought stress and ABA have also been identified. Mundy and Chua (1988) isolated a gene that is induced in rice plants under ABA treatment or water stress. This gene, named *Rab21* (Responsive to ABA), encodes a basic glycine-rich protein with a deduced two-domain structure with a predicted molecular weight of 16kDa. Transcripts for this gene can be induced in a variety of tissues upon treatment with 200-mM NaCl or 10-μM ABA or both. The effects of these two different stimuli are not cumulative, suggesting that they share a common response pathway. Induction of *Rab21* mRNA is rapid (less than 15 min) and does not depend on *de novo* protein synthesis, indicating that preformed nuclear–cytosolic factors mediate its responsiveness. An ABA-induced cDNA clone from barely aleurone layer cells encodes a protein of predicted mass of 22,000 with an apparent molecular weight of 27 kDa. The protein is lysine rich and, therefore, very basic (Hong *et al.*, 1988); its function is unknown.

Another ABA-inducible gene isolated from maize embryos encodes a mRNA that accumulates in epidermal cells. The gene encodes a 157 amino acid polypeptide with a predicted mass of 15,423 (Gomez *et al.*, 1988). The ABA inducibility of this gene is thought to reflect the normal pathway of water stress during seed dormancy accompanying seed desiccation. This increase in ABA levels in embryos before desiccation is part of a normal developmental program that ensures the embryo's survival by the expression of a series of genes whose products are thought to protect subcellular structures.

Normally during seed maturation, as ABA levels increase, the expression of yet another class of proteins increases. These are known as late embryogenesis abundant (LEA) proteins of unknown function. Several cDNAs for these genes have been isolated and characterized from developing seeds from cotton (Barker et al., 1988), wheat (Litts et al., 1987), and rape (Harada et al., 1989). Products of LEA genes are highly soluble. Considering that these proteins exhibit amphiphilic regions, computer modeling resulted in structure predictions that support self-assembly of (at least some) LEA proteins to form long rods, which may be important in mediating protection (Dure et al., 1989). However, the proposed mechanisms for protecting embryos against conditions of desiccation stress are not understood in detail. More recently, a set of cDNAs were isolated from barley and corn seedling aleuron undergoing dehydration (Close et al., 1989). These dehydration-induced genes encode proteins, termed *dehydrins,* which have conserved structural features among themselves and share features with *Rab21* from rice (Mundy and Chua, 1988) and with one of the cotton LEA proteins (D11) (Baker et al., 1988). In contrast, the dehydrins appear not to be closely related to the ABA-induced gene from barley aleurone (Hong et al., 1988), to osmotin from tobacco or tomato (Singh et al., 1989; King et al., 1988), or to most of the other LEA proteins from cotton embryos (Baker et al., 1988).

Loss of turgor via water deficit leads to ABA induction of gene expression that is dependent on transcription (Guerrero and Mullet, 1986). Loss of turgor leads to decreased growth rates and to increases and decreases in a number of specific *in vitro* translation products as analyzed by two-dimensional polyacrylamide gel electrophoresis (2D-PAGE) (Bensen et al., 1988). Abscisic acid-treated plants exhibited lower polysome content and reduced growth inhibition preceding the decline in polysome content (Bensen et al., 1988).

E.   Energy Metabolism, Growth, and Development

Only defined aspects of metabolic pathways needing adjustment have been targeted in some detail. Energy metabolism and especially cell growth, and the relation of overall plant growth with physical characteristics of the cell wall, have apparently been too complex to be amenable to biochemical analysis up to now.

The increased energy consumed by metabolic adjustments in reaction to salt stress may be deduced from the increases in respiration that have been documented (Veen, 1980) and from the decreases in yield that accompany stress conditions of crop plants (Flowers and Yeo, 1989). However, information relating metabolic cost, tolerance reactions, and the adaptation of energy-producing pathways is sketchy at best. Only certain aspects of costs, such as the changes in plant amino acid metabolism as related to stress-induced proline accumulation (Rhodes *et al.*, 1986; Rhodes and Handa, 1989), have been studied in depth. This aspect of salt stress requires much more study.

Growth is affected by salt stress (and water deficits) almost instantaneously (see a collection of articles in Cosgrove and Kniefel, 1987). Most crop plants exhibit drastically reduced growth rates. This was initially related to the loss of turgor and its accompanying relaxation of cell wall tension. However, the relationship appears to be more complex. Reductions in growth rate occur even without loss of turgor (Bressan *et al.*, 1982; Boyer *et al.*, 1985), and it appears likely that growth is actively controlled, independently of the sensing of turgor pressure, via changes in cell wall extensibility (Hasegawa *et al.*, 1984; Singh *et al.*, 1989). Such changes appear to be related to changes in the protein composition of cell walls (Bozarth *et al.*, 1987; Singh *et al.*, 1989; N. C. Carpita, pers. comm.).

Maintenance of turgor by osmotic adjustment is not the limiting factor in growth rates of salt-adapted tobacco suspension cells grown in NaCl (Binzel *et al.*, 1985). Maximum turgor occurs approximately at the onset of exponential fresh weight accumulation. Water content and fresh weight decrease with increasing levels of NaCl adaptation. Dry weight remains unchanged. The decrease in fresh weight was a result of the decrease in cell volume. Salt-adapted cells were better at gaining fresh weight and dry weight than unadapted cells in the same medium. The tolerance of unadapted cells and adapted cells was influenced by growth stage; the highest degree of tolerance was exhibited by cells in the exponential growth phase. The overall reduction in cell expansion with increasing adaptation to salt concentration was not a result of the cells' failure to maintain turgor, since some cells adapted to NaCl underwent osmotic adjustment in excess of the change in water potential caused by the addition of NaCl to the medium. Tolerance did not increase proportionately with an increase in turgor. Adaptation of these glycophytic cells to NaCl appears to involve mechanisms that result in an altered relationship between turgor and cell expansion.

Even less research has been targeted toward understanding the stress response during the course of plant development. Only certain aspects of plant development, such as seed imbibition, seedling emergence and establishment, induction of flowering, or timing of senescence, are known to be sensitive to osmotic stresses. In-depth research should especially be directed toward understanding a plant's changing response to stress during ontogeny.

To a varying degree, all metabolic parameters in a salt-stressed plant appear to require at least fine-tuning as the plant adapts to the environment. In no instance, however, has there been a report that conclusively indicates causes for tolerance or sensitivity, whether it is the primary stress perception, signal recognition and transfer, or the initial responses on the level of gene expression. What has been studied to date addresses what one might term consecutive *opportunistic* consequences of the tolerant plant or terminal consequences of the sensitive phenotype.

F. Patterns of Transcription and Translation

Plants change their patterns of transcription and protein synthesis during water stress due to either transcriptional changes in the utilization of genes or translational switches in the utilization of mRNAs. Numerous studies have suggested that gene expression is closely related to and controls or mediates the processes of turgor adjustment, cessation of cell elongation, and growth during water deficits. Recently, Mason *et al.,* (1988) found that water deficit caused changes in polysome composition and that the *in vitro* translation products from isolated polysomal mRNA changed as assayed by two-dimensional polyacrylamide gel electrophoresis (2D-PAGE). They postulated that the observed polysome disaggregation during water deficit could limit growth by reducing the rate of synthesis of a population of rapidly depleted labile "growth-limiting proteins." This work established that gene regulation is involved in both the maintenance of growth inhibition and recovery of growth. It is not known whether there are fundamental differences between glycophytes and halophytes with respect to such characters as mRNA utilization.

IV. EXPERIMENTAL APPROACHES TO STUDYING SALT-SPECIFIC GENE EXPRESSION OF PLANTS

A. Cells versus Whole Plants

As it became clear that the response of plants to high salinity and water stress was multifactorial, some researchers opted to work with cell suspensions cultures to simplify complex physiological mechanisms that could be difficult to study in organized tissues. It is argued that suspension cultures provide the opportunity to work with a relatively homogenous population of cells. This approach will limit the number of responses to only those that operate at the cellular level, but it may offer a means to focus on the physiological and biochemical processes inherent to the cell that contribute to salinity tolerance. In some instances it has been reported that salt-tolerant plants have been regenerated from cells that had been adapted to NaCl in culture (Nabors *et al.*, 1980; R. Bressan, pers. comm.). Ignoring whole plant processes involved in salt tolerance, there are many salt-tolerance mechanisms that can be considered at the cellular level alone. These include such processes as sequestering ions into the vacuole (Flowers *et al.*, 1977; Greenway and Munns, 1980; Wyn Jones and Gorham, 1983), synthesis and accumulation of osmotically compatible solutes (Watad *et al.*, 1983; Wyn Jones and Gorham, 1983), and the active exclusion of Na^+ and Cl^- from the cell (Ben-Hayyim and Kochba, 1983; Greenway and Munns, 1980). Many cell lines with enhanced tolerance to NaCl have been isolated from many glycophytic species (Rhodes *et al.*, 1986; Tal, 1983; Binzel *et al.*, 1985; Warren and Gould, 1982). Various physiological processes appear to contribute to cell adaptation to salinity; for example, NaCl-tolerant cells of tobacco accumulate Na^+ and Cl^- (Heyser and Nabors, 1981; Watad *et al.*, 1983), whereas salt-tolerant citrus cells accumulate less Na^+ and Cl^- than do salt-sensitive cells (Ben-Hayyim and Kochba, 1983). Such cellular mechanisms may be especially important to the study of nonhalophytic species, which lack some of the highly specialized anatomical structures found in many halophytes, such as salt glands, salt hairs, and epidermal bladder cells.

One argument for concentrating on the mechanisms of NaCl tolerance at the cellular level is the assumption that such mechanisms will be more amenable to genetic manipulation than the more complex mechanisms involved in cell–tissue interactions or unique morphological structures (Wyn Jones and Gorham, 1983). However, this argument appears questionable given that most cellular processes are complicated enough to remain beyond the grasp of genetic manipulation for the near future. Another argument for using suspension cells for studying the salt-tolerant phenomenon is that it can be difficult to separate cellular mechanisms of tolerance from those based on the use of anatomical structures or physiological specialization requiring the cell and tissue organization that exists in intact plants (Warren and Gould, 1982). A disadvantage of using salt-tolerant cells is that, while many reports claim stability of the salt-tolerant trait through prolonged periods of culture in the absence of NaCl (Nabors *et al.*, 1980), there are some reports where tolerance was only stable in some lines (Bressan *et al.*, 1985) or was unstable (Hasegawa *et al.*, 1980).

B. Protein Analysis

One approach to understanding the genetic basis of salt tolerance is to examine changes in the patterns of proteins expressed in plants or cell cultures that are exposed or adapted to high levels of NaCl. By comparing these patterns from unstressed–stressed or unadapted–adapted plants or cell lines in culture, proteins may be detected whose steady-state levels are modulated under stress. The appearance of proteins that are not present in the unstressed state or enhanced levels of preexisting proteins are measured and analyzed by one- or two-dimensional polyacrylamide gel electrophoresis.

One of the earlier experiments that attempted to identify proteins associated with salt stress compared the protein patterns of NaCl-adapted cultured tobacco cells to unadapted controls (Ericson and Alfinito, 1984). Three major hydroxyproline-containing proteins (20,26, and 32 kDa) were identified as being salt induced. The 20- and 32-kDa proteins were enhanced; the 26-kDa protein was apparently induced. In another study, salt- or PEG-adapted cultured tobacco cells produced several new or enhanced protein bands and other bands with reduced intensities. A total of eight proteins were induced, including a 26-kDa protein, which increased until it constituted about 10% of the total cell protein. During a later stage of culture growth beginning at midlog phase in unadapted cells and during adaptation to NaCl, a major 26-kDa (osmotin) protein accumulated that was suggested to be involved in adaptation of cultured tobacco cells to NaCl and water stress (Singh *et al.*, 1985).

Another group also identified a 26-kDa protein (osmotin) that accumulated in a *N. tabacum* suspension cell culture grown in NaCl containing a medium developed by Hasegawa *et al.* (1980). Using immunoblot analysis, it could be shown that this protein is also present in suspension cell cultures without salt as the cells approach stationary phase, but it does not accumulate to as high a level as in salt-adapted cells (King *et al.*, 1986). The protein accumulates in a variety of members of the *Solanaceae* family specifically in response to osmotic stress by various substances (PEG, KCl, NaCl) that lower the water potential (King *et al.*, 1986). Tomato species with different tolerances to NaCl do not show any difference in the amount of the 26-kDa protein compared to salt-sensitive species.

Given that investigations into the "molecular effects" of salt stress were scant at the time, Ramagopal (1987a,b) investigated changes in gene expression at the mRNA level in salt-tolerant and salt-sensitive lines of barley. Seedlings were exposed to NaCl, and mRNA was isolated from roots and shoots. The barley mRNA was used in a reticulocyte cell-free translation system and the *in vitro* translation products were resolved by 2-D PAGE. Stress triggered the differential transcription of mRNA specific to genotype and tissue. Unique root transcripts appeared preferentially in the salt-tolerant genotype, whereas accumulation of some other proteins appeared to be restricted to mRNA from shoots of the salt-sensitive genotype. These results suggested that both transcriptional and posttranscriptional mechanisms were at work regulating gene expression during salt stress. This was the first demonstration of a genotype-specific regulation of gene expression by environmental stress.

Hurkman and Tanaka (1987; Hurkman *et al.*, 1988) investigated the *in vivo* patterns of protein synthesis in barley roots due to 200-mM NaCl stress. Barley was used because it is the most NaCl tolerant of the major agricultural crops. Roots from intact plants were labeled *in vivo* and analyzed by 2D-PAGE. Although the patterns between control and NaCl-stressed plants were qualitatively similar, a number of proteins changed quantitatively. In particular, a pair of proteins of 26 and 27 kDa increased in roots: The 27-kDa protein was more heavily labeled in the microsomal fraction, whereas the 26-kDa protein was enriched in the soluble fraction. Both proteins returned to their normal control levels when plants were transferred to nonsaline conditions, suggesting that they may be involved in the salt-tolerant phenotype of barley. Antibodies raised to the barley root 26- and 27-kDa proteins did not cross-react with the 26-kDa protein from salt-adapted tobacco cell lines (Ericson and Alfinito, 1984; Singh *et al.*, 1985). In addition to the two major proteins found to increase in barley roots under salt stress, there were more than 46 visible differences (32 in the membrane fraction alone) between control and salt-treated plants (Hurkman and Tanaka, 1987; Hurkman *et al.*, 1988).

Even in salt-tolerant species, stress reduces growth rates. The apparent metabolic cost involved in the expression of the tolerance phenotype lets us assume that genes controlling tolerance may be tightly regulated on demand. Gulick and Dvorak (1987) compared the 2D-PAGE patterns of *in vitro* translation products from isolated mRNA to compare the mRNA populations present in roots, expanding leaves, and old leaves from the highly salt-tolerant artificial amphiploid arising from a cross between *T. aestivum* cv. Chinese Spring (salt sensitive) and *Elytrigia elongata* (salt tolerant) with those of the salt-sensitive parent before and after acclimation to high salt levels. Various mRNAs were induced, enhanced, or repressed, some differentially between the two cultivars, but only in root tissue. No differences were seen in gene expression between control and NaCl-treated plants in leaves, meristems, and unexpanded leaves of the amphiploid. It was postulated that the roots appeared to be the major site of change in gene expression in response to salt stress. Failure to detect any changes in expression in the shoots of both genotypes agrees with the findings of Storey *et al.* (1985) that these plants exclude salt from the shoots. It is possible that induced-shoot mRNAs were of such low abundance that they could escape detection on the autoradiograms. These types of experiments cannot answer the question of whether the gene products shown to be affected by salt stress are causally related to the amphiploid's ability to tolerate higher NaCl levels. They do establish that a certain limited number of genes are induced by salt stress and that these genes are differentially expressed in the two genotypes.

The main drawback to this type of approach is that the correlation of *in vitro* and *in vivo* profiles is difficult due to preferential utilization of mRNAs in the translation assays and due to differences in posttranslational maturation or modification. The most severe problem, however, may be that claims for induction of a protein, as opposed to mere enhancement of a protein that is present in the unstressed sample, are largely dependent on the quality of separation and the bias of the experimenter.

The techniques of *in vitro* translations and computer-assisted two-dimensional PAGE protein analysis have often been used to measure the extent of a stress response in terms of gross changes in mRNA populations. One of the best examples of this approach compared the effect of salt stress on the polypeptides and mRNAs from roots of two different barley cultivars that differed in NaCl tolerance (Hurkman *et al.*, 1989). The patterns of *in vivo*-labeled polypeptides or *in vitro* translation products were computer-analyzed to identify and quantitate changes in gene expression resulting from salt stress. There were no qualitative changes and the quantitative changes that did occur were relatively minor. The abundance of most mRNAs was not affected. Another category of transcripts changed transiently, either by diminishing for some time or by being enhanced. In a totally adapted plant, the abundance of these transcripts appears unchanged from the unstressed state. Finally, transcripts can be found that either increase or decrease with the stress and assume a new level of relative abundance. In this instance, because of the similarity in the two cultivars used, no specific polypeptides of translation products could be identified that related specifically to salt tolerance.

This technique, however, has some inherent flaws. First, heterologous translation systems such as wheat germ or rabbit reticulocyte lysates are being used. It is becoming increasingly obvious that the structure of the untranslated 5'-leaders and the untranslated 3'-ends of mRNAs are of crucial importance for the translatability of messages. It appears possible that the actual two-dimensional distributions of polypeptides are influenced by such subtle differences in the untranslated regions of mRNAs in various species. Another concern is mRNA abundance. One might expect that stress-induced changes in transcript abundance, which are detectable by *in vitro* protein analysis, are irrelevant to salt tolerance as they most likely involve abundant "housekeeping" genes, while the transcripts in which we might be interested might not be abundant. We have estimated from our own observations that we may be able to see *in vitro* translation products of mRNAs that are represented as 0.1–0.01% of all mRNAs. Rare mRNAs, even when their abundance is increased by a factor of 100, for example, may not be detected in such analysis. Another point to be emphasized is that the salt stress response of cells or plants is complex. One should not assume that reactions are only related to stress or tolerance adaptation mechanisms. Changes may be elicited that are related not to stress but to growth, such as changes in cell-cycle regulation. In fact, we have recently found that in cell suspension cultures of a halophytic species, *Distichlis spicata*, one response to increased NaCl (Zhao *et al.*, 1989) is the expression of genes that are normally tightly controlled during the cell cycle, for example, histone 2B (Z. Zhao, pers. comm.).

In summary, two-dimensional protein analysis is a valuable tool to observe gross changes in mRNA or protein abundance in response to salt stress. It is valuable to obtain a time frame for specific analysis of stress treatments, but it has the inherent limitation of daunting complexity and providing very little real knowledge about the role or function of the spots described on the gels.

C. Transcript Analysis

Protein analysis is increasingly being replaced by transcript analysis because the cloning of genes induced by salt stress appears to be the best strategy to learn more about the mechanisms of induction and to determine the molecular homology of the proteins induced between different organisms. cDNA libraries can be established from stressed and unstressed plants with ease. Such transcript libraries can be analyzed by differential screening, i.e., by probing replica filters of representative portions of cDNA libraries with isolated mRNA from both stressed and unstressed plants. In this way cDNAs have been isolated that represent mRNAs whose steady-state amount is either up-regulated, or down-regulated, or unchanged by the stress treatment. It is, however, important to remember that such gene identification is associated with similar problems encountered with the protein screening approach. The mRNAs are cloned according to their abundance. Furthermore, mRNAs might be present in the unstressed cells, but they are only translated upon stress treatment. Finally, mRNAs representing transcripts of different members of a gene family, where one transcript is specific for the stressed state, may not be distinguishable during screening.

A more promising approach is the use of so-called subtraction cDNA libraries (Davis *et al.*, 1984; Travis and Sutcliffe, 1988). Such libraries enrich for transcripts that are specific to a particular state. Excess mRNA from an unstressed state is used to subtract the first-strand cDNAs from mRNA from the stressed state by hybridization. In this way mRNAs that are present in both states can be separated as DNA–RNA hybrids from the single-strand cDNA by hydroxyapatite chromotography. The advantage of this approach is that a sublibrary is generated in which specific transcripts of the stressed state are enriched, thus allowing the isolation of rare mRNA molecules.

Yet another approach is direct genomic screening. Gene libraries may be screened randomly using mRNA from the two states to reveal differentially expressed transcripts. While this approach usually circumvents the problem of mRNA abundance, it is restricted to gene libraries of plants that are distinguished by relatively small genome sizes, such as *Arabidopsis thaliana* or *M. crystallinum*. With the ice plant, we are using this approach (G. Meyer, pers. comm.). The results, after screening a representative number of clones containing single-copy (genomic) DNA of the plant, suggest that in total more than 100 induced transcripts might be found, i.e., transcripts that cannot be detected in mRNA from unstressed plants.

D. Gene Identification and Functional Analysis

As more genes that respond to ABA, salt, or desiccation stress become known, comparisons of the primary and secondary structure of the proteins reveal similarities or partial identities between proteins of diverse origin. A high degree of evolutionary conservation between regions of such proteins suggests functional significance. One example is osmotin, which displays similarity to thaumatin (Edens *et al.*, 1982), to the TMV-induced pathogenesis-related protein of tobacco (Cornelissen *et al.*, 1986), and to the maize trypsin/ α-amylase inhibitor (Richardson *et al.*, 1987). The partial identities of these three proteins, which are all synthesized in response to stress-related conditions, suggest a common function for at least one domain of the protein. Although the exact function of some of the proteins remains to be defined, it was argued that they may play a role in plant defense strategies against a variety of environmental injuries including mechanical wounding (Cleveland and Black, 1982), microbial infections (Peng and Black, 1976), and insect attack (Green and Ryan, 1972). Such proteins have been described as a plant's "primary immune response," contributing to the overall defense of the plant by reducing the rates of endogenous mobilization of stored reserves (proteinase inhibitors) or by providing the enzyme systems for inhibiting invading pathogens (Ryan, 1973). Likewise, osmotin may function as a means of adjusting turnover rates of stored carbon compounds (e.g., organic acids in the vacuole). Given its localization to the vacuole and the fact that it is inducible by ABA, it may be a new type of desiccation-specific storage protein whose accumulation is linked to reduced growth and stress adaptation.

A comparable context appears to apply to the similarities between some other ABA and drought- or salt-induced proteins. LEA proteins and dehydrins exhibit highly conserved regions (Dure *et al.*, 1989). A subset of LEA proteins may have a fundamental role in plant responses to dehydration, although there is still no direct evidence that these proteins possess any biochemical or biophysical significance during desiccation of the cells in the maturing seed. Similarities with α-amylase and proteinase inhibitors suggest that these proteins may function in defense of the seed, preventing degradation of stored reserves by secreted enzymes from bacteria or fungi.

Yet another type of information can be obtained from the analysis of genes that are induced by the same (set of) environmental signals. Such genes must contain similarities in their structures with respect to the elements that control their expression. Common structures in promoters of functionally related or unrelated genes that respond to the same signal in different plants have been identified by sequence comparisons. Sequence identity has, in a few cases, been demonstrated to include functional identity. For example, the specificity of the promoter for *RbcS* genes is conferred by a small sequence element (Kuhlemeier *et al.*, 1987). When the *RbcS* promoter was supplied with a control element from a heat shock gene, this element imposed heat inducibility to the modified promoter (Strittmatter and Chua, 1987). Control elements of gene expression in plants have been implied for light induction, heat shock, anaerobiosis, tissue specificity and cell type, developmental programs, and hormone action such as auxin, gibberellins, and ABA. Although none of these control elements are fully functionally analyzed, some generalizations can be made.

Such genes are characterized by the presence of short-sequence elements, often in repetition, in their promoter regions. "Consensus" sequence elements have emerged for heat shock, light response, anaerobiosis, and auxin and ABA responses (McClure *et al.*, 1989; Mundy and Chua, 1988; for reviews, see Goldberg *et al.*, 1989; Okamura and Goldberg, 1989). The study of *cis*-acting elements that confer cell and tissue specificity (e.g., in the expression of chalcone synthase [CHS] [Schulze-Lefert *et al.*, 1989 and developmental specificity [Goldberg *et al.*, 1989]) is a topic of intense study. In other systems the function of these control regions of gene expression has in some cases been demonstrated experimentally. It appears that such elements are required for the recognition and binding of *trans*-acting factors, which then modulate expression by either repression or enhancement of transcription. Transgenic plants that express the modified gene or promoter elements of the normal context have been essential for these analyses.

V. MOLECULAR GENETICS OF SALT STRESS

In the previous sections we have attempted to outline important metabolic responses in the adaptation to salt, and we have tried to review the several lines of research that are taken to approach the problem with techniques of molecular biology. In the following section we present the areas to which more research should be devoted.

A.   Metabolic Control

Physiological analyses have indicated that a limited number of pathways are involved in the stress responses of plants. Most likely, however, many other pathways need to be adapted during stress. With what is known today, understanding such adaptive changes can be considered a problem of both quantitative genetics and comparative physiology.

The pathways that warrant attention at the molecular level are properties of the plasmalemma that control NaCl uptake in the roots and salt transport in the xylem; the nature and location of ion pumps in the plasmalemma and tonoplast and maintenance of ion selectivity and pH; control and pathways for synthesis of organic osmolytes; maintenance of photosynthesis and energy charge; and maintenance of growth, involving cell cycle control and the biosynthesis and regulation of cell wall components.

B.   Signal Transduction Mechanisms

One of the major questions of salt adaptation and tolerance is how plants perceive osmotic changes in their environments and what mechanisms they use to transfer signals from perception to initiate changes in gene expression. Recent experiments have shown that changes in turgor or extracellular osmotic state bring about physical and chemical changes in the plasmalemma, which result in the rapid activation of specific genes. Rapid loss of turgor brought about by exposing excised pea shoots to a stream of air results in the increase in several poly-A$^+$ mRNAs in wilted plants. These were not induced by heat shock or exogenously applied ABA even though ABA levels do increase 50 times 30 min after wilting (Guerrero and Mullet, 1988). Differential screening of a Lambda gt10 cDNA library, generated with mRNA from wilted shoots, identified several clones corresponding to rapidly induced poly A$^+$ mRNAs (Guerrero and Mullet, 1988). The link between the perception of changes in turgor and the activation of these genes is unknown at present.

Some evidence has been provided for the types of signal transduction mechanisms that may be operating when plants are exposed to salinity or desiccation stresses. In maize root protoplasts, it was demonstrated indirectly that salinity stress disturbs the intracellular Ca^{2+} content (Lynch *et al.*, 1987; Lynch and Läuchli, 1988). By measuring cytoplasmic Ca-activity with a fluorescent probe (Indo-1), Lynch *et al.* (1989) found that high NaCl concentrations caused immediate elevation of intracellular cytoplasmic Ca^{2+}-activity and presented evidence for the existence of a phosphoinositol regulatory system (Lynch *et al.*, 1989). Hypothetically, such a system would be an integral part of a Ca^{2+}–based signal transduction pathway that would link stress signals to the activation of gene expression via protein phosphorylation. There is direct evidence that inositoltriphosphate may serve as a second messenger for the mobilization of intracellular Ca^{2+} in higher plants (Rincon and Boss, 1987; Rincon *et al.*, 1989). Phosphatidyl serine activation of a plant protein kinase C (Elliot *et al.*, 1988) strongly suggests that a Ca^{2+} second messenger system operates in plant membranes and that it is important to plant hormone action (Elliot, 1986). The recent isolation of plant protein kinases (Lawton *et al.*, 1989) and the cloning of calmodulin from barley (Ling and Zielinski, 1989) provide further support for the existence of such a signal transduction pathway.

As discussed above, the role of secondary messengers, particularly Ca^{2+}, in regulating plant growth and development is established. It is also clear that changing parameters of growth and development must be preceded by alterations in the expression of genes. The relationship between environmental signals, changes in plant hormones, and subsequent changes in gene expression have been reported in only a few instances (Guilfoyle, 1986; Etlinger and Lehle, 1988; Morse *et al.*, 1987; McClure *et al.* 1989). Interpreting these experiments suggests that changes in light quality (Morse *et al.*, 1987) or changes in intracellular pH (Felle, 1989) may involve Ca^{2+} as a second messenger. However, these correlations up to now remain hypothetical. The relation between second messengers and gene expression has recently been reviewed (Guilfoyle, 1989). The possible candidates for such messengers in plants, such as Ca^{2+}, or cAMP, or phorbol esters, or signaling processes mediated by protein kinase C, are explored. Guilfoyle (1989) pointed to the need for focused research mainly in two areas: signal-transduction pathways elicited by environmental stress and *in vitro* soluble transcription systems that would provide functional assays (which have been essential in animal systems).

C. Regulatory Mechanisms of Gene Expression

A different though interrelated aspect is the study of molecular mechanisms that control gene expression, which finally result in metabolic adjustments. Metabolic responses have to be used to trace gene expression that had previously been difficult to trace because many of the biochemical reactions have not been studied. What information we have about the induced expression of genes under salt stress is largely molecular phenomenology.

It is presumed that salt stress acts via a chain of events: perception, signal transfer, and gene expression. Such a chain is conjecture, although examples from other biological systems, such as the action of glucocorticoid hormones in animals (Evans, 1988) whose effects are similar in many respects to salt stress responses make it likely that such a chain exists. The number of regulatory mechanisms is not known, nor is the sequence of mechanisms that leads from stress perception to, for example, the biosynthesis of osmo-lytes or changes in carbon metabolism.

D. Transcriptional Induction or Translational Control

While transcriptional activation of genes by environmental stresses has been shown, or deduced from the increases in steady-state levels of processed transcripts, analysis is far from sufficient or complete. For example, in-creases in the rate of transcription alone do not necessarily imply that the resulting mRNA is translationally utilized. In plants, chloroplast transcrip-tion and translation have been shown to be largely uncoupled (Gruissem, 1989) and similar relations have been found in animal systems. A number of control circuits appear to be operative between transcription and the appearance of mRNA in the cytoplasm: controls that affect maturation of primary transcripts, transport of transcripts through nuclear pores, and the formation of cytoplasmic ribonucleoprotein particles. Finally, translational arrest of ribosomes, or specific sequences in the untranslated regions of a mRNA have been found to affect mRNA half-life and, hence, protein syn-thesis. Control processes that occur in the nucleus (in plant and animal sys-tems) have been refractory to precise analysis because of technical difficult-ies. Callis et al. (1987) showed that the presence of introns and the type of intron present in plants influence gene expression. Since introns should be removed before mRNAs appear outside the nucleus, this result points to the realization of an intranuclear control mechanism. Control mechanisms that act on mRNA half-life in the cytoplasm have only recently received atten-tion. The influence of 5'- and 3'-leader sequences on mRNA stability or on translational efficiency, which ultimately determines plant gene expression (Okamura and Goldberg, 1989), has been reported (Gallie et al., 1989; An et al., 1989).

E. Plant Models

From the published data, a number of metabolic reactions have emerged that are modulated under stress. Their in-depth biochemical study is far from complete, and the involvement of gene expression requires much more attention. One is left however, with the impression that monitoring bio-chemical behavior and outlining response chains in gene expression, while necessary, is just phenomenology on a different level than in the past. What is needed is a plant model on which to concentrate in order to accumulate enough data to allow the targeting of mechanisms. We do not wish to imply that we can point out the ultimate model plant. Extensive work on salt and drought stress has been done with many plants. The plants most often used are probably *Atriplex,* barley, *Distichlis,* ice plant, rice, tobacco, tomato and wheat. However, certain aspects of salt tolerance have been looked at in virtually any crop plant.

The following requirements for a plant model appeared desirable for our own work. The plant should have a short lifetime, exhibit stress-specific physiological and biochemical reactions, and possess morphological or developmental stress markers that would aid in molecular analysis. Finally, the plant (or a close relative) should be salt tolerant. This latter characteristic was considered essential since it would ensure that physiological reactions to salt were analyzed, rather than senescence or prolonged dying. One plant emerged as a good candidate: the *facultative* halophyte, *M. crystallinum* (ice plant). The previously published work using this plant as an experimental system has been summarized (Bohnert *et al.*, 1988). Its physiological behavior under salt stress has been well studied (Winter and von Willert, 1972; Holtum and Winter, 1982; Monson *et al.*, 1983; Winter, 1985) with respect to carbon metabolism, accumulation of osmolytes (Demmig and Winter, 1986; Treichel, 1986), and tonoplast biochemistry (Struve and Lüttge, 1987; Bremberger *et al.*, 1988). Ecological studies have been performed in which the normal life cycle of the plant was described (Winter *et al.*, 1978). Recently, studies using NMR have focused on water mobility in the tissues (Walter *et al.*, 1989).

When stressed by addition of salt to the medium, by drought, or by cold, *Mesembryanthemum* plants reproducibly change their primary mode of carbon assimilation from C3 to CAM (Ting, 1985). We have used this switch and the concomitant changes in the activity and expression of specific enzymes as a yardstick to dissect the CAM induction process (Bohnert *et al.*, 1988). During the course of our work, we realized that the plants follow a developmental program that determines their stress response. The plants are characterized by a distinct growth pattern that is developmentally programmed. This pattern provides an excellent marker for recognizing the plant's physiological state.

The onset of stress and the buildup of the CAM pathway has been documented in detail. The enzymes indicative of CAM—namely phosphoenolpyruvate carboxylase (PEPCase), pyruvate, orthophosphate dikinase, and NADP malic enzyme—are induced *de novo* (Höfner *et al.*, 1987; Schmitt *et al.*, 1988; R. Höfner, per. comm.) within hours after the plants are stressed by NaCl. This is due to a significant increase in the steady-state amounts of mRNAs for these enzymes (Schmitt *et al.*, 1988). Phosphoenolpyruvate carboxylase mRNAs and genes have been studied in depth (Ostrem *et al.*, 1987; Vernon *et al.*, 1988; Michalowski *et al.*, 1989a; Cushman *et al.*, 1989). We have shown that PEPCase isogenes are differentially expressed in stressed and unstressed plants (Cushman *et al.*, 1989).

One gene, *Ppc2*, encodes a nonphotosynthetic PEPCase, which appears to carry housekeeping functions in roots and leaves, while *Ppc1* is the induced gene that accumulates predominantly in leaves and stems upon stress. Transcription assays in isolated nuclei from stressed and unstressed leaves indicated that the stress-induced transcription from the gene *Ppc1* is enhanced approximately 8-fold within 3 days of stress (Cushman *et al.*, 1989), demonstrating transcriptional enhancement. However, this has to be compared with the increase in steady-state amount of PEPCase mRNA, which is approximately 40-fold within this period. Consideration of the kinetics of reactions in series leads to the conclusion that the increase in steady-state mRNA by a factor greater than the increase in transcription rate can only be accounted for by a change in mRNA stability. In addition, transcription of *Ppc1* in unstressed plants is as high as that of all four (E. J. DeRocher, unpubl.) genes for the small subunit of Rubisco, *RbcS*, taken together, and yet the mRNA level for *Ppcl* is close to the limit of detection, while *RbcS* mRNAs amount to approximately 1% of all mRNAs. We deduce from these results that the posttranscriptional modification of primary transcripts may be as important in determining gene expression in response to stress as control mechanisms acting at the level of transcriptional activation. Some preliminary data have already been obtained (Vernon *et al.*, 1988) by measuring PEPCase mRNA and protein half-life after removal of salt from previously stressed plants. These experiments indicated a mRNA half-life of approximately 2.5 hr, while the protein half-life was 3 days. We used these "de-stress" experiments to show that CAM induction was under environmental control. However, similar experiments with plants stressed for shorter periods followed by removal of salt should provide data about posttranscriptional processes.

VI. PERSPECTIVES

Salt stress and, especially, drought already affect large portions of the agriculturally essential land on our planet. Although estimates of the actual number of acres affected vary widely (Kelley *et al.*, 1979; Epstein *et al.*, 1980), the necessity to farm land of marginal quality, mainly in arid climate zones, and the salinization of irrigated land (Norlyn, 1980; Jefferies, 1981) pose increasing problems. As we approach the next millenium, even moderately pessimistic projections assume that the deterioration of arable land and the competition for fresh water between urban populations and farmers will increase. Global changes in the climate, also brought about by human activity, are projected to compound the problem. This statement is true, irrespective of whether climatic changes due to the greenhouse effect have already started or whether these effects are still in the future. We need to know what the basic targets of desiccation stresses (salinity and drought) in plants are. We need to find out how naturally salt- or drought-tolerant species cope with such stresses. We need to study the molecular and genetic mechanisms that have evolved in adapted plants and to find out which mechanistic differences distinguish halophytes and drought-resistant plants from salt- and desiccation-sensitive species. Remnants of identified halophytic traits have to be studied in salt-sensitive plants, and if possible, their expression in crop plants has to be enhanced.

The question of what limits photosynthesis is probably best answered by the statement that it is water. Limits restraining photosynthesis affect productivity. In this respect it may be significant that in the ice plant model, photosynthesis appears to be unaffected by salt stress since we have not found any changes in the expression of mRNA and protein for ferredoxin-NADP$^+$-oxidoreductase (Michalowski *et al.*, 1989b). Although physiological measurements also point out that photosynthetic capacities are little affected in *Mesembryanthemum* (Winter *et al.*, 1978; Köster and Anderson, 1988) during stress, more studies are needed to corroborate or generalized this result to include all halophytes.

Molecular biology of salt stress is a relatively new research area. Only during the last decade have experimental expertise and the concepts matured enough to allow us to formulate hypotheses that can be tested. We list a few of those concepts, and we give—in the form of questions—a compilation of topics that we think should attract more groups to join this field in order to gain new insights. Salt stress elicits a plethora of responses that may be attempts to achieve a new metabolic equilibrium that confers tolerance when approached or reached. Equally likely, however, is that these responses are indicators of accelerated development, injury, or senescence. The sum of the responses and the speed with which the new equilibrium is achieved contribute to the complexity of the tolerance phenotype, much of which appears to be a problem of quantitative genetic contributions with synergistic effects. Yet halophytes exist and thrive under conditions of high salinity, which suggests that life under salt stress is not necessarily life on the edge.

1. Are there principal differences in metabolism between glycophytes and halophytes? Why are some plants dependent on high levels of salt?

2. What is the significance of accumulation of osmotically active substances? Are the genes that determine this metabolism induced or are they constitutively expressed or novel in some plants? How amenable to genetic engineering are these pathways? Do they provide an ecological advantage by being the basis for salt tolerance? Proline synthesis under stress, after

compensation for catabolism, for example, is linearly correlated with the solute potential (Rhodes and Handa, 1989), suggesting that solute sensing may be a signal.

3. How is energy charge and the ability of the plant to maintain high rates of electron transport and carbon dioxide assimilation correlated with (1) the tolerant phenotype, and specifically with (2) altered biochemical pathways, (3) specific proteins, (4) signaling and response mechanisms that are not expressed in sensitive plants?

4. To what extent is the ability of plants to cope with salt correlated with a particular hormonal response? Which hormones or ratios of hormones are involved? Do specific receptors in the response to salt share receptors for hormone recognition? What is the relation of salt perception with a hypothetical signal transfer chain and hormones?

5. Is salt perception a membrane-bound event, and if so, how is perception transmitted to chromosomal genes? Is this perception associated with a physicochemical event, such as salt concentration, pH, or energy charge, or is it a substantial entity, a protein, a hormone, carbohydrate, or ion as such?

6. Finally, we propose that it will be most important to determine at which levels gene expression is controlled—transcriptional or posttranscriptional. This will involve more study of the functional characterization of *cis*-elements and corresponding *trans*-acting factors and of the mechanisms of their activation. We hypothesize that the study of intranuclear events of the maturation of primary transcripts and their fate on transport into the cytoplasm might even be more important.

ACKNOWLEDGMENTS

Supported by USDA-CRGP-8712475 (HJB), Arizona Agricultural Experiment Station (#17441, HJB), and a NSF Postdoctoral Fellowship (DMB 8710662) awarded to JCC.

REFERENCES

An, G., Mitra, A., Choi, H. K., Costa, M. A., An, K., Thornburg, R. W., and Ryan, C. A. (1989). *Plant Cell* **1**, 115–122.
Baker, J., Steel, C., and Dure, L. (1988). *Plant Mol. Biol.* **11**, 277–291.
Ben-Hayyim, G., and Kochba, J. (1983). *Plant Physiol.* **72**, 685–690.
Bensen, R. J., Boyer, and J. S., Mullet, J. E. (1988). *Plant Physiol.* **88**, 289–294.
Binzel, M. L., Hasegawa, P. M., Handa, A. K., and Bressan, R. A. (1985). *Plant Physiol.* **79**, 118–125.

Binzel, M. L., Hess, F. D., Bressan, R. A., and Hasegawa, P. M. (1988). *Plant Physiol.* **86**, 607–614.

Blumwald, E., Cragoe, E. J., and Poole, R. J. (1987). *Plant Physiol.* **85**, 30–33.

Bohnert, H. J., Ostrem, J. A., Cushman, J. C., Michalowski, C. B., Rickers, J., Meyer, G., DeRocher, E. J., Vernon, D. M., Krueger, M., Vazquez-Moreno, L., Velten, J., Höfner, R., and Schmitt J. M. (1988). *Plant Mol. Biol. Rep.* **6**, 10–28.

Bowman, E. J., Tenney, K., and Bowman, B. J. (1988). *J. Biol. Chem.* **263**, 13994–14001.

Boyer, J. S., Cavalieri, A. J., and Schulze, E. D. (1985). *Planta* **163**, 527–543.

Bozarth, C. S., Mullet, J. E., and Boyer, J. S. (1987). *Plant Physiol.* **85**, 261–267.

Braun, Y., Hassidim, M., Lerner, H. R., and Reinhold, L. (1986). *Plant Physiol.* **81**, 1050–1056.

Bremberger, C., Haschke, H. P., and Lüttge, U. (1988). *Planta* **175**, 465–470.

Bressan, R. A., Handa, A. K., Handa, S., and Hasegawa, P. M. (1982). *Plant Physiol.* **70**, 1303–1309.

Bressan, R. A., Singh, N. K., Handa, A. K., Kononowicz, A., and Hasegawa P. M. (1985). *In* "Plant Genetics" (M. Freeling, ed.), pp. 755–769. Liss, New York.

Brulfert, J., Müller, D., Kluge, M., and Queiroz, O. (1982a). *Planta* **154**, 326–331.

Brulfert, J., Guerrier, D., and Queiroz, O. (1982b). *Planta* **154**, 332–338.

Callis, J., Fromm, M., and Walbot, V. (1987). *Genes Dev.* **1**, 1183–1200.

Chandler, S. F., and Thorpe, T. A. (1987). *Plant Physiol.* **84**, 106–111.

Cheeseman, J. M. (1988). *Plant Physiol.* **87**, 547–550.

Cherry, J. H., ed. (1989). "Environmental Stress in Plants." Springer-Verlag, Berlin.

Cleveland, T. E., and Black, L. L. (1982). *Plant Physiol.* **69**, 537–542.

Close, T. J., Kortt, A. A., and Chandler, P. M. (1989). *Plant Mol. Biol.* **13**, 95–108.

Cornelissen, B. J. C., van Huijsduijnen, R. A. M., and Bol, J. F. (1986). *Nature (London)* **321**, 531–532.

Cosgrove, D. J., and Kniefel, D. P., eds. (1987). "Physiology of Cell Expansion during Plant Growth." Am. Soc. Plant Physiol., Rockville, Maryland.

Csonka, L. N. (1989). *Microbiol. Rev.* **53**, 121–147.

Cushman, J. C., Meyer, G., Michalowski, C. B., Schmitt, J. M., and Bohnert, H. J. (1989). *Plant Cell* **1**, 715–725.

Davies, M. J., and Mansfield, T. (1983). *In* "Abscisic Acid" (F. T. Addicott, ed.), pp. 237–268. Praeger, New York.

Davis, M. M., Cohen, D. I., Neilsen, E. A., Steinmetz, M., Paul, W. E., and Hood, L. (1984). *Proc. Natl. Acad. Sci. U.S.A.* **81**, 2194–2198.

Demmig, B., and Winter, K. (1986). *Planta* **168**, 421–426.

Dure, L., Crouch, M., Harada, J., Ho, T. H. D., Mundy, J., Quatrano, R., Thomas, T., and Sung, Z. R. (1989). *Plant Mol. Biol.* **12**, 475–486.

Dvorak, J., Ross, Z., and Mendlinger, S. (1985). *Crop. Sci.* **25**, 306–309.

Edens, L., Helslinga, L., Klok, R., Ledeboer, A. M., Maat, J., Toonen, M. Y., Visser, C., and Verrips, C. T. (1982). *Gene* **18**, 1–12.

Elliot, D. C. (1986). *In* "Molecular and Cellular Aspects of Calcium in Plant Development" (A. J. Trewavas, ed.), pp. 285–300. Plenum, New York.

Elliot, D. C., Fournier, A., and Kokke, Y. S. (1988). *Phytochemistry* **12**, 3725–3730.

Epstein, E., and Rains, D. W. (1987). *Plant Soil* **99**, 17–29.

Epstein, E., Norlyn, J. D., Rush, D. W., Kingsbury, R. W., Kelley, D. B., Cunningham, G. A., and Wrona, A. F. (1980). *Science* **210**, 399–404.

Ericson, M. C., and Alfinito, S. H. (1984). *Plant Physiol.* **74**, 506–509.

Ettlinger, C., Lehle, L. (1988) *Nature* **331**, 176–178.

Evans, R. M. (1988). *Science* **240**, 889–895.

Felle, H. (1989). *In* "Second Messengers in Plant Growth and Development" (F. Boss and D. J. Morré, eds.), pp. 145–166. Liss, New York.

Finkelstein, R. R., and Crouch, M. L. (1986). *Plant Physiol.* **81**, 907–912.

Flores, H. E., and Galston, A. W. (1982). *Science* **217**, 1259–1261.

Flowers T. J., and Yeo, A. R. (1989). *In* "Environmental Stress in Plants" (J. H. Cherry, ed.), pp. 101–120. Springer-Verlag, Berlin.

Flowers, T. J., Troke, P. F., and Yeo, A. R. (1977). *Annu. Rev. Plant Physiol.* **28**, 89–121.

Gallie, D. R., Lucas, W. J., and Walbot, V. (1989). *Plant Cell* **1**, 301–311.

Garabarino, J., and DuPont, F. M. (1988). *Plant Physiol.* **86**, 231–236.

Goldberg, R. B., Barker, S. J., and Perez-Grau, L. (1989). *Cell (Cambridge, Mass.)* **56**, 149–160.

Gomez, J., Sanchez-Martinez, D., Stiefel, V., Rigau, J., Puigdomenech, P., and Pages, M. (1988). *Nature (London)* **334**, 262–264.

Green, T. R., and Ryan, C. A. (1972). *Science* **175**, 776–777.

Greenway, H., and Munns, R. (1980). *Annu, Rev. Plant Physiol.* **31**, 149–190.

Gruissem, W. (1989). *Cell (Cambridge, Mass.)* **56**, 161–170.

Guerrero, F. D., and Mullet, J. E. (1986). *Plant Physiol.* **80**, 588–591.

Guerrero, F. D., and Mullet, J. E. (1988). *Plant Physiol.* **88**, 401–408.

Guilfoyle, T. (1986). *CRC Crit. Rev. Plant Sci.* **4**, 247–276.

Guilfoyle, T. (1989). *In* Second Messengers in Plant Growth and Development (W. F. Boss and D. J. Morré, eds.), pp. 315–326. Liss, New York.

Gulick, P., and Dvorak, J. (1987). *Proc. Natl. Acad. Sci. U.S.A.* **84**, 99–103.

Handa, S., Handa, A. K., Hasegawa, P. M., and Bressan, R. A. (1986). *Plant Physiol.* **80**, 938–945.

Hanson, A. D., and Hitz, W. D. (1982). *Annu. Rev. Plant Physiol.* **33**, 163–203.

Harada, J., DeLisle, A., Baden, C., and Crouch, M. (1989). *Plant Mol. Biol.* **12**, 395–401.

Harper, J. F., Surowy, T. K., Sussman, M. R. (1989). *Proc. Natl. Acad. Sci. U.S.A.* **86**, 1234–1238.

Hasegawa, P. M., Bressan, R. A., and Handa, A. K. (1980). *Plant Cell Physiol.* **21**, 1347–1355.

Hasegawa, P. M., Bressan, R. A., Handa, S., and Handa, A. K. (1984). *HortScience* **19**, 371–376.

Hassidim, M., Braun, Y., Lerner, H. R., and Reinhold, L. (1986). *Plant Physiol.* **81**, 1057–1061.

Hedrich, R., Barbier-Brygoo, H., Felle, H., Flügge, U. I., Lüttge, U., Maathuis, F. J. M., Marx, S., Prins, H. B. A., Raschke, K., Schnabl, H., Struve, I., Taiz, L., and Ziegler, P. (1988). *Bot. Acta* **101**, 7–13.

Heyser, J. W., and Nabors, M. W. (1981). *Plant Physiol.* **67**, 720–727.

Higgins, C. F., Cairney, J., Stirling, D. A., Sutherland, L., and Booth, I. R. (1987). *Trends Biochem Sci.* **12**, 339–343.

Higgins, C. F., Dorman, C. J., Stirling, D. A., Waddell, L., Booth, I. R., May, G., and Bremer, E. (1988). *Cell (Cambridge, Mass.)* **52**, 569–584.

Höfner, R., Vasquez-Moreno, L., Winter, K., Bohnert, H. J., and Schmitt, J. M. (1987). *Plant Physiol.* **83**, 915–919.

Holtum, J. A. M., and Winter, K. (1982). *Planta* **155**, 8–16.

Hong, B., Uknes, S. J., and Ho, T. H. D. (1988). *Plant Mol. Biol.* **11**, 495–506.

Hurkman W. J., and Tanaka, C. K. (1987). *Plant Physiol.* **83**, 517–524.

Hurkman W. J., Tanaka, C. K., Dupont, F. M. (1988). *Plant Physiol.* **88**, 1263–1273.

Hurkman, W. J., Fornari, C. S., and Tanaka, C. K. (1989). *Plant Physiol.* **90**, 1444–1456.

Jefferies, R. L. (1981). *BioScience* **31**, 42–46.

Jones, H., Leigh, R. A., Tomos, A. D., and Jones, R. G. W. (1987). *Planta* **170**, 257–262.

Jones, M. B. (1975). *Planta* **123**, 91–96.

Kelley, D. B., Norlyn, J. D., and Epstein, E. (1979). *In* "Arid Land Plant Resources" (J. R. Goodin and D. K. Northington, eds.), pp. 320–334. Texas Tech Univ., Lubbock.

King, G. J., Hussey, C. E., and Turner, V. A. (1986). *Plant Mol. Biol.* **7**, 441–449.

King, G. J., Turner, V. A., Hussey, C. E., Wurtele, E. S., and Lee, S. M. (1988). *Plant Mol. Biol.* **10**, 401–412.

Kluge, M., and Ting, I. P. (1978). "Crassulacean Acid Metabolism" Ecol. Stud., Vol. 30. Springer-Verlag, Berlin.

Köster, S., and Anderson, J. M. (1988). *Photosynth. Res.* **19**, 251–264.

Kuhlemeier, C., Green, P. J., and Chua, N. H. (1987) *Annu. Rev. Plant Physiol.* **38**, 221–257.

Lai, S. P., Randall, S. K., and Sze, H. (1988). *J. Biol. Chem.* **263**, 16731–16737.

LaRosa, P. C., Handa, A. K., Hasegawa, P. M., Bressan, R. A. (1985). *Plant Physiol.* **79**, 138–142.

LaRosa, P. C., Hasegawa, P. M., Rhodes, D., Clithero, J. M., Watad, A. A., and Bressan, R. A. (1987). *Plant Physiol.* **85**, 174–181.

Lawton, M. A., Yamamoto, R. T., Hanks, S. K., and Lamb, C. J. (1989). *Proc. Natl. Acad. Sci. U.S.A.* **86**, 3140–3144.

Lazof, D., and Cheeseman, J. M. (1986). *Plant Physiol.* **81**, 742–747.

Ling, V., and Zielinski, R. E. (1989). *Plant Physiol.* **90**, 714–719.

Litts, J. C., Colwell, G. W., Chakerian, R. L., and Quatrano, R. S. (1987). *Nucleic Acids Res.* **15**, 3607–3618.

Lynch, J., and Läuchli, A. (1988). *Plant Physiol.* **87**, 351–356.

Lynch, J., Cramer, G. R., and Läuchli, A. (1987). *Plant Physiol.* **83**, 390–394.

Lynch, J., Polito, V. S., and Läuchli, A. (1989). *Plant Physiol.* **90**, 1271–1274.

Mason, H. S., Mullet, J. E., and Boyer, J. S. (1988). *Plant Physiol.* **86**, 725–733.

McClure, B. A., Hagen, G., Brown, C. S., Gee, M. A., and Guilfoyle, T. (1989). *Plant Cell* **1**, 229–239.

Michalowski, C. B., Olson, S. W., Piepenbrock, M., Schmitt, J. M., and Bohnert, H. J. (1989a). *Plant Physiol.* **89**, 811–817.

Michalowksi, C. B., Schmitt, J. M., and Bohnert, H. J. (1989b). *Plant Physiol.* **89**, 817–823.

Moftah, A. E., and Michel, B. E. (1987). *Plant Physiol.* **83**, 238–240.

Monson, R. K., Rumpho, M. E., and Edwards, G. E. (1983). *Planta* **159**, 97–104.

Morgan, J. (1984). *Annu. Rev. Plant Physiol.* **35**, 299–319.

Morse, M. J., Crain, R. C., and Satter, R. L. (1987). *Proc. Natl. Acad. Sci. U.S.A.* **84**, 7075–7078.

Mundy, J., and Chua, N. H. (1988). *EMBO J.* **7**, 2279–2286.

Nabors, M. W., Gibbs, S. E., Bernstein, C. S., and Meis, N. E. (1980). *Z. Pflanzenphysiol.* **97**, 13–17.

Nelson, T., and Langsdale, J. A. (1989). *Plant Cell* **1**, 3–13.

Norlyn, J. (1980). *In* "Genetic Engineering of Osmoregulation." (D. W. Rains, R. C. Valentine, and A. Hollaender, eds.), pp. 293–309. Plenum, New York.

Okamura, J. K., and Goldberg, R. B. (1989). *In* "The Biochemistry of Plants" (A. Marcus, ed.), Vol. 15, pp. 1–82. Academic Press, San Diego, California.

Oross, J. W., and Thomson, W. W. (1982). *Am. J. Bot.* **69**(6), 939–949.

Osmond, C. B., and Holtum, J. A. M. (1981). *In* "The Biochemistry of Plants" (M. D. Hatch and N. K. Boardman, eds.), Vol. 8, pp. 283–328. Academic Press, New York.

Ostrem, J. A., Olson, S. W., Schmitt, J. M., and Bohnert, H. J. (1987). *Plant Physiol.* **84**, 1270–1275.

Pardo, J. M., and Serrano, R. (1989). *J. Biol. Chem.* **264**, 8557–8562.

Peng, J. H., and Black, L. L. (1976). *Phytopathology* **66**, 958–963.

Quatrano, R. S. (1986). *Oxford Surv. Plant Mol. Cell Biol.* **3**, 467–477.

Rains, D. (1979). *In* "The Biosaline Concept" (A Hollaender, ed.), pp. 47–57. Plenum, New York.

Ramage, R. T. (1988). *In* "Chromosome Engineering in Plant Genetics and Breeding" (T. Tsuchiya, and R. K. Gupta, eds.). Elsevier, Amsterdam, (in press).

Ramagopal, S. (1987a). *Plant Physiol.* **84**, 324–331.

Ramagopal, S. (1987b). *Proc. Natl. Acad. Sci. U.S.A.* **84**, 94–98.

Randall, S. K., and Sze, H. (1987). *Plant Physiol.* **89**, 1292–1298.

Reinhold, L., Braun, Y., Hasidim, M., and Lerner, H. R. (1989). *In* "Environmental Stress in Plants" (J. H. Cherry, ed.), pp. 121–130. Springer-Verlag, Berlin.

Rhodes, D. (1987). *In* "The Biochemistry of Plants" (D. D. Davies, ed.), Vol. 12, pp. 201–241. Academic Press, San Diego, California.

Rhodes, D., and Handa, S. (1989). *In* "Environmental Stress in Plants" (J. Cherry, ed.), pp. 41–62. Springer-Verlag, Berlin.

Rhodes, D., Handa, S., and Bressan, R. A. (1986). *Plant Physiol.* **82**, 890–903.

Richardson, M., Valdes-Rodriguez, S., and Blanco-Labra, A. (1987). *Nature (London)* **327**, 432–434.

Rincon, M., and Boss, W. F. (1987). *Plant Physiol.* **83**, 375–398.

Rincon, M., Chen, Q., and Boss, W. F. (1989). *Plant Physiol.* **89**, 126–132.

Rodriguez, M., and Heyser, J. W. (1988). *Plant Cell Rep.* **7**, 305–308.

Ryan, C. A. (1973). *Annu. Rev. Plant Physiol.* **24**, 173–196.

Schmitt, J. M., Michalowski, C. B., and Bohnert, H. J. (1988). *Photosynth. Res.* **17**, 159–171.

Schroeder, J. I., and Hedrich, R. (1989). *Trends Biochem. Sci.* **14**, 187–192.

Schulze-Lefert, P., Becker-Andre, M., Schulz, W., Hahlbrock, K., and Dangl, J. L. (1989). *Plant Cell* **1**, 707–714.

Shannon, M. C. (1985) *Plant Soil* **89**, 227–241.

Singh, N. K., Handa, A. K., Hasegawa, P. M., and Bressan R. A. (1985). *Plant Physiol.* **79**, 126–137.

Singh, N. K., LaRosa, P. C., Handa, A. K., Hasegawa, P. M., and Bressan, R. A. (1987a). *Proc. Natl. Acad. Sci. U.S.A.* **84**, 739–743.

Singh, N. K., Bracker, C. A., Hasegawa, P. M., Handa, A. K., Buckel, S., Hermodson, M. A., Pfankoch, E., Regnier, F. E., and Bressan, R. A. (1987b). *Plant Physiol.* **85**, 529–536.

Singh, N. K., Nelson, D. E., Kuhn, D., Hasegawa, P. M., and Bressan, R. A., (1989). *Plant Physiol.* **90**, 1096–1101.

Sipes, D. L., and Ting, I. P. (1985). *Plant Physiol.* **77**, 59–63.

Storey, R., Graham, R. D., and Shepherd, K. W. (1985). *Plant Soil* **83**, 327–330.

Strittmatter, G., and Chua, N. H. (1987). *Proc. Natl. Acad. Sci. U.S.A* **84**, 8986–8990.

Struve, I., and Lüttge, U. (1987). *Planta* **170**, 111–120.

Tal, M. (1983). *In* "Handbook of Plant Cell Culture" Vol. 1 (D. A. Evans, W. R. Sharp, P. V. Ammirato, and Y. Yamada, eds.) Vol. 1, pp. 461–488. Macmillan, New York.

Ting, I. P. (1985). *Annu. Rev. Plant Physiol.* **36**, 595–622.

Ting, I. P., and Rayder, L. (1982). *In* "Crassulacean Acid Metabolism" (I. P. Ting and M. Gibbs, eds.), pp. 193–207. Am. Soc. Plant. Physiol., Rockville, Maryland.

Travis, G. H., and Sutcliffe, J. G. (1988). *Proc. Natl. Acad. Sci. U.S.A.* **85**, 1696–1700.

Treichel, S. (1986). *Physiol. Plant.* **67**, 173–181.

Veen, B. W. (1980). *In* "Genetic Engineering of Osmoregulation" (D. W. Rains, R. C. Valentine, and A. Hollaender, eds.), pp. 187–202. Plenum, New York.

Vernon, D. M., Ostrem, J. A., Schmitt, J. M., and Bohnert, H. J. (1988). *Plant Physiol.* **86**, 1002–1004.

von Willert, D. J., Treichel, S., Kirst, G. O., and Curdts, E. (1976a). *Phytochemistry* **15**, 1435–1436.

von Willert, D. J., Kirst, G. O., Treichel, S., and von Willert, D. J. (1976b). *Plant Sci. Lett.* 7, 341–346.

Walter, L., Balling, A., Zimmermann, U., Haase, A., and Kuhn, W. (1989). *Planta* 178, 524–530.

Wang, M. Y., Lin, Y. H., Chou, W. M., Chung, T. P., and Pan, R. L. (1989). *Plant Physiol.* 90, 475–481.

Warren, R., and Gould, A. (1982). *Z. Pflanzenphysiol.* 107, 347–356.

Watad, A. E. A., Reinhold, L., and Lerner, H. R. (1983). *Plant Physiol.* 73, 624–629.

Winter, K. (1985). *In* "Photosynthetic Mechanisms and the Environment" (J. Barber and N. R. Baker, eds.), pp. 329–387. Elsevier, Amsterdam.

Winter, K., and von Willert, D. J. (1972). *Z. Pflanzenphysiol.* 67, 166–170.

Winter, K., Lüttge, U., and Winter, E. (1978). *Oecologia* 34, 225–237.

Wyn Jones, R. G., and Gorham, J. (1983). *Encycl. Plant Physiol New Ser.* 12C, 35–58.

Yancey, P. H., Clark, M. E., Hand, S. C., Bowlus, R. S., and Somero, G. N. (1982). *Science* 217, 1214–1222.

Zhao, Z., Heyser, J. W., and Bohnert, H. J. (1989). *Plant Cell Physiol.* 30 861–867.

CHAPTER 9

Use of Two-Dimensional Gel Electrophoresis to Characterize Changes in Gene Expression Associated with Salt Stress of Barley

William J. Hurkman

U.S. Department of Agriculture
Agricultural Research Service
Western Regional Research Center
Albany, Calfornia

1. INTRODUCTION

Understanding the mechanisms of salt tolerance in plants is increasingly important as soil salinity becomes a greater problem in countries where agricultural land is dependent on irrigation. A high concentration of salt in the soil solution has adverse effects on plants because of reduced water potential, specific ion stress or toxicity, and ion imbalance or nutrient deficiency (Yeo, 1983), all of which reduce growth and decrease crop

Environmental Injury to Plants
Copyright © 1990 by Academic Press, Inc. All rights of reproduction in any form reserved.

productivity. Worldwide, approximately 230 million hectares are currently under irrigation and one-third of this number is affected by salinity (Epstein *et al.*, 1980, and references therein). Attempts to alleviate the progressive salt buildup in crop lands include the implementation of reclamation, drainage, and irrigation practices that minimize the salinity levels to which crops are exposed. However, even with highly sophisticated agrotechnology, increasing soil salinity remains a problem. For example, in California, 20% of the irrigated agricultural land is affected to some extent by salinity (Croughan and Rains, 1982). The need for costly engineering solutions to combat salinity may be reduced by the development of crop plants that are more salt tolerant, particularly since many major crop plants have relatively low salt tolerance. Epstein and associates (Epstein and Norlyn, 1977; Epstein *et al.*, 1980; Norlyn and Epstein, 1982) have expressed optimistic views on the efficient use of saline soils through the development of crop plants with improved salt tolerance used in conjunction with agrotechnology. To develop salt-tolerant crops, it is important to understand the mechanisms that plants use to cope with a saline environment and still maintain productivity. This entails understanding the effects of salt at the physiological, biochemical, and molecular levels. The problems associated with salt stress and the mechanisms that plants use to tolerate increased salinity have been extensively reviewed (Cheeseman, 1988; Downton, 1984; Epstein, 1980; Flowers *et al.*, 1977; Greenway and Munns, 1980; Hasegawa *et al.*, 1986; Hellebust, 1980; Levitt, 1980; Munns *et al.*, 1983; Pasternak, 1987; Pasternak and San Pietro, 1985; Poljakoff-Mayber, 1982; Poljakoff-Mayber and Gale, 1975; Rains, 1972; Staples and Toenniessen, 1984; Wainwright, 1980; Wyn Jones, 1981; Yeo, 1983). Rather than present a similar appraisal of the literature, this chapter focuses on one aspect of salt stress—its affect on gene products.

One approach to studying the mechanisms involved in salt tolerance is to use two-dimensional polyacrylamide gel electrophoresis (2D-PAGE) to identify polypeptides whose levels increase in salt-tolerant plants in response to salt stress. The assumption is that these polypeptides have a role in the plant's adaptation to salt stress. Protein synthesis responds dramatically to environmental stresses such as heat shock (Cooper and Ho, 1983; Key *et al.*, 1981) and anaerobiosis (Sachs *et al.*, 1980), where the synthesis of most proteins ceases and the synthesis of a new set of proteins is induced. A similar effect on protein synthesis had been observed for excision shock (Theillet *et al.*, 1982). For other environmental stresses the response is more subtle; water stress (Bewley *et al.*, 1983; Dasgupta and Bewley, 1984; Heikkila *et al.*, 1984), osmotic shock (Fleck *et al.*, 1982), wounding (Schuster and Davies, 1983b; Shirras and Northcote, 1984), cold acclimation (Cloutier, 1983; Guy and Haskell, 1988; Guy *et al.*, 1985), and salt stress (Gulick and Dvořak, 1987; Hurkman and Tanaka, 1987, 1988; Hurkman *et al.*, 1988, 1989; Ramagopal, 1987a,b, 1988) result in an increase in the synthesis of some proteins and a decrease in the synthesis of others, with or without an induction of unique stress proteins. This chapter is concerned with the use of 2D gels to identify polypeptides that increase with salt stress in barley and the use of related technologies to obtain the corresponding genes.

II. BARLEY AS AN EXPERIMENTAL SYSTEM

Barley is the most salt-tolerant grain of agricultural importance and has been grown successfully in fields salted out by previous irrigation practices (Epstein *et al.*, 1980; Norlyn and Epstein, 1982). Barley genetics and physiology have been extensively studied (Briggs, 1978), as has the relationship between salt tolerance and ion transport on both the cellular and whole-plant levels (reviewed: Greenway and Munns, 1980; Jeschke, 1984; Pitman, 1984). Improving the salt tolerance of crop plants depends on biological variability for this trait within the species of interest. Evaluation of 22,000 accessions of barley revealed wide genetic diversity for salt tolerance (Norlyn, 1980), indicating that well-chosen composite crosses may constitute a potential genetic resource for salt tolerance. Based on this information, Hurkman and Tanaka (1987) initiated studies to analyze the effect of salt stress on gene regulation in barley.

The effect of salt on polypeptide levels was examined in the roots (Hurkman and Tanaka, 1987), because they are the first plant organ that copes with increased soil salinity. Ion transport studies (Jeschke, 1984; Pitman, 1984) show that mechanisms are present in the roots that may contribute to the salt tolerance of barley. In order to study changes caused by salt in a system where results are not complicated by physiological processes such as photosynthesis and transpiration or by changes in light intensity and relative humidity, barley seedlings were grown in the dark at 100% relative humidity. To facilitate salt treatment, plants were grown hydroponically in a standard nutrient solution (Hurkman and Tanaka, 1987). In preliminary experiments (Hurkman and Tanaka, 1987), germination and growth of CM 72, a relatively salt-tolerant cultivar, was examined as a function of increasing salt concentration (0- to 400-mM NaCl). Germination decreased substantially, as did root and shoot growth, when seeds were sown at NaCl concentrations above 200 mM. All NaCl concentrations tested inhibited shoot growth, but root growth was stimulated by 50- and 100-mM NaCl.

Because 200-mM NaCl did not greatly reduce growth, this concentration was used to study the effect of salt on protein synthesis in barley roots (Hurkman and Tanaka, 1987).

III. TWO-DIMENSIONAL POLYACRYLAMIDE GEL ELECTROPHORESIS

A. Value in the Study of Salt Stress

In preliminary studies, no changes in polypeptide levels were found when polypeptides from the roots of control plants and plants treated with NaCl were examined on Coomassie-stained sodium dodecyl sulfate (SDS) gels (W. J. Hurkman and C. K. Tanaka, unpubl. obs.). Some changes were observed, however, when roots were labeled with [^{35}S]-methionine following salt treatment for 24 hr or 6 days (Fig. 9.1). The labeling increased in two bands (bands 3 and 4) and decreased in several others (bands 1, 5, 6, 8, 12). There were also some differences between the 6-day (lane 2) and 24-hr (lane 3) treatments. For example, the labeling of band 2 decreased at 6 days and increased at 24 hr relative to the control; the labeling of bands 7 and 9–11 did not change at 6 days but decreased at 24 hr. These results indicated that the changes in polypeptide labeling caused by salt were both quantitative and temporal. Because of the subtlety of the changes and the fact that a single band on a SDS gel can contain a number of polypeptides, an electrophoretic technique with higher resolution is essential for analysis of the changes in polypeptide levels caused by salt.

One powerful analytical tool for separating and detecting individual polypeptides is 2D-PAGE, a method where polypeptides are separated in the first dimension by differences in isoelectric points and in the second dimension by variations in molecular weight. The important advantage of 2D-PAGE is the ability to resolve simultaneously large numbers of polypeptides in complex protein mixtures; SDS-PAGE can resolve 100–200 polypeptide bands, whereas 2D-PAGE has the potential to resolve 1000–2000 polypeptides. It is a useful technique in various fields of plant biology such as genetic studies of seeds (Garcia-Olmedo *et al.*, 1988; Krochko and Bewley, 1988; Shewry *et al.*, 1988), analysis of gene expression during development (Bassett *et al.*, 1988; Goday *et al.*, 1988a,b; Hirsch *et al.*, 1989; Sáchez-Martinez *et al.*, 1986; Stafstrom and Sussex, 1988), hormone treatment (Broglie *et al.*, 1986; Chen and Leisner, 1985; Chory *et al.*, 1987; Christoffersen and Laties, 1982; Meyer *et al.*, 1984a,b; Schuster and Davies, 1983c; Theologis and Ray, 1982; Van der Zaal *et al.*, 1987; Zurfluh and Guilfoyle, 1980, 1982a,b,c and aging (Malik, 1987; Schuster and Davies, 1983a; Skadsen and Cherry, 1983). This technique is also useful in studies of plants' responses to biological (Cramer *et al.*, 1985; Somssich *et*

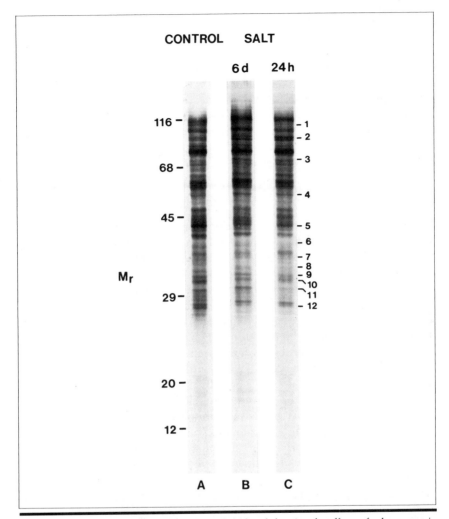

Figure 9.1. Fluorograph of an SDS polyacrylamide gel showing the effects of salt on protein synthesis. (A) Polypeptides in the roots of control plants grown for 6 days in the absence of salt. (B) Polypeptides in the roots of the plants grown for 6 days in the presence of 200-mM NaCl salt. (C) Polypeptides in the roots of plants grown for 5 days in the absence of salt and then transferred to a nutrient solution containing 200-mM salt for 24 hr. The polypeptides were labeled *in vivo* with [^{35}S]-methionine following the treatments. The numbered polypeptides are those whose levels changed with the salt treatments.

al., 1986; Traylor *et al.*, 1987; Wolpert and Dunkle, 1983) or environmental stresses such as wounding (Schuster and Davies, 1983b; Theillet *et al.*, 1982), heat shock (Cooper and Ho, 1983; Key *et al.*, 1981), cold stress (Guy and Haskell, 1988; Guy *et al.*, 1985; Mohapatra *et al.*, 1988; Perras and Sarhan, 1989; Robertson *et al.*, 1987; Sarhan and Perras, 1987), water stress (Bewley *et al.*, 1983; Bray, 1988; Damerval *et al.*, 1988; Dasgupta and Bewley, 1984; Heikkila *et al.*, 1984; Lalonde and Bewley, 1986; Mason *et al.*, 1988), and salt stress (Gulick and Dvořák, 1987; Hurkman and Tanaka, 1987, 1988; Hurkman *et al.*, 1988, 1989; Ramagopal, 1987a,b, 1988). The sensitivity of 2D-PAGE permits the identification of specific differences in the levels of individual polypeptides. Because 2D-PAGE can separate polypeptides with high resolution, another advantage is that it can be used to purify specific polypeptides from whole cell extracts for use in characterization studies (Aebersold *et al.*, 1987a; Kennedy *et al.*, 1988b).

B. Importance of Sample Solubilization

Despite its advantages, poor separation of plant proteins on two-dimensional (2D) gels has often prevented utilization of 2D-PAGE to its full potential. One reason is that 2D-PAGE involves a number of steps and there are numerous opportunities for technical errors. For example, the quality of reagents and ampholytes, sample preparation, equilibration of isoelectric focusing (IEF) gels and transfer of proteins between dimensions, the acrylamide composition of the second dimension, and the method of polypeptide detection can influence the resolution of 2D gels. Detailed descriptions of many of the potential problems have been reported previously (Dunbar, 1987a,b; Duncan and Hershey, 1984; Dunn and Burghes, 1983a,b; Garrels, 1979; Marlow *et al.*, 1988; O'Farrell, 1975). The inherent composition of plant tissue has been the greatest obstacle in obtaining 2D gels of high resolution. Plant cells have a relatively low protein content and contain a number of components such as phenolics, tannins, phytic acid, polynucleotides, polysaccharides, organic acids, pigments, and proteases (Granier, 1988; Marlow *et al.*, 1988) that interfere with protein extraction, solubilization, and IEF. Once the technical aspects of 2D-PAGE have been optimized, the single most important step in the procedure becomes sample solubilization.

Silver-stained 2D gels were used to evaluate the relationship of the solubilization method to gel quality. When lysis buffer (O'Farrell, 1975) was used to solubilize barley roots directly, the polypeptides were obscured by streaking, smearing, and background staining in the 2D gels. In addition, few polypeptides greater than 85 kDa were present in the gels, an indication that proteolysis had occurred during sample preparation. Roots were therefore homogenized in an extraction buffer containing SDS and phenylmethylsulfonyl fluoride (PMSF). The proteins were then precipitated and rinsed with acetone to remove the SDS prior to solubilizing the proteins in a urea buffer. Not only were a greater number of high-molecular-weight polypeptides present in the gels, but polypeptide spots were better resolved and background staining was reduced. The increase in high-molecular-weight polypeptides may have been due to the use of PMSF or the presence of SDS in the extraction buffer, or both. Initial extraction of wheat leaf proteins in an SDS buffer for 2D-PAGE significantly decreased proteolysis of the large subunit of ribulose biphosphate (bisP) carboxylase/oxygenase (Colas des Francs *et al.*, 1985).

One remaining problem was that although the SDS procedure reduced the background of 2D gels considerably, the gels still had an unacceptably high level of background staining. Recovery of proteins from a phenol phase following initial extraction with a high ionic strength buffer had been used to prepare plant proteins for analysis by 2D-PAGE (Schuster and Davies, 1983a; Wolpert and Dunkle, 1983). The 2D gels of barley polypeptides that were extracted and partitioned into phenol had improved resolution due in large part to low background staining. The polypeptide pattern obtained on 2D gels following the phenol method was similar to that obtained on gels of polypeptides that had initially been extracted with SDS. High-molecular-weight polypeptides were present in the gels, indicating that the phenol method prevented proteolysis to the same extent as SDS. The phenol extraction method also worked well in the analysis of membrane polypeptides by 2D-PAGE (Hurkman and Tanaka, 1986). Solubilization of plant membrane proteins for 2D-gel analysis has been particularly difficult (Hurkman and Tanaka, 1986, and references therein). The improved solubilization and enhanced resolution on IEF gels obtained when proteins were extracted by the phenol method resulted in the resolution of far more polypeptides in 2D gels of membrane fractions enriched in endoplasmic reticulum, tonoplast membrane, and plasma membrane from barley (Hurkman *et al.*, 1988) than was previously obtained for membrane preparations of other plants (Booz and Travis, 1980; Oleski and Bennett, 1987; Uemura and Yoshida, 1984, 1985). The phenol partitioning method has also been used to separate polypeptides of etioplasts, chloroplasts, and thylakoids of pea leaves (Dietz and Bogorad, 1987) and tonoplast of primary mary leaves of barley (Dietz *et al.*, 1988).

One conclusion that can be drawn from the solubilization studies is that when plant proteins are solubilized and loaded directly onto the first-dimension IEF gels, the second-dimension SDS gels often have low resolution. This is true (Hurkman and Tanaka, 1986) with O'Farrell's method (1975) and also with the method devised by Ames and Nikaido (1976), where proteins are first solubilized with SDS, then diluted with a buffer containing urea, NP-40, and mercaptoethanol, and loaded onto the IEF gels (e.g., Booz and Travis, 1980, 1981, 1983; Oleski and Bennett, 1987; Uemura and Yoshida, 1984, 1985). However, if the proteins are solubilized, precipitated out of solution, and then resolubilized with the urea buffer, as done in the SDS and phenol extraction procedures, high-resolution 2D gels can be obtained. This strategy worked well with green leaf tissue, a particularly difficult tissue from which to prepare proteins for 2D-PAGE. Leaves were ground to a powder in the presence of liquid nitrogen, extracted, and proteins precipitated with acetone (Colas des Francs *et al.*, 1985; Flengsrud and Kobro, 1989; Grainer, 1988; Marlow *et al.*, 1988) or ammonium sulfate (Cremer and van de Walle, 1985) prior to solubilization for IEF. Precipitation of proteins with TCA in acetone (Granier, 1988) also resulted in good 2D gels of leaf proteins. Apparently, these precipitation methods inactivate proteases and remove nonprotein components that interfere with the resolution of proteins during IEF.

C.   Computer Analysis of Two-Dimensional Gels

When 2D gels are used to examine polypeptide changes, two or more complex patterns must be compared and analyzed. Differences can be quantitative, where spot intensity varies, or qualitative, where a spot is absent from one pattern and present in another. Analyses of these quantitative and qualitative differences in polypeptide patterns are most often done by visual inspection of the 2D gels. This can be a formidable task considering the number of polypeptides per gel, the necessity for coordinating the results for different experimental conditions, and for autoradiographs, the need for multiple film exposures to compensate for the wide range of polypeptide levels. Visual analysis is useful for observations of major polypeptide changes, but computer analysis of 2D gels is a more accurate and objective method for quantitation of changes in polypeptide levels. The complex data of 2D gels can be analyzed to provide reliable information on synthesis, turnover, regulation, and modification of polypeptides. Computer analysis of 2D gels was used to identify and quantitate changes caused by salt stress at the polypeptide and messenger RNA (mRNA) levels in barley.

IV.   SALT STRESS IN BARLEY ROOTS

A.   Changes in Polypeptides and Translatable mRNAs

The effect of salt on polypeptide levels was examined in roots of CM 72, a relatively salt-tolerant cultivar. Polypeptides were labeled with [^{35}S]-methionine in roots of intact plants to eliminate the complications from a wound response that would occur if excised root segments were used. Polypeptides were labeled in control roots, in roots of salt-grown plants (grown with 200-mM NaCl for 6 days), in roots of salt-grown plants that were taken off salt (grown with 200-mM NaCl for 6 days, then transferred to a nutrient solution without salt for 24 hr), and in roots of salt-shocked plants (grown for 5 days without salt, then treated with 200-mM NaCl for 24 hr). Solubilized samples were sent to Protein Databases, Inc.[1] (Huntington Station, New York) for separation of polypeptides on 2D gels, fluorography, and subsequent matching and quantitation. Each of the fluorographs of the 2D gels contained over 800 polypeptides that had molecular weights ranging from 18,000 to 120,000 and isoelectric points (pIs) ranging from less than 4.4 to greater than 7.0 (Figs. 9.2 and 9.3). The computer analysis revealed that salt caused only quantitative changes and did not induce the synthesis of new polypeptides or cause polypeptides to disappear. Following salt treatment for 6 days, over 80 polypeptides increased and 500 decreased 1.5-fold or more. Following salt treatment for 24 hr, over 90 polypeptides increased and 480 decreased 1.5-fold or more. For the data in Figs. 9.2 and 9.3 and Tables 9.1 and 9.2, the analyses were limited to paired sets of polypeptides in which one polypeptide in each pair had 70 disintegrations per minute (dpm) or more and those polypeptides that increased or decreased 5-fold or more. Approximately 20 polypeptides met these criteria. Of these polypeptides, 3 polypeptides (numbers 3–5) increased and 5 polypeptides (numbers 14–18) decreased, whether plants were grown in the presence of salt for 6 days or treated with salt for 24 hr. When salt-grown plants were taken off salt for 24 hr, polypeptide labeling returned to control levels (Table 9.1, numbers 1, 4, 5, 9, 13) or remained depressed (Table 9.1, numbers 16, 17) or elevated (Table 1, numbers 2, 3).

Polypeptides with a molecular weight of 26 kDa and pIs of 6.3 and 6.5 (numbers 4 and 5 in Figs. 9.2 and 9.3) increased noticeably and consistently with salt treatment. They increased when plants were treated with salt for 6 days or for only 24 hr, and they decreased when salt-grown plants were transferred to a nutrient solution without salt, indicating that regulation of the genes corresponding to these polypeptides is influenced by salt stress. The pI 6.3 and 6.5 polypeptides were further characterized with respect to intracellular location and response to other stresses. Cell fractionation studies revealed that these polypeptides were present in a soluble and a microsomal fraction (Hurkman and Tanaka, 1987) and also in a cell wall fraction (Hurkman *et al.*, 1988). In the microsomal fraction, they appeared to be predominantly associated with and increased the most in the tonoplast fraction (Hurkman *et al.*, 1988). In the course of these fractionation studies, it was found that they were resistant to proteases, including chymotrypsin, trypsin, and protease K (Hurkman *et al.*, 1988). The increases in the pI 6.3 and 6.5 polypeptides were due specifically to NaCl treatment and were not a result of a general response to a decrease in water potential. Although the levels of a number of polypeptides increased or decreased with osmotic stresses induced by water deficit or treatment with PEG, the levels of the pI 6.3 and 6.5 polypeptides did not change (Hurkman and Tanaka, 1988). A heat shock treatment of 42°C for 1 hr did not cause increases in the labeling levels of these polypeptides (W. J. Hurkman and C. K. Tanaka, unpubl. obs.).

---

[1]Mention of a specific product name by the U.S. Department of Agriculture does not constitute an endorsement and does not imply a recommendation over other suitable products.

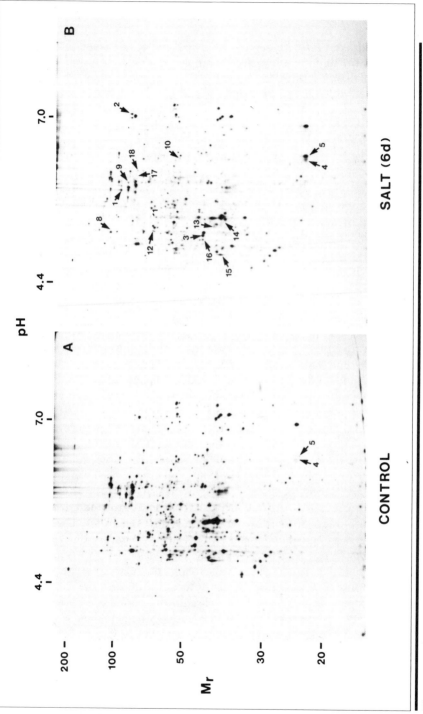

Figure 9.2. Fluorographs of 2D polyacrylamide gels showing the effects of salt on protein synthesis. (A) Polypeptides in the roots of control plants grown for 6 days in the absence of salt. (B) Polypeptides in the roots of plants grown for 6 days in the presence of 200-mM NaCl salt. The polypeptides were labeled *in vivo* with [^{35}S]-methionine following the treatments. The numbered polypeptides are those whose levels changed with the salt treatment; numbers 1–5 increased and 8–18 decreased fivefold or more. The quantitative data for these changes are listed in Table 9.1.

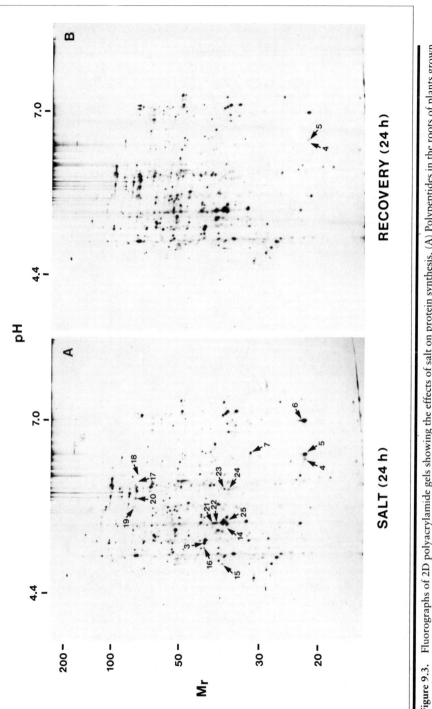

Figure 9.3. Fluorographs of 2D polyacrylamide gels showing the effects of salt on protein synthesis. (A) Polypeptides in the roots of plants grown for 5 days in the absence of salt and transferred to a nutrient solution containing 200-m*M* salt for 24 hr. (B) Polypeptides in the roots of plants grown for 6 days in the presence of 200-m*M* NaCl salt and transferred to a nutrient solution without salt for 24 hr. The polypeptides were labeled *in vivo* with [³⁵S]-methionine following the treatments. The numbered polypeptides are those whose levels changed with the salt treatment; numbers 3–7 increased and 14–25 decreased. Note also that numbers 3–5 increased and 14–18 decreased in salt-grown roots (see Fig. 9.2A). The quantitative data for the changes in the fluorographs are listed in Tables 9.1 and 9.2.

Table 9.1   Changes in Polypeptide Levels in Barley Roots Grown in Salt for Six Days[a]

Polypeptide number	Molecular weight	Isoelectric point	Disintegrations per minute		
			Control[b]	Salt[c]	Recovery[d]
		Polypeptides that increase			
1	93.0	6.06	28	173	48
2	78.6	7.19	38	190	244
3	45.4	5.49	48	345	328
4	26.0	6.30	25	170	67
5	26.0	6.50	11	287	1

Polypeptide number	Molecular weight	Isoelectric point	Disintegrations per minute		
			Control	Salt	Recovery
		Polypeptides that decrease			
8	107.0	5.53	120	18	75
9	80.2	6.18	248	1	192
10	51.2	6.56	96	11	59
11	55.5	5.78	120	1	53
12	67.9	5.53	194	1	71
13	42.5	5.62	180	1	221
14	39.6	5.68	101	1	30
15	40.4	5.16	73	1	30
16	45.6	5.43	674	123	118
17	77.5	6.27	72	5	2
18	77.0	6.35	135	1	19

[a]Plants were labeled for 3 hr *in vivo*, the polypeptides separated on 2D gels, and the fluorograph patterns matched and quantitated using computer-assisted analysis. Polypeptides are those that have increased or decreased fivefold or more; they are shown in the fluorographs depicted in Figs. 9.2A and B and 9.3B.
[b]Polypeptides in the roots of control plants grown for 6 days in the absence of salt.
[c]Polypeptides in the roots of plants grown for 6 days in the presence of salt.
[d]Polypeptides in the roots of plants grown for 6 days in the presence of salt and then transferred to a nutrient solution without salt for 24 hr.

The analysis of barley cultivars that differ in salt tolerance provides the opportunity to determine if polypeptides whose levels change with salt treatment are related genetically to salt tolerance. The response of CM 72, the relatively salt-tolerant cultivar, was compared with that of Prato, a salt-sensitive cultivar. In contrast to CM 72, where germination was not significantly inhibited by NaCl concentrations up to 300 mM (Hurkman and Tanaka, 1987), germination of Prato was significantly inhibited by concentrations of NaCl above 50 mM (Hurkman et al., 1989). Despite the substantial effect of NaCl on germination of Prato seeds, growth of the surviving plants was similar to that of CM 72 (Hurkman et al., 1989). Polypeptides were labeled in the roots of control and salt-grown plants, solubilized, and separated by 2D-PAGE. The dried gels were sent to Protein Databases, Inc. for matching and quantitation (Hurkman et al., 1989). The polypeptide patterns were qualitatively the same for CM 72 and Prato, a finding not

Table 9.2.   Changes in Polypeptide Levels in Roots of Salt-Shocked Barley Plants[a]

Polypeptides that increase

Polypeptide number	Molecular weight	Isoelectric point	Disintegrations per minute	
			Control[b]	Salt[c]
3	45.4	5.49	48	384
4	26.0	6.30	25	149
5	26.0	6.50	11	358
6	24.0	7.07	3	89
7	33.5	6.64	14	154

Polypeptides that decrease

Polypeptide number	Molecular weight	Isoelectric point	Disintegrations per minute	
			Control	Salt
14	39.6	5.68	101	1
15	40.4	5.16	73	1
16	45.6	5.43	674	66
17	77.5	6.27	72	1
18	77.0	6.35	135	5
19	81.6	5.96	147	28
20	81.1	6.06	313	53
21	42.8	5.74	540	101
22	42.1	5.74	445	65
23	40.1	6.15	108	1
24	39.2	6.16	190	34
25	39.7	5.80	152	1

[a]Plants were labeled for 3 hr. *in vivo*, the polypeptides separated on 2D gels, and the fluorograph patterns matched and quantitated using computer-assisted analysis. Polypeptides are those that have increased or decreased fivefold or more; they are shown in the fluorographs depicted in Figs. 9.2A and 9.3A.
[b]Polypeptides in the roots of control plants grown for 6 days in the absence of salt.
[c]Polypeptides in the roots of plants grown for 5 days in the absence of salt and then transferred to a nutrient solution with salt for 24 hr.

entirely surprising since both cultivars have common progenitors in their genetic backgrounds (Schaller *et al.*, 1977, 1979). The findings were complex (see Hurkman *et al.*, 1989) but can be summarized easily. Over 950 polypeptides had the same mobilities on the fluorographs of CM 72 and Prato. Following salt treatment, over 220 increased and 320 decreased in CM 72 and over 240 increased and 260 decreased 1.5-fold or more in Prato. The majority of polypeptide changes were specific for CM 72 or Prato, indicating that each cultivar had a different response to salt, a finding in keeping with the differences in their genetic backgrounds. No newly induced polypeptides were detected nor were there any polypeptides that disappeared with salt treatment in either cultivar. The pI 6.3 and 6.5 polypeptides that increased with salt treatment in CM 72 did not increase in Prato.

However, it is premature to speculate that the changes in the pI 6.3 and 6.5 polypeptides are correlated with salt tolerance, particularly since the incorporation of label into these polypeptides was greater in the control roots of Prato than in the control roots of CM 72.

Poly(A)$^+$ RNA was isolated from roots of control and salt-treated plants of CM 72 and Prato, and changes in translatable mRNA levels were assayed by *in vitro* translation of the poly(A)$^+$ RNA in a rabbit reticulocyte lysate system (Hurkman *et al.*, 1989). Computer analysis of the translation products showed that over 850 polypeptides for CM 72 and Prato had the same mobilities on the fluorographs. As was found *in vivo*, a large number of these polypeptides increased or decreased in each cultivar with salt treatment. In CM 72, over 140 increased and 460 decreased and in Prato, over 300 increased and 200 decreased 1.5-fold or more. The majority of the translation products whose levels changed with salt treatment were specific for CM 72 or Prato. No polypeptides were newly induced by salt and no polypeptides disappeared.

B.   Interpretations of Changes

The computer analyses illustrate the complexity of the salt stress response in barley roots. Salt treatment causes a large number of quantitative changes in polypeptide and translatable mRNA levels in CM 72 and Prato; overall, the changes are relatively small. The numerous changes in polypeptide and mRNA levels no doubt reflect changes in plant growth, water relations, ion transport, and cell metabolism that are induced by salt stress (Bohnert *et al.*, 1988; Greenway and Munns, 1980; Jeschke, 1984; Munns *et al.*, 1983). It will be difficult to determine which changes caused by salt represent primary or secondary responses and whether specific changes are related genetically to the trait of salt tolerance in barley. The comparison of CM 72 and Prato demonstrates the similarity of these two cultivars with respect to salt tolerance and also illustrates the problem of relying on germination as the criterion for salt tolerance at the seedling stage of development. Although salt substantially inhibited seed germination of Prato, the plants that survived were nearly as salt tolerant as those of CM 72. This demonstrates that the seed population of Prato is heterogeneous for the trait of salt tolerance and that salt treatment selected for seedlings of Prato as salt tolerant as those of CM 72. In addition, comparison of the salt response in the two cultivars is probably complicated by the age and water status of the roots. Salt treatment could alter the physiological age of the two cultivars to the extent that the observed changes could be due, in part, to differences in developmental states. The possibility that the specificity of the responses of the two cultivars to salt could be due to differences in intracellular water potentials cannot be overlooked.

The changes in polypeptide and translatable mRNA levels indicate that gene regulation is altered by salt stress. However, the specific mechanisms are not known and cannot be determined by 2D-PAGE alone. Whether the alterations in protein synthesis are due to changes in the efficiency of mRNA translation (initiation and elongation rates, amount of mRNA associated with ribosomes or polysomes) or to regulation of mRNA transcription, processing, transport, or stability or to altered rates of protein degradation requires information from additional experiments.

The possibility that salt inhibits or stimulates the translation of mRNAs to varying degrees was studied *in vitro*. Translation of mRNAs showed that protein synthesis was inhibited by Na^+ and Cl^- and that amino acid incorporation was inhibited to the same extent for mRNAs isolated from salt-tolerant and salt-sensitive plants (Gibson *et al.*, 1984). It was also found for the mRNAs isolated from the salt-tolerant plants that a specific population of cytoplasmic mRNAs was not preferentially translated in the presence of Na^+ and Cl^-. Gibson *et al.* (1984) concluded that the abundant mRNAs from salt-tolerant and salt-sensitive species are equally sensitive to the limitations imposed by ions on the initiation reactions of protein synthesis. It is interesting to note that the activities of many enzymes from glycophytes or halophytes are inhibited to a similar extent *in vitro* (Flowers, 1972; Flowers *et al.*, 1977; Greenway and Osmond, 1972; Pollard and Wyn Jones, 1979). Whether levels of Na^+ and Cl^- similar to those tested *in vitro* inhibit protein synthesis and enzyme activity *in vivo* is not known. There are mechanisms, including the nonspecific buffering model postulated by Richey *et al.* (1987), that allow protein-DNA interactions *in vivo* even in the presence of high levels of salt. These processes depend on the nature and concentration of electrolyte ions and occur *in vitro* only within a narrow range of salt concentrations and pH (Richey *et al.*, 1987, and references therein). In studies on the interactions of *E. coli* lac repressor with lac operon mutants and on the interactions of RNA polymerase with various promoters, Richey *et al.* (1987) found that formation of functional protein-DNA complexes *in vivo* was weakly dependent, if at all, on intracellular ion concentration. Further, they showed, by calculations based on a thermodynamic model of the lac operon, that protein bound to nonspecific binding sites for lac repressor could reduce the salt sensitivity of specific binding at the operon site.

C. Comparison to Other Studies

Relatively few studies have investigated the effect of salt stress at the protein level. Considerable information has been obtained on a 26-kDa polypeptide that is associated with the adaptation of cultured tobacco cells to a medium containing high levels of NaCl (Singh *et al.*, 1985, 1987a,b). This polypeptide was named *osmotin* because it is synthesized and accumulates in cells undergoing gradual osmotic adjustment (Singh *et al.,* 1987b). Osmotin is homologous in molecular weight and amino acid composition with thaumatin, a protein with potential use as a low-calorie sweetener; it is also homologous with a tobacco pathogenesis-related (PR) protein and with maize trypsin/α-amylase inhibitor (Singh *et al.,* 1987b). Osmotin accounts for 10 to 12% of the total cellular protein (Singh *et al.,* 1985), is synthesized as a precursor protein that contains a signal polypeptide, and is concentrated in dense inclusion bodies within vacuoles in adapted cells (Singh *et al.,* 1987b). Osmotin is synthesized in unadapted cells, but has a lower pI than the polypeptide of the adapted cells and is not accumulated under normal growth conditions (Singh *et al.,* 1985). The synthesis of osmotin is induced by ABA, but accumulation is dependent on the presence of NaCl (Singh *et al.,* 1987a). A 26 kDa polypeptide, NP24, which is immunologically related to osmotin and homologous with thaumatin, has been found in cultured tomato cells adapted to medium containing NaCl (King *et al.,* 1986, 1988). Accumulation of the 26 kDa polypeptide in intact tobacco plants appears to be tissue specific, but there is some question about which tissue. King *et al.* (1986) found a greater amount in roots than in stems or leaves, whereas Singh *et al.* (1987a) found more in the outer stem than in roots or leaves. When immunoblots of polypeptides of barely roots were probed with antibodies to osmotin, no cross-reaction was detected (Hurkman and Tanaka, 1987). This result was not unexpected because the barley polypeptides have more basic pIs and slightly greater molecular weights than osmotin (Hurkman and Tanaka, 1987).

Much of the work on the effect of salt on polypeptide and translatable mRNA levels has been done using barley (Hurkman and Tanaka, 1987, 1988; Ramagopal, 1987a,b, 1988), and similar studies have been reported for wheat (Gulick and Dvořàk, 1987). Overall, these studies show that certain polypeptides and translatable mRNAs are enhanced, repressed, or induced by salt stress. The specific details of the response to salt stress differ in the reports for barley. For example, in an *in vivo*–labeling experiment, Ramagopal (1987b) found that the levels of relatively few polypeptides changed with salt treatment, new polypeptides were induced by salt, and the same changes occurred in both CM 72 and Prato. In contrast, the results of Hurkman and co-workers (1989) showed a far greater number of salt-induced changes in CM 72 and Prato, no unique polypeptides were induced, and the majority of changes were cultivar specific. Ramagopal (1987a) also reported that the levels of relatively few translation products changed with salt treatment, the patterns of CM 72 and Prato were qualitatively different and, different products were newly induced by salt in CM 72 and Prato. In contrast, Hurkman and co-workers (1989) showed that there were far more salt-induced changes in the translation products, the majority of changes were cultivar specific, and the translation products of CM 72 and Prato poly(A)$^+$ RNAs were qualitatively the same. In addition, no unique salt-induced translation products were found for either cultivar. A major difference between the studies is the technical quality of the 2D gels. Because of the low resolution of Ramagopal's fluorographs (1987a,b), comparisons of the data are particularly difficult for polypeptides of mid to high molecular weight. In addition, experimental treatment of the plants was different. Hurkman and co-workers (Hurkman and Tanaka, 1987, 1988; Hurkman *et al.*, 1988, 1989) labeled intact plants and treated them with 200-mM NaCl, whereas Ramagopal (1987a,b, 1988) labeled excised plant pieces and used 342-mM NaCl.

Wounding responses have been reported at both the polypeptide and the mRNA levels (Schuster and Davies, 1983b; Shirras and Northcote, 1984; Theillet *et al.*, 1982). In pea epicotyls, wounding causes an increase in protein synthesis and prompted Davies and Schuster (1981) to warn that the properties of wounded tissues are markedly different from those of intact plants. Wounding has been demonstrated to promote the synthesis of certain heat shock proteins in spinach leaves (Guy *et al.*, 1985) and maize mesocotyls (Heikkila *et al.*, 1984). In a study of low-molecular-weight heat shock proteins in a variety of crop plants, Mansfield and Key (1987) detected 24 low-molecular-weight heat shock proteins in maize; previous studies had reported only one (Baszczynski *et al.*, 1982) or a few (Cooper *et al.*, 1984). Mansfield and Key (1987) suggested that these differences in the data were the result of using excised tissues in the earlier studies. It is possible that the data reported by Ramagopal (1987a,b, 1988) are also complicated by a wounding response.

Greenway and Munns (1980) recommended that salt stress should be investigated at a NaCl concentration where growth is not reduced. They suggested that the physiological and biochemical changes observed under such conditions are more likely to be of adaptive value. Hurkman and co-workers (Hurkman and Tanaka, 1987; Hurkman *et al.*, 1989) have shown for CM 72 and Prato that growth is substantially inhibited by NaCl concentrations above 200 mM. The differences in the response observed by Ramagopal (1987a,b, 1988) could be due to the higher concentration of salt used and may reflect salt damage rather than an adaptive response to salt.

Several conclusions can be drawn from the comparisons of the data on the response of polypeptides and translatable mRNAs to salt stress of barley. Changes observed in 2D gels are often interesting, but are highly dependent on high quality 2D gels and on experimental design. For example, Robinson *et al.* (1989) demonstrated that the changes in the levels of certain polypeptides and translation products differ with length of salt treatment.

Since 2D gels are a "snapshot" of the polypeptides in cells at one instant in time, a number of experimental comparisons (eg., time of exposure, various salt concentrations, recovery from salt stress, different tissues and developmental stages, other stresses) must be done to prove that changes occur and that they are a reproducible and specific response to salt stress. Even with such comparisons, 2D-PAGE can only indicate that specific changes occur at the polypeptide and, indirectly, mRNA levels. Of primary importance is the identity and function of the polypeptides whose levels change with salt stress. Two-dimensional PAGE should be considered the initial step in an experimental approach aimed at obtaining this information.

V.   TECHNIQUES THAT COUPLE TWO-DIMENSIONAL GEL ELECTROPHORESIS WITH ANTIBODY PRODUCTION, SEQUENCE ANALYSIS, AND GENE ISOLATION

Although the underlying mechanisms for salt tolerance remain elusive, the increase in specific polypeptides during salt stress could be important in the adaptation of plants to saline conditions. In barley, 2D-PAGE revealed that polypeptides with pIs of 6.3 and 6.5 accumulate with salt treatment and their synthesis increases noticeably over a wide range of treatment times and NaCl concentrations. Whether these polypeptides have a role in the salt tolerance of barley remains to be proved. In order to determine the role of these polypeptides in salt stress, it will be necessary to establish their identities and functions. Several approaches can be taken. The salt-responsive polypeptides can be isolated by 2D-PAGE and used for preparation of antibodies and to obtain amino acid sequences. The antibodies can be used to determine *in situ* localization of these polypeptides as well as to examine the possibility of posttranslational modifications. The amino acid sequences are valuable in the identification of unknown polypeptides through homologies to known proteins. Corresponding recombinants can be isolated from expression libraries using antibodies and/or from cDNA libraries by oligonucleotide probes. The recombinants can then be used in experiments designed to examine their expression during salt stress. These experimental approaches are covered briefly in this section.

A.   Antibodies to Polypeptides in Two-Dimensional Gels

1.   Preparation of Antibodies

Immunogens obtained by SDS-PAGE (Boulard and Lecroisey, 1982; Knudsen, 1985) have been used to produce antisera. The use of polypeptides separated by 2D-PAGE as immunogens increases the probability of obtaining monospecific antibodies, but one limitation is the low levels of polypeptides in the 2D gels. When IEF gels were loaded with 60 μg of protein from barley roots, 10–50 ng of the pI 6.3 and 6.5 polypeptides were present in the second-dimension analytical gels (W. J. Hurkman and C. K. Tanaka, unpubl. results). An increased amount of protein can be loaded for preparative 2D gels, where the diameter of the IEF gels and the thickness of the SDS gels is increased; polypeptides in a sample containing 750 μg could be separated with good resolution in such gels. The efficiency in obtaining suitable quantities of immunogen can also be improved by running only those regions of the focusing gels that contained the pI 6.3 and 6.5 polypeptides into the second-dimension gel. Polypeptides from several IEF gels can then be run into the same second-dimension gel, a technique that has been used in other studies (Aebersold *et al.*, 1987a; Meyer *et al.*, 1988).

Antibodies to the pI 6.3 and 6.5 polypeptides were raised using standard immunological procedures. The portion of the gel containing the polypeptides was injected directly (Meyer *et al.*, 1988; Tracy *et al.*, 1983; Wood and Dunbar, 1981), rather than eluting the polypeptides from the gel prior to injection (Bravo *et al.*, 1983) or blotting them to nitrocellulose for implantation as solid-base immunogens (Chiles *et al.*, 1987). Because of the low amount of protein, immunogens were injected into mice and, following injection of a sarcoma, the antibody-rich ascites fluid was collected (W. J. Hurkman, C. K. Tanaka and M. W. Bigelow, A. E. Karu, College of Natural Resources Hybridoma Center, University of California-Berkeley, unpubl. results). This procedure provides a greater amount of antibody-rich fluid than sera alone.

2. Use of Antibodies

Antibodies can provide important information on the *in situ* localization of intracellular molecules. The immunocolloidal gold method for transmission electron microscopy, where an antibody–colloidal gold particle complex is reacted with thin sections of embedded tissue, allows the detection of antigenic material with high resolution and high specificity (reviewed: Herman, 1988). Immunocytochemistry was used to show that osmotin (discussed in Section IV.C) was localized in vacuolar inclusion bodies in cultured tobacco cells adapted to grow on high levels of NaCl (Singh *et al.*, 1987b). Osmotin has been hypothesized to be involved in the osmotic adjustment of the tobacco cells during adaptation to salt stress (Singh *et al.*, 1987b).

Antibodies can also be used to identify *in vitro* translation products that correspond to *in vivo* polypeptides and to aid in determining the nature of posttranslational modifications. In barley roots, only a few *in vitro* translation products have identical mobilities to polypeptides labeled *in vivo* and none of the translation products had identical mobilities to the pI 6.3 and 6.5 polypeptides (Hurkman *et al.*, 1989). The differences in mobilities of the polypeptides labeled *in vivo* and the *in vitro* translation products can be due to posttranslational modifications (including cleavage of signal peptides, glycosylation, or phosphorylation) that occur *in vivo*. When confirmed, such features can provide information on the site of synthesis and possible functions of the polypeptides.

Finally, antibodies can be used to identify and isolate particular gene sequences from recombinant DNA expression libraries constructed using either plasmid (Helfman and Hughes, 1987) or bacteriophage vectors (Jendrisak *et al.*, 1987; Snyder *et al.*, 1987). These expression vectors are capable of producing a fusion protein with a portion of the amino acid sequence specified by the DNA insert fragment. The bacteriophage, λgt11, is often used as a recombinant expression vector because of the high efficiency of introducing recombinant DNA into *E. coli*, the stability of the β-galactosidase fusion protein, and the efficiency of detection of antigen at high plaque densities (Huynh *et al.*, 1985; Jendrisak *et al.*, 1987; Young and Davis, 1983). Polyclonal antibodies have generally been used as probes for screening λgt11 libraries because they recognize a number of different epitopes in the fusion polypeptides (Huynh *et al.*, 1985; Mierendorf *et al.*, 1987). Work is in progress to screen a λgt11 library with the polyclonal antibodies obtained from mice.

B. Microsequencing Polypeptides from Two-Dimensional Gels

The use of 2D-PAGE as a preparative method for isolating polypeptides has been coupled with the improved sensitivity of protein sequencing instrumentation, making it possible to obtain N-terminal sequence information from 20 pmol of protein (Aebersold *et al.*, 1986). Electroblotting allows the small amounts of polypeptides in 2D gels to be bound directly onto an inert support such as glass-fiber filters (Aebersold *et al.*, 1986; Bauw *et al.*, 1987; Vandekerckhove *et al.*, 1985), nitrocellulose (Aebersold *et al.*, 1987b), or polyvinylidene difluoride membranes (Bauw *et al.*, 1987; Matsudaira, 1987) in a suitable form for gas-phase sequence analysis. Not only can the N-terminal sequences be obtained, but methods have been developed for the generation and isolation of peptide fragments for internal amino acid sequence analysis (Aebersold *et al.*, 1986; Kennedy *et al.*, 1988a,b). The N-terminal sequences were obtained for the pI 6.3 and 6.5 polypeptides of barley. These sequences differed significantly from that for osmotin, the salt-induced protein in cultured tobacco cells (W. J. Hurkman, C. K. Tanaka, and H. P. Tao, unpubl. results). The sequences for the barley polypeptides can be used to design and synthesize oligonucleotides that can be used as hybridization probes to isolate recombinants from cDNA libraries. Probes consisting of single oligonucleotides (Lathe, 1985) or mixtures of oligonucleotides (Suggs *et al.*, 1981) can be used but because of the redundancy of the genetic code, a mixture of oligonucleotides chosen to cover all the coding possibilities of a particular amino acid sequence is most often used (Lathe, 1985).

VI. CONCLUSIONS

Two-dimensional PAGE is a powerful analytical technique for examining changes in individual polypeptide levels that are caused by perturbations in the cellular environment. In barley roots, 2D-PAGE revealed that salt stress alters the levels of polypeptides and the populations of translatable mRNAs. It can be speculated that these changes may be a reflection of numerous biochemical and structural adjustments that enable barley to tolerate salt stress. However, data from other techniques are needed to support this hypothesis. Future studies must be aimed at identifying some of the proteins whose levels change with salt stress and at determining their functions. Because of the number and complexity of the changes caused by salt, one approach is to study initially those polypeptides that increase significantly and specifically with salt stress, such as the pI 6.3 and 6.5 polypeptides. Techniques that take advantage of the resolution of 2D gels can be utilized to isolate these polypeptides in order to make antibodies against them and to microsequence them for procedures aimed at isolating the corresponding complementary DNAs (cDNAs). Once the cDNAs have been obtained and verified, they can be sequenced and, through comparisons with sequences in a data base, used to determine homologies with known genes and proteins. The cDNAs can be used in experiments designed to characterize parameters that influence their expression and to determine whether changes in expression are correlated with differences in the salt tolerance of barley cultivars. These cDNAs can also be used as probes to screen a genomic library. The genomic clones can then be used to study the structure and regulation of salt-responsive genes. The isolation of these genes will provide an opportunity to assess their expression more accurately during salt stress and to determine in what way they contribute to salt tolerance in barley.

REFERENCES

Aebersold, R. H., Teplow, D. B., Hood, L. E., and Kent, S. B. H. (1986). *J. Biol. Chem.* **261**, 4229–4238.
Aebersold, R. H., Leavitt, J., Hood, L., and Kent, S. (1987a). *In* "Methods in Protein Sequence Analysis—1986" (K. A. Walsh, ed.), pp. 277–294, Humana Press, Clifton, New Jersey.
Aebersold, R. H., Leavitt, J., Saavedra, R. A., and Hood, L. E. (1987b). *Proc. Natl. Acad. Sci. U.S.A.* **84**, 6970–6974.

Ames, G. F. L., and Nikaido, K. (1976). *Biochemistry* 15, 616–623.

Bassett, C. L., Mothershed, C. P., and Galau, G. A. (1988). *Planta* 175, 221–228.

Baszczynski, C. L., Walden, D. B., and Atkinson, B. G. (1982). *Can. J. Biochem.* 60, 569–579.

Bauw, G., De Loose, M., Inzé, D., Van Montagu, M., and Vandekerckhove, J. (1987). *Proc. Natl. Acad. Sci. U.S.A.* 84, 4806–4810.

Bewley, J. D., Larsen, K. M., and Papp, J. E. T. (1983). *J. Exp. Bot.* 34, 1126–1133.

Bohnert, H. J., Ostrem, J. A., Cushman, J. C., Michalowski, C. B., Rickers, J., Meyer, G., Derocher, E. J., Vernon, D. M., Krueger, M., Vasquez-Moreno, L., Velten, J., Hoefner, R., and Schmitt, J. M. (1988). *Plant Mol. Biol. Rep.* 6, 10–28.

Booz, M. L., and Travis, R. L. (1980). *Plant Physiol.* 66, 1037–1043.

Booz, M. L., and Travis, R. L. (1981). *Phytochemistry* 20, 1773–1779.

Booz, M. L., and Travis, R. L. (1983). *Phytochemistry* 22, 353–358.

Boulard, C., and Lecroisey, A. (1982). *J. Immunol. Methods* 50, 221–226.

Bravo, R., Fey, S. J., Larsen, P. M., Coppard, N., and Celis, J. E. (1983). *J. Cell Biol.* 96, 416–423.

Bray, E. A. (1988). *Plant Physiol.* 88, 1210–1214.

Briggs, D. E. (1978). "Barley." Chapman & Hall, London.

Broglie, K. E., Gaynor, J. J., and Broglie, R. M. (1986). *Proc. Natl. Acad. Sci. U.S.A.* 83, 6820–6824.

Cheeseman, J. M. (1988). *Plant Physiol.* 87, 547–550.

Chen, C.-M., and Leisner, S. M. (1985). *Plant Physiol.* 77, 99–103.

Chiles, T. C., O'Brien, T. W., and Kilberg, M. S. (1987). *Anal. Biochem.* 163, 136–142.

Chory, J., Voytas, D. F., Olszewski, N. E., and Ausubel, F. M. (1987). *Plant Physiol.* 83, 15–23.

Christoffersen, R. E., and Laties, G. G. (1982). *Proc. Natl. Acad. Sci. U.S.A.* 79, 4060–4063.

Cloutier, Y. (1983). *Plant Physiol.* 71, 400–403.

Colas des Francs, C., Thiellement, H., and de Vienne, D. (1985). *Plant Physiol.* 78, 178–182.

Cooper, P., and Ho, T.-H. D. (1983). *Plant Physiol.* 71, 215–222.

Cooper, P., Ho, T.-H. D., and Hauptman, R. M. (1984). *Plant Physiol.* 75, 431–441.

Cramer, C. L., Ryder, T. B., Bell, J. N., and Lamb, C. J. (1985). *Science* 227, 1240–1243.

Cremer, F., and Van de Walle, C. (1985). *Anal. Biochem.* 147, 22–26.

Croughan, T. P., and Rains, D. W. (1982). *In* "Handbook of Biosolar Resources" (A. Mitsui and C. C. Black, eds.), Vol.1, Part 2, pp. 245–255. CRC Press, Boca Raton, Florida.

Damerval, C., Vartanian, N., and de Vienne, D. (1988). *Plant Physiol.* 86, 1304–1309.

Dasgupta, J., and Bewley, J. D. (1984). *J. Exp. Bot.* 35, 1450–1459.

Davies, E., and Schuster, A. (1981). *Proc. Natl. Acad. Sci. U.S.A.* 78, 2422–2426.

Dietz, K.-J., and Bogorad, L. (1987). *Plant Physiol.* 85, 808–815.

Dietz, K.-J., Kaiser, G., and Martinoia, E. (1988). *Planta* 176, 326–367.

Downtown, W. J. S. (1984). *CRC Crit. Rev. Plant Sci.* 1, 183–201.

Dunbar, B. S. (1987a). *BioTechniques* 5, 218–226.

Dunbar, B. S. (1987b). "Two-Dimensional Electrophoresis and Immunological Techniques." Plenum, New York.

Duncan, R., and Hershey, J. W. B. (1984). *Anal. Biochem.* 138, 144–155.

Dunn, M. J., and Burghes, A. H. M. (1983a). *Electrophoresis* 4, 97–116.

Dunn, M. J., and Burghes, A. H. M. (1983b). *Electrophoresis* 4, 173–189.

Epstein, E. (1980). *In* "Genetic Engineering of Osmoregulation" (D. W. Rains, R. C. Valentine, and A. Hollaender, eds.), pp, 7–21. Plenum, New York.

Epstein, E., and Norlyn, J. D. (1977). *Science* 197, 249–251.

Epstein, E., Norlyn, J. D., Rush, D. W., Kingsbury, R. W., Kelley, D. B., Cunningham, G. A., and Wrona, A. F. (1980). *Science* 210, 399–404.

Shewry, P. R., Parmar, S., and Field, J. M. (1988). *Electrophoresis* **9**, 727–737.

Shirras, A. D., and Northcote, D. H. (1984). *Planta* **162**, 353–360.

Singh, N. K., Handa, A. K., Hasegawa, P. M., and Bressan, R. A. (1985). *Plant Physiol.* **79**, 126–137.

Singh, N. K., LaRosa, P. C., Handa, A. K., Hasegawa, P. M., and Bressan, R. A. (1987a). *Proc. Natl. Acad. Sci. U.S.A.* **84**, 739–743.

Singh, N. K., Bracker, C. A., Hasegawa, P. M., Handa, A. K., Buckel, S., Hermodson, M. A., Pfankock, E., Regnier, F. E., and Bressan, R. A. (1987b). *Plant Physiol.* **85**, 529–536.

Skadsen, R. W., and Cherry, J. H. (1983). *Plant Physiol.* **71**, 861–868.

Snyder, M., Elledge, S., Sweetser, D., Young, R. A., and Davis, R. W. (1987). *In* "Methods in Enzymology" (R. Wu, ed.), Vol. 154, pp. 107–128. Academic Press, San Diego, California.

Somssich, I. E., Schmelzer, E., Bollmann, J., and Hahlbrock, K. (1986). *Proc. Natl. Acad. Sci. U.S.A.* **83**, 2427–2430.

Stafstrom, J. P., and Sussex, I. M. (1988). *Planta* **176**, 497–505.

Staples, R. C., and Toenniessen, G. H. (1984). "Salinity Tolerance in Plants, Strategies for Crop Improvement." Wiley, New York.

Suggs, S. V., Wallace, R. B., Hirose, T., Kawashima, E. H., and Itakura, K. (1981). *Proc. Natl. Acad. Sci. U.S.A.* **78**, 6613–6617.

Theillet, C., Delpeyroux, F., Fiszman, M., Reigner, P., and Esnault, R. (1982). *Planta* **155**, 478–485.

Theologis, A., and Ray, P. M. (1982). *Proc. Natl. Acad. Sci. U.S.A.* **79**, 418–421.

Tracy, R. P., Katzmann, J. A., Kimilinger, T. K., Hurst, G. A., and Young, D. S. (1983). *J. Immunol. Methods* **65**, 97–107.

Traylor, E. A., Shore, S. H., Ransom, R. F., and Dunkle, L. D. (1987). *Plant Physiol.* **84**, 975–978.

Uemura, M., and Yoshida, S. (1984). *Plant Physiol.* **75**, 818–826.

Uemura, M., and Yoshida, S. (1985). *Plant Cell Physiol.* **26**, 35–46.

Vandekerckhove, J., Bauw, G., Puype, M., Van Damme, J., and Van Montagu, M. (1985). *Eur. J. Biochem.* **152**, 9–19.

Van der Zaal, E. J., Mennes, A. M., and Libbenga, K. R. (1987). *Planta* **172**, 514–519.

Wainwright, S. J. (1980). *Adv. Bot. Res.* **8**, 221–261.

Wolpert, T. J., and Dunkle, L. D. (1983). *Proc. Natl. Acad. Sci. U.S.A.* **80**, 6576–6580.

Wood, D. M., and Dunbar, B. S. (1981). *J. Exp. Zool.* **217**, 423–433.

Wyn Jones, R. G. (1981). *In* "Physiological Processes Limiting Plant Productivity" (C. B. Johnson, ed.), pp. 271–284. Butterworth, London.

Yeo, A. R. (1983). *Physiol. Plant.* **58**, 214–222.

Young, R. A., and Davis, R. W. (1983). *Proc. Natl. Acad. Sci. U.S.A.* **80**, 1194–1198.

Zurfluh, L. L., and Guilfoyle, T. J. (1980). *Proc. Natl. Acad. Sci. U.S.A.* **77**, 357–361.

Zurfluh, L. L., and Guilfoyle, T. J. (1982a). *Plant Physiol.* **69**, 332–337.

Zurfluh, L. L., and Guilfoyle, T. J. (1982b). *Plant Physiol.* **69**, 338–340.

Zurfluh, L. L., and Guilfoyle, T. J. (1982c). *Planta* **156**, 525–527.

CHAPTER 10

Mechanisms of Trace Metal Tolerance in Plants

Paul J. Jackson*
Pat J. Unkefer**
Emmanuel Delhaize+
Nigel J. Robinson†

*Genetics Group, Life Sciences Division
Los Alamos National Laboratory
Los Alamos, New Mexico

**Isotope and Structural Chemistry Group

Isotope and Nuclear Chemistry Division
Los Alamos National Laboratory
Los Alamos, New Mexico

+CSIRO, Division of Plant Research
Canberra, Australia

†Department of Biological Sciences
University of Durham Science
Laboratories
Durham, England

I. INTRODUCTION

A. Sources of Metal Contamination

Surface deposits containing high concentrations of specific metals occur in natural geological formations, which can be identified by their distinctive flora (Agricola, 1556). Localized deposits of metals have also resulted from anthropogenic change. Toxic trace metal ions can enter the soil from stack emissions of fossil fuel-burning power plants (Natusch *et al.*, 1974), from the mining of coal, shale, and oil, and from metallurgical mining and processing (Takijima *et al.*, 1973a,b); through additions of phosphate fertilizers (Van Bruwaene *et al.*, 1984; Street *et al.*, 1978); from automobile emissions (Lagerwerff and Specht, 1970); and through addition of metal-containing pesticides and industrial sewage sludge to the soil (Street *et al.*, 1978; Van Bruwaene *et al.*, 1984; Vlamis *et al.*, 1985). Toxic trace metal ions can also be released from soils in free ionic form by changes in soil acidity (Godbold *et al.*, 1988; Løbersli and Steinnes, 1988; Unsworth and Harrison, 1985). Mining has led to the contamination of soil and groundwater in certain isolated sites. This contamination with toxic metal ions is often limited to the site of mining activities. However, due to airborne distribution, emissions from power plants, smelters, and automobiles have resulted in a more widespread deposition of contaminants within soils (Løbersli and Steinnes, 1988; Garty, 1988, for examples). Deposition of trace metals is often accompanied by a change in soil acidity as emissions result in a deposition of acidic compounds that far exceeds the soil's buffering capacity (Johnson *et al.*, 1981; Folster, 1985). This change in acidity induces the release of normally insoluble metal ions and allows for their uptake by plants (Ulrich *et al.*, 1979; Foy *et al.*, 1978). Perhaps the most detrimental deposition of these metals occurs as the result of sewage disposal operations. Sludge generated from all major industrial cities contains high concentrations of toxic metal ions; concentrations of ions can approach 3 g/kg of dry sludge (Street *et al.*, 1978). Different methods of disposal of this sludge result in release of these toxic ions into aquatic and marine environments and into agriculturally active soils.

B. Uptake of Toxic Metal Ions by Humans and Other Animals

Drinking water and ambient air contribute relatively little to toxic metal intake in humans and other animals. For populations not subjected to occupational exposure, uptake of toxic metal ions is normally through ingestion of food (Van Bruwaene *et al.*, 1984; Wagner *et al.*, 1984). Grains and cereals constitute a primary dietary source of trace metal ions. Smoking of tobacco products also results in a significant accumulation of these metals (Van Bruwaene *et al.*, 1984; Wagner *et al.*, 1984). Therefore, accumulation of metal ions from contaminated or modified soils in higher plants, especially those plants consumed either by humans or as part of ecological food webs, results in a significant impact on humans and other consuming species. It also has a direct effect on plant ecosystems.

C. Ecological Aspects of Metal Toxicity

Contamination of different sites with toxic metal ions and acidification of certain soils as a result of deposition by acid rain and the subsequent solubilization of metal ions have had an adverse effect on many environments (Faulstich and Stournasas, 1985; Garty, 1988; Godbold *et al.*, 1988; Løbersli and Steinnes, 1987; Shortle and Smith, 1988; Unsworth and Harrison, 1985; Watabe *et al.*, 1984). However, populations of higher plants have rapidly colonized environments contaminated with high concentrations of certain metal ions. This colonization is perhaps the best-known example of evolution through natural selection in higher plants (Bradshaw, 1984). Metal-tolerant cultivars of some common grasses are commercially available for the revegetation of contaminated sites (Gemmel, 1977). Plants grown in such environments may accumulate trace metal ions, subject to the genetic characteristics of the plant (Foy *et al.*, 1978; Turner, 1969) and environmental factors such as soil pH, cation exchange capacity of the soil, and different concentrations of competing metal ions. Accumulation of toxic trace metal ions in plants affects many metabolic processes (Prasad and Prasad, 1987a; Petolino and Collins, 1985; Reese and Roberts, 1985, for examples), yet some plant species and some selected ecotypes of other species can grow and reproduce in the presence of normally toxic concentrations of these ions (Page *et al.*, 1972; Rauser and Curvetto, 1980; Poulter *et al.*, 1985; Robinson and Thurman, 1986a,b; Taylor and Foy, 1985; Rauser and Winterhalder, 1985; Van Steveninck *et al.*, 1987; Gilissen and Van Staveren, 1986, for examples). Much research has focused on ecological and genetic aspects of metal tolerance in higher plants but, until recently, little information concerning the biochemical and molecular aspects of tolerance had been elucidated. This recent work is the subject of this chapter.

II. MECHANISMS OF TOLERANCE TO DIFFERENT METAL IONS

Metal ions have been placed into different categories based on their affinities for different ligands found within organic molecules (Nieboer and Richardson, 1980). Class A metals, which include Al^{3+}, form stable complexes with ligands containing oxygen. Class B metals, which include Cu^{2+} and Cd^{2+}, form more stable complexes with ligands containing sulfur and nitrogen centers. Many molecules synthesized by higher plants contain one or more of these ligands; some molecules contain an abundance of ligands. It is therefore probable that interaction with critical organic molecules is often related to the mechanism of the metal ions' toxicity. However, the production of large amounts of certain other molecules containing these ligands probably also plays an important role in metal ion detoxification in higher plants.

A. Aluminum

Although aluminum is not a trace metal ion, it is considered here because changes in soil pH result in small amounts of free ion being transported into plants. Approximately 15% of the earth's crust is made up of Al_2O_3

(Hunt, 1972). However, aluminum toxicity occurs only in soils where the pH is below 5.5 and increases in severity when the pH drops below 5.0 (Foy, 1974). While this toxicity is often a problem in soils that are naturally acidic, toxic effects of this ion are also evident in areas where soil acidity has increased as a result of deposition of airborne contaminants emitted by industrial processes. The hypothesis that aluminum toxicity may be a primary factor in forest decline has been suggested (Ulrich *et al.*, 1979).

Despite a large body of literature documenting aluminum toxicity and tolerance, their physiological bases remain elusive. The identity of the phytotoxic species is not known. Several monomeric species have been implicated as the primary toxic species, and the sum of all monomeric species has also been correlated with phytotoxicity (Blamey *et al.*, 1983). Wagatsuma and Ezoe (1985) concluded that aluminum polymers might be phytotoxic. Toxicity appears to be the result of several interactions, and there is clearly no consensus on the mechanisms of aluminum toxicity in higher plants. To some extent, aluminum clearly has a deleterious effect on numerous aspects of the affected species' physiology. The inability to use phosphate in the presence of aluminum appears to contribute to toxicity (Helyar and Anderson, 1971; Foy and Campbell, 1984; Awad *et al.*, 1976), probably as a result of the formation of aluminum phosphate complexes within the tissue. Exposure to aluminum also results in the reduction of free magnesium and calcium in plant tissue. It has been shown in different conifer species that magnesium levels are reduced below those critical for tissue survival in plants exposed to aluminum (Godbold *et al.*, 1988); calcium uptake was reduced by as much as 90%. It has been speculated that aluminum may sufficiently reduce calcium and magnesium uptake to levels that are critical for survival (Ulrich *et al.*, 1979). It has also been hypothesized that aluminum rhizotoxicity is related to disruption of membrane function. Several of the biochemical effects are probably due to changes in the structure and function of the root cell plasmalemma (Zhao *et al.*, 1987; Hecht-Buchholz and Foy, 1981). Depending on the pH, aluminum can bind to membrane proteins and lipids (Foy and Campbell, 1984). It can also participate in the formation of cross-links between proteins and pectins within the cell wall, making it more rigid. These interactions should have a deleterious effect on membrane integrity. However, wheat roots that exhibit severe symptoms of toxicity have an undiminished capacity to extrude protons, suggesting that these membranes are intact and ATP synthesis is sufficient to supply the proton-translocating ATPases (Kinraide, 1988). It has been reported that abnormal root growth is the result of disturbed mitotic processes (Foy, 1982, Foy and Campbell 1984; Morimura *et al.*, 1978; Neitzke and Runge, 1985). Aluminum is particularly concentrated in the nucleus (DeBoni *et al.*, 1974; Foy *et al.*, 1978). The cell cycle is inhibited, probably at the level of DNA replication where aluminum inhibits the synthesis of DNA (Foy *et al.*, 1978; Foy, 1974, 1982). It therefore appears that aluminum probably acts at a number of critical sites and that toxicity can be explained by the impact of this metal ion on a number of critical physiological processes.

Just as the mechanisms of toxicity are not entirely clear, the mechanisms of tolerance are not yet well defined either. Tolerance occurs naturally in certain species and within selected ecotypes. Therefore the mechanism of tolerance is probably not the result of changes within the physiology or biochemistry related to the specific site of aluminum toxicity, since numerous changes would be required. Rather, mechanisms that reduce or eliminate the uptake of this ion or reduce its impact after uptake are most likely to be part of the tolerance mechanism. A number of different mechanisms that confer tolerance have been hypothesized. These mechanisms include aluminum binding to specific sites within cell walls of the epiderm and mesophyll or inside the cytoplasm, thus preventing the toxic ion from reaching sensitive metabolic sites (Foy, 1974, 1983), and mechanisms that facilitate the uptake of mineral nutrients in the presence of the aluminum ion or involve low requirements for certain critical nutrients (Bieleski, 1970; Medappa and Dana, 1970). Some plants increase the pH of their rhizosphere, lowering the solubility of aluminum (Mugwira *et al.*, 1976; Foy *et al.*, 1978; Foy and Fleming, 1982). Changes in ammonium versus nitrate ion uptake result in a change in rhizosphere acidity. Aluminum-tolerant cultivars of *Triticum aestivum* take up less ammonium ion than their sensitive counterparts, resulting in a higher solution pH and less solubility of the toxic ion (Taylor and Foy, 1985).

Aluminum-tolerant plants may also produce different chelating agents, causing the formation of metal complexes that are no longer toxic to the plant. Since this metal ion has a high affinity for organic acids, it is not surprising that some tolerant plants produce large amounts of organic acids (Mengel and Kirkby, 1982). Aluminum-tolerant barley maintains higher concentrations of organic acids in roots exposed to aluminum (Foy *et al.*, 1987). In maize roots, there is a large increase in organic acids, especially malate, citrate, and *trans*-aconitate (Suhayda and Haug, 1986). In suspension cultures, the toxic effect of aluminum decreased greatly after addition of an acid fraction of the conditioned media from tolerant cells. This fraction contained four different organic acids, primarily citrate. Moreover, the toxic effects of this metal ion could also be reduced by addition of citric or malic acid to the cells. Therefore, at least in cell culture, tolerance may be related to the production and release of organic acids by the cells (Ojima *et al.*, 1984). Chelation of the metal ion then occurs externally.

Aluminum-tolerant cells cultures of tomato have been selected (Meredith, 1978). These cells do not exclude the metal ion. Therefore, it is likely that it is somehow chelated in a less toxic form within the cells. Tolerant plants that accumulate aluminum accumulate it in the roots rather than in the shoots (Foy and Campbell, 1984; Thornton et al., 1986). In barley and maize, tolerance is correlated with the ability to maintain higher concentrations of organic acids within root cells in the presence of aluminum (Foy et al., 1987; Suhayda and Haug, 1986). It therefore appears that chelation with organic acids may occur within the tissue or within the rhizosphere of the tolerant plants.

The relative overproduction of organic acids in aluminum-tolerant ecotypes suggests that they clearly play a role in tolerance. However, the production of large amounts of organic acids changes the physiological and biochemical conditions within the cells and also results in the chelation of a number of necessary micronutrients. It remains to be demonstrated that changes in organic acid concentrations alone are sufficient to explain aluminum tolerance in most species. Certainly, methods of removal or storage of the metal in a nontoxic state after chelation must be required. There are a number of other physiological aspects of aluminum tolerance that are not yet understood, and more research is required before these mechanisms and the underlying molecular aspects of their regulation will be understood. A review by Andersson (1988) provides more information on aluminum toxicity and tolerance in higher plants.

B. Cadmium

Cadmium can enter the environment from stack emissions of fossil fuel–burning power plants, mining activity, additions of phosphate fertilizers, and addition of domestic and industrial sewage sludge to the soil. It can also be introduced by the discharge of contaminated industrial effluent into the environment. This metal ion, a group IIB trace metal, is known to be harmful to human health. Cadmium has been implicated in hypertension, emphysema, chronic bronchitis, and "itai-itai" disease (Ellis et al., 1979; Yasumura et al., 1980; Yost et al., 1980).

The mobility of cadmium in the soil and its availability for uptake by plants depends on the chemical form of cadmium present. This, in turn, is related to the amount of cadmium present; the pH of the soil; the occurrence of other elements, notably competing metal ions; the presence of complexing ligands and adsorption sites associated with the soil; and environmental factors such as soil temperature and moisture content. Cadmium concentrations and soil pH are the two main factors influencing uptake by food crops (Van Bruwaene et al., 1984).

The toxic species of this metal ion is Cd^{2+}. Cadmium toxicity is the result of a complex set of interactions between this nonessential ion and different critical aspects of cellular and whole-plant physiology. This ion interferes with respiratory carbohydrate metabolism in plant cells, probably by substituting irreversibly for another micronutrient in critical enzymes (Reese and Roberts, 1985). It also inhibits the formation of chlorophyll by interfering with protochlorophyllide reduction and the synthesis of aminoevulinic acid (Stobart *et al.*, 1985). While cadmium does not appear to inhibit photochemical reactions, it does interfere with different steps of the Calvin cycle, resulting in the inhibition of photosynthetic CO_2-fixation (Weigel, 1985). This toxic ion also inhibits uptake of other metal ions by roots (Lindberg and Wingstrand, 1985). Cadmium is known to irreversibly replace copper and zinc in critical metalloenzymes. Many of these enzymes are involved in RNA, DNA, and protein metabolism. Therefore, cadmium has a deleterious effect on much of the biochemical machinery required for cell survival.

Like aluminum, cadmium has numerous sites of action within the plant. Therefore it is unlikely that tolerance will occur as the result of changes at these susceptible sites. It is more likely that tolerance will be associated with a mechanism that either excludes this toxic ion from plant tissues or sequesters it in a less toxic form.

In *Euglene gracilis* tolerance has been associated with a lower accumulation of cadmium (Bariaud *et al.*, 1985). However, in higher plants most tolerant ecotypes do not exclude this ion, although high quantities of cadmium and other toxic metal ions have been localized within the cell walls of the shoots and roots of *Azolla* (Sela *et al.*, 1988). This has also been shown for rice (Dabin *et al.*, 1978) and *Zea mays* (Klan *et al.*, 1984). However, in most species the majority of the cadmium absorbed is found associated with the soluble cytoplasmic fractions of leaves and roots (see Petit *et al.*, 1978; Weigel and Jäger, 1980; Casterline and Barnett, 1982; Wagner and Trotter, 1982; Cataldo *et al.*, 1981; Rauser, 1984, for examples).

In mammals, cadmium tolerance is associated with the ability to produce large amounts of metallothionein—small, heat-stable cysteine-rich proteins that tightly bind cadmium, zinc, copper, and several other metal ions (Nordberg and Kojima, 1979). Cytoplasmic thiol-rich cadmium complexes have been known in plants for over a decade (Bartolf *et al.*, 1980; Bennetzen and Adams, 1984; Casterline and Barnett, 1982; Jackson *et al.*, 1984; Rauser *et al.*, 1983, for examples). It was thought that these complexes contained metallothionein. However, in all plants whose structure has been elucidated, these complexes do not contain this protein (Robinson and Jackson, 1986). They contain cadmium and poly(γ-glutamylcysteinyl)glycines (Fujita, 1985; Grill *et al.*, 1985, 1987; Jackson *et al.*, 1985a, 1987; Steffens *et al.*, 1986). These compounds—also known as cadystins, phytochelatins, class III metallothioneins (Fowler *et al.*, 1987) or γ-glutamyl binding peptides—were first discovered in the fission yeast *Schiazosaccharomyces pombe* (Kondo *et al.*, 1984). They appear to be ubiquitous in plants (Grill *et al.*, 1988). Some complexes also contain sulfide; they have a higher affinity for cadmium than those that do not contain sulfide (Steffens *et al.*, 1986; Reese *et al.*, 1988). The presence of γ-carboxamide linkages between glutamate and adjacent cysteine residues suggests that these polypeptides are not encoded by a structural gene. In fact, they are the product of a biosynthetic pathway (Robinson *et al.*, 1988) that consumes glutathione (Berger *et al.*, 1989; Scheller *et al.*, 1987; Grill *et al.*, 1987; Huang *et al.*, 1987; Reese and Wagner, 1987a). It has been suggested that merely the ability to produce large amounts of these polypeptides may confer tolerance to cadmium (Jackson *et al.*, 1987; Grill *et al.*, 1988). Growth of plants and plant cells in the presence of buthionine sulfoximine inhibits the synthesis of γ-glutamylcysteine, glutathione, and these metal-binding polypeptides (Huang *et al.*, 1987; Reese and Wagner, 1987a) and also leads to cadmium sensitivity in normally tolerant cell cultures. However, both cadmium-tolerant and cadmium-sensitive plants can produce these compounds (Grill *et al.*, 1988). This suggests that while these polypeptides definitely play a role in metal tolerance, the ability to produce them does not in itself confer tolerance to cadmium in a cell line of *Datura innoxia*. Research demonstrates that cadmium tolerance is correlated with the ability to form cadmium:polypeptide complexes and not with the ability to synthesize the polypeptides (Delhaize *et al.*, 1989a; Jackson *et al.*, 1989). Clearly, tolerance is not simply related to the ability to produce large amounts of these polypeptides.

Cadmium in many plants is found in soluble complexes containing metal-binding polypeptides. However, the fate of the cadmium once it is taken into the plant is not well understood. Cadmium-tolerant plant cell cultures accumulate cadmium as long as they are exposed to it and do not release it into the media in a chelated form. On removal of cadmium from the culture media, the average cellular content of cadmium is reduced at a rate consistent with dilution by cell division (Jackson *et al.* 1985b). There is no indication that plants excrete bound cadmium. In beans as much as 70% of the plant cadmium is stored in the cytoplasm of roots and leaves (Weigel and Jäger, 1980). Calcium and phosphate have been implicated in the precipitation of cadmium in tomato, where cadmium may be stored as an insoluble salt of cadmium phosphate in the epidermal layers of the stem. An insoluble form of cadmium has also been demonstrated inside root parenchyma cells of *Agrostis giganta* and *Z. mays* (Rauser and Ackerley, 1987). Perhaps metal-binding polypeptides bind cadmium as it enters the cytoplasm, then the metal ion is deposited as crystals of cadmium phosphate or within crystals of calcium oxalate. It is possible that the polypeptides are stored as part of these crystals since thiols have been found in stored granules within plants. Clearly, cadmium is maintained in a less toxic form within the plant and is not excreted into the environment.

Exposure of plants and plant cells to cadmium results in a large number of physiological, biochemical, and molecular changes, in addition to the synthesis of metal-binding polypeptides. Some of these responses are similar to those elicited by other abiotic stresses (Weinstein *et al.*, 1986; Edelman *et al.*, 1988; Huang and Goldsbrough, 1988; Delhaize *et al.*, 1989b). Some responses may result from the requirement for the synthesis of the polypeptide precursors, but many others have no clear relationship to the production of these molecules. The mechanisms of cadmium tolerance and the response of plants to this toxic metal are quite complex and cannot be explained by the production of only one class of molecules.

C. Copper

Copper enters the environment as the result of many of the same activities that are associated with the introduction of cadmium. Mining and processing of copper result in contamination of soils (Folkeson and Andersson-Bringmark, 1988; Løbersli and Steinnes, 1988 for examples). Copper is also found in agrichemicals that are applied to different crop species (Magalhães *et al.*, 1985). Like cadmium, copper is found in industrial effluent.

While copper is a toxic metal ion at high concentrations, it is also a necessary micronutrient (Lipman and Mackinney, 1931). This complicates the study of toxicity since plants must contend with the problem of maintaining a minimal concentration to meet their needs while, simultaneously, preventing the occurrence of a toxic level. Copper is required for several important biochemical and physiological processes (Bussler, 1981). It is an essential component of the plant metalloenzymes diamine oxidase, ascorbate oxidase, o-diphenol oxidase, *cytochrome c* oxidase, superoxide dismutase, plastocyanin oxidase, and quinol oxidase (Delhaize *et al.*, 1985). Removal of this ion results in their inactivation (Vallee and Wacker, 1970). Copper toxicity is primarily the result of its high affinity for sulfhydryl groups, causing the inactivation of sulfhydryl-containing enzymes or altering their catalytic specificity or control. Copper ions may also be responsible for stimulating the peroxidative degradation of the lipids of chloroplast membranes (Woolhouse, 1983). Copper is required for photosynthesis but high concentrations adversely affect the primary charge separation of the photosystem II reaction center, resulting in the release of photosynthetic excitation energy as heat (Hsu and Lee, 1988). These results suggest that there must be some mechanism of copper homeostasis in plants that provides the necessary copper ions at the required enzymatic sites, while reducing or eliminating their deleterious interaction with sulfhydryl groups required for other normal cell functions.

In animals metallothionein proteins are thought to be involved in copper and zinc homeostasis. Recently, a metallothionein has been found in higher plants (submitted). This protein binds both Cd^{2+} and Cu^{2+}. However, there is no evidence that synthesis of these proteins is induced in either copper-tolerant plants or sensitive plants exposed to copper. Rather, synthesis of this protein appears to be constitutive and is not greatly elevated in plants or plant cells exposed to high concentrations of metal ions. It is not known whether or not this protein plays a role in metal-tolerance in higher plants. Many reports of the presence of metallothioneins in plants have resulted from the incomplete characterization of other copper-binding compounds or the incorrect interpretation of results. There are many reports of thiol-rich copper-binding proteins occurring in copper-tolerant and copper-sensitive plants and plant cells, but many of these proteins have not been completely characterized. Available information suggests that these "proteins" are often copper:poly(γ-glutamylcysteinyl)glycine complexes (Robinson and Jackson, 1986; Rauser and Curvetto, 1980). These polypeptides have a higher affinity for copper than for cadmium (Jackson et al., 1985a; Reese et al., 1988). Production of metal-binding polypeptides appears to be involved in the tolerance mechanism in some species. All plants tested appear to contain the enzymes required for polypeptide synthesis (Grill et al., 1988), and these enzymes are present constitutively in D. innoxia cell cultures (Robinson et al., 1988). Copper-tolerant and copper-sensitive Mimulus guttatus both synthesize these polypeptides in response to copper exposure (Robinson and Thurman, 1986b). Moreover, only 6% of the copper in the roots is bound by this complex after growth of the plants in high copper concentrations. This implies that the mechanism of tolerance does not simply involve the sequestration of all excess copper by these copper:-polypeptide complexes. It may be that in copper-tolerant M. guttatus the polypeptides transiently bind copper in the cytoplasm prior to accumulating it in the vacuole or at some other site. This is consistent with X-ray microanalysis of root sections, which revealed most of the copper localized within electron-dense bodies associated with sulfur and encapsulated within bodies thought to be vacuoles (Mullins et al., 1985). Such entrapment of relatively high concentrations of copper in such nonmetabolic centers would allow the plant to avoid toxicity (Woolhouse and Walker, 1981).

While copper tolerance in some species involves the synthesis of thiol-rich polypeptides, this is clearly not always the case. There is no evidence of the production of such peptides in response to copper in tolerant spinach (Tukendorf et al., 1984). Copper-tolerant Deschampsia caespitosa produces less of a thiol-rich copper complex than nontolerant plants of the same species (Schultz and Hutchinson, 1985), and while sulfur deficiency reduced the amount of thiol-rich compounds produced, it did not interfere with copper tolerance. Copper is stored in the walls of some species while uptake is slower in others (Lolkema and Vooijs, 1986). In some species, copper is apparently excluded from chloroplasts (Lolkema and Vooijs, 1986). Copper-tolerant *Silene cucubalus* may excrete a chelating compound into the rhizosphere. Copper-tolerant cell cultures of this species excrete such a compound into the medium, which can then be used to protect sensitive cells from toxicity. Such a result suggests that the tolerant variety would have a very high demand for copper; this appears to be the case (Lolkema *et al.*, 1986). Some tolerant cells produce large amounts of citrate and malate when exposed to copper. The addition of these organic acids to sensitive cell cultures prevents copper toxicity (Kishinami and Widholm, 1987). It therefore appears that the accumulation of these compounds is at least partially responsible for tolerance.

Several different biochemical mechanisms appear to be involved in copper tolerance in higher plants. Most research programs have only focused on one putative mechanism of tolerance. Therefore it is not clear whether the reports cited here represent different mechanisms of copper tolerance of different aspects of complex mechanisms. It is apparent that certain chelating molecules are involved in copper tolerance in plants. However, the role these molecules play in the tolerance mechanism is not as well understood.

D. Manganese

Manganese, like copper, is a necessary micronutrient for plants. Although manganese forms relatively insoluble compounds at high pH (Rufty *et al.*, 1979), manganese toxicity is a problem where the soil pH is characteristically low and where high rates of acid-forming nitrogenous fertilizers are applied. The effects of manganese toxicity are difficult to differentiate from other factors because soils containing free manganese often also contain significant amounts of free aluminum ion. Some effects may also be directly related to the acidity of the soil.

An early physiological symptom of manganese toxicity is a decrease in the rate of net photosynthesis and an increase in leaf polyphenol oxidase activity. This may be the result of interaction between this ion and enzymes involved in CO_2 fixation (Houtz *et al.*, 1988). Continued exposure to this ion results in further decreases in the rate of photosynthesis, the onset of visible foliar symptoms, and a decrease in chlorophyll and leaf total protein (Nable *et al.*, 1988). Visual symptoms include leaf chlorosis and necrosis, altered leaf morphology, and discoloration of the roots. These symptoms are the result of a number of physiological disorders. Long-term exposure to excess manganese results in increased peroxidative destruction of indole-3-acetic acid (IAA) and increased synthesis of ethylene (Fowler and Morgan, 1972; Morgan *et al.*, 1976; Sirkar and Amin, 1974). Manganese may also interact with iron, leading to disorders as a consequence of limited availability of this ion (Foy *et al.*, 1978). The long-term effects of manganese toxicity on enzymic activities are an increase in peroxidase activity (Fowler and Morgan, 1972; Morgan *et al.*, 1976) and diminished activities in iron-containing enzymes such as catalase and *cytochrome c* oxidase (Sirkar and Amin, 1974). There is an increase in the copper-containing enzyme polyphenol oxidase, and cellular ATP concentrations are also lowered (Sirkar and Amin, 1979).

Little is known of the mechanisms of tolerance to supraoptimal concentrations of this ion, but it is unlikely that it will be associated with a change in enzyme affinities for manganese since that would require modifications of numerous enzymes within the plant. It is more likely that this ion is chelated in a less toxic form or accumulated in nonmetabolic sites. In some instances roots have been shown to reduce manganese toxicity by precipitating manganese as MnO_2 on root surfaces (Foy *et al.*, 1978). Maize avoids toxicity by entrapment of relatively high concentrations of this metal in vacuoles (Foy, 1973). Manganese has been found associated with organic complexes in xylem and phloem exudates (Bradfield, 1976), and the organic components of these complexes may be involved in mechanisms of tolerance to this ion.

E. Nickel

Nickel is an essential micronutrient for the growth of plants (Brown *et al.*, 1987); it is a component of the enzyme urease (Dixon *et al.*, 1975). However, the roles of nickel in plant metabolism remain largely unknown. The broad range of effects attributed to nickel deficiency suggests that it may be involved in several physiological processes (Brown *et al.*, 1987). Once nickel is absorbed into root cells, mechanisms to prevent the sorption and interaction of this ion with critical biochemical processes must be present. This metal ion is associated with polypeptides or amino acids transported within the xylem. Nickel accumulates in leaves of plants containing high concentrations of citrate (Lee *et al.*, 1978), suggesting that this organic acid may be involved in nickel detoxification. Few studies address either the toxicity of nickel or mechanisms of tolerance to this ion.

F. Selenium

Much of our knowledge about selenium toxicity and selenium tolerance in plants has been obtained during the past two decades. Brown and Shrift (1982) wrote an excellent review that contains detailed information on this subject. The major aspects of current knowledge of selenium toxicity and plant tolerance are summarized here. We have attempted to provide an update on selenium as a micronutrient because this role is a newly appreciated one that is presently being studied.

A substantial portion of the selenium found in soils is present as selenate (SeO_4^-) and selenite (SeO_3^-). Only these two selenium-containing ions are toxic to plants. Soils that contain the selenate or selenite anions are termed *seleniferous*, whereas soils that contain the nonextractable selenium forms are termed *nonseleniferous*.

Seleniferous soils are found in a broad section of the central part of the United States bounded on the north by the Canadian borders of Montana and North Dakota, then south along an area just east of the Mississippi River to the Gulf of Mexico at the panhandle of Florida; the western boundary is defined by the western edges of the Rocky Mountains and widens in the Southwest to include most of Arizona and the southern tip of California. Most information about selenium toxicity in plants comes from studies of U.S. flora because the seleniferous soils include a major livestock grazing range. In fact, investigation of two livestock-poisoning syndromes referred to as "alkali disease" and "blind staggers" revealed that ingestion of plants containing high levels of selenium by the grazing livestock caused these poisonings. Related investigations led to the discovery of selenium-accumulating and selenium-tolerant plants.

Most of the selenium-accumulating plants belong to the genus *Astragalus* (milk vetch), although two Australian accumulators, unrelated to *Astragalus*, have been identified as *Neptunia amplexicaulis* Domin. (Leguminosae, Peterson and Butler, 1971) and the shrub *Morinda reticulata* Benth. (Rubiaceae, Peterson and Butler, 1971). Seeds of a Venezuelan deciduous tree *Lecythis ollaria* Linn. contain the highest levels of selenium ever measured in plant material (18,000 µg/g) (Aronow and Kerdel-Vegas, 1965). The North American flora can be divided into three groups: (1) primary selenium accumulators, or plants that accumulate several thousand micrograms of selenium per gram of tissue, including milk vetch, goldenweed *(Oonopsis)*, and prince's plume *(Stanleya)*; (2) secondary accumulators, or plants that absorb up to 1000 µg selenium per gram, including wild aster *(Aster)*, salt bush *(Atriplex)*, paintbrush *(Castilleja)*, and matchweed *(Gutierrezia)*; and (3) nonaccumulators, or plants that generally do not contain more than 25 µg selenium per gram, including forage and crop plants. The primary accumulator plants are also indicators of seleniferous soil because they require this soil for growth. This fact suggests that these plants have a metabolic requirement for selenium.

A small, although growing, body of knowledge is beginning to define a metabolic role for selenium in prokaryotes and eukaryotes. Two good reviews on this subject (Stadtman, 1980, 1987) are available. A select group of enzymes contains selenocysteine. These enzymes include glycine reductase (Sliwkowski and Stadtman, 1987), formate dehydrogenase (Zinoni et al., 1987), and glutathione peroxidase (Forström et al., 1978). In these enzymes there is evidence that the selenol group, which is largely ionized at physiological pH, functions as a redox center. Selenocysteine is also part of a 75-kDa protein found in liver, kidney, and serum. Selenium has been found as a dissociable component of nicotinic acid hydroxylase (Dilworth, 1982) and xanthine dehydrogenase of Clostridia. The UGA stop codon specifies selenocysteine incorporation into these proteins. Several bacterial tRNAs contain selenium as 5-methylaminomethyl-2-selenouridine in the "wobble" position of the anticodons of certain transfer RNAs (tRNAs) for glutamate and lysine. The anticodon-codon interaction is significantly influenced by the selenouridine (Ching et al., 1985). The physiological consequences of these seleno-tRNAs is unknown. Glutathione peroxidase is induced by the addition of sodium selenite to the growth media for Chlamydomonas reinhardii (Yokota et al., 1988). Glutathione peroxidase from the marine diatom Thalassiosira pseudonana is a selenoprotein. This is consistent with the observation that this organism requires selenium for growth (Price and Harrison, 1988). The requirement for selenium appears to be an indicator of selenoenzymes or seleno-tRNAs in these organisms. If this is indeed the case, the requirement that selenium-tolerant and selenium-accumulating plants have for selenium suggests that these plants also contain selenobiopolymers. This area of plant research is essentially unexplored and in serious need of investigation.

A basic property of selenium that contributes to its toxicity to plants is its structural similarity to sulfur and its capability of mimicking and therefore replacing sulfur in certain biological compounds. It is not a perfect mimic, however, and cannot carry out the functions that sulfur provides to living systems. This failure is considered to be the biological basis for the toxic effects selenium produces in most plants and in the grazing livestock that feed on selenium accumulators.

In order for a metal to be toxic, it must first be taken up by the plant. Only the inorganic anions selenate and selenite can readily enter and be translocated to other parts of the plant (Shrift and Ulrich, 1969; Ulrich and Shrift, 1968; Asher et al., 1977). Selenate is accumulated in cells against a concentration gradient by what is probably an active transport process. This anion inhibits sulfur transport and may move into the cell via the sulfur transport mechanism (Persson, 1969; Pettersson, 1961, 1966). Although the uptake and translocation of selenite are not as well understood, they are clearly distinct from the mechanisms that move selenate. The first step in the metabolism of selenate is thought to be the reduction to selenite. The enzyme ATP sulfurylase catalyzes the reduction of sulfate to sulfite in cells and can also catalyze the reduction of selenate to selenite *in vitro* (Burnell, 1981a; Ellis, 1969; Shaw and Anderson, 1974), although this function has not yet been demonstrated *in vivo*. Selenite is reduced to selenide. This reduction has been examined in pea chloroplasts, and a scheme has been proposed (Ng and Anderson, 1978a,b, 1979). Selenide is combined with O-acetylserine to form selenocysteine in a reaction catalyzed by cysteine synthetase (Ng and Anderson, 1978a,b, 1979). Again, selenium mimics sulfur because cysteine synthetase catalyzes the normal biosynthesis of cysteine in the absence of selenium. It is not known if selenocysteine is further metabolized by the enzymes of the biosynthetic pathway that uses cysteine as a substrate for methionine synthesis. Selenocysteine and selenomethionine are found in selenium-sensitive plants.

The nonaccumulating, selenium-sensitive plants readily incorporate selenocysteine and selenomethionine into their proteins. The differences between sulfur and selenium are responsible for the toxicity of selenium. The elements differ in their radii, selenium is much larger than sulfur, and a bond between two selenium atoms is longer and weaker than the analogous disulfide bond. Furthermore, the ionization properties of selenol (-SeH) and sulfydryl (-SH) are vastly different. At pH 7.0 to 7.5 the selenol group of selenocysteine is almost completely ionized ($-Se^+ + H^+$), whereas the sulfydryl group of cysteine is about 80% protonated (-SH) (Huber and Criddle, 1967). These differences translate into significant alterations in the tertiary structure of proteins that have selenocysteine substituted for cysteine. Cysteine plays an important role in the tertiary structure of many proteins because of the formation of disulfide bonds between cysteine residues within a protein and because of the role of the sulfydryl group in hydrogen bonding to other amino acid residues. The sulfur atoms of cysteine also form iron-binding sites in enzymes carrying out electron transfers. Methionine is much less critical to protein folding and conformation than cysteine, and as expected, substitution with selenomethionine is much less disruptive to protein structure.

Sulfur amino acids participate in other cellular functions that could be disrupted by selenium, although most of our understanding of this comes from studies of bacteria. One such cellular function is the transfer of the sulfur of cysteine to uracil to form 4-thiouracil, a component of specific types of tRNA (Ching *et al.*, 1985). The extent of the cellular disruption and toxicity caused by this substitution is not understood. A second possible toxic substitution is the formation of Se-adenosylmethionine in place of the common methyl donor, S-adenosylmethionine (Colombani *et al.*, 1975).

Selenium-insensitive plants differ from selenium-sensitive plants in two recognized properties. First, the insensitive plants, which are selenium accumulators, accumulate nonprotein selenoamino acids. These compounds usually contain most of the selenium found in the accumulators and have been suggested to act in selenium detoxification. The major compound identified is Se-methylselenocysteine, which has been detected in several *Astragalus* accumulators and in a few other accumulators (Trelease *et al.*, 1960). Sulfur-methylcysteine is also found in *Astragalus* accumulators (Dunnill and Fowden, 1967). These two compounds are apparently absent in nonaccumulator species (Dunnill and Fowden, 1967). A dipeptide, γ-glutamyl-Se-methylselenocysteine, is also unique to accumulators and has been detected in *Astragalus* (Nigam *et al.*, 1969). The two Australian selenium accumulators contain large amounts of selenocystathione (Peterson and Robinson, 1972). The enzyme β-cystathionase, which cleaves cystathionine, is unable to cleave the selenium analogue and the analogue therefore accumulates. The selenoanalogues are apparently synthesized in the same way as the sulfur-containing compounds they resemble (Chow *et al.*, 1973).

Second, selenium accumulators are capable of excluding selenoamino acids from their proteins (Peterson and Butler, 1967; Brown and Shrift, 1981). Exclusion of the selenoamino acids from cellular protein was first suggested for *N. amplexicaulis* (Peterson and Butler, 1967), and the existence and function of the mechanism was later confirmed in *Astragalus* by Brown and Shrift (1981). Exclusion occurs by one of two mechanisms. Accumulator species either limit the intracellular concentration of the selenoamino acids or discriminate against them and prevent their incorporation into proteins. The intracellular concentration of selenoamino acids might be managed by synthesis of significant amounts of the nonprotein selenoamino acids discussed above. Alternatively, the inability of cysteinyl-tRNA synthetase to attach Se-cysteine to the cysteine tRNA results in the exclusion of selenocysteine from the cellular proteins of *A. bisulcatus* (Burnell and Shrift, 1979; Burnell, 1981a,b). This mechanism is not universal among accumulator species or within *Astragalus*. Other accumulator plants may discriminate against selenocysteine at a later stage of protein synthesis (Burnell and Shrift, 1979; Burnell, 1981a,b).

G. Zinc

Zinc enters the soil as the result of many of the same activities associated with the introduction of other trace metal ions. Mining and processing of zinc result in contamination of soils (Folkeson and Andersson-Bringmark, 1988; Løbersli and Steinnes, 1988, for examples). Zinc has recently been introduced into certain phytopharmaceutical products that are applied to different crop species (Magalhães *et al.*, 1985). Like cadmium and copper, zinc is a component of industrial waste and sewage sludge.

This metal ion is also required by plants in small amounts. Metalloenzymes containing zinc include many of the enzymes involved in DNA and RNA synthesis and processing and protein synthesis and metabolism. Indeed, zinc is known to be a cofactor to over 200 enzymes, and of the elements encountered at the active site of enzymes, zinc is the only one that participates with representation of all six classes of enzymes (oxidoreductases, transferases, hydrolases, lyases, isomerases, and ligases). However, in higher concentrations, this ion causes the blockage of xylem elements (Robb *et al.*, 1980) and inhibits photosynthesis by inhibiting electron transport and the carboxylase capacity of ribulose-1,5-bisphosphatase carboxylase/oxygenase (Van Assche and Clijsters, 1986a,b).

There is restricted uptake of zinc by zinc-tolerant *Agrostis tenuis* (Mathys, 1973). However, this appears to be an uncommon mechanism of trace metal tolerance in higher plants. Zinc normally is not excluded from either tolerant or sensitive plants. On entry into the plant, zinc must be complexed in order to avoid nonspecific interactions that might be deleterious to different biochemical processes. The cell wall does not appear to be involved in this process (Poulter *et al.*, 1985). Small organic molecules are probably involved in this chelation. Molecules that might chelate zinc include organic acids and poly(γ-glutamylcysteinyl)glycines. Induction of synthesis of these polypeptides in *Rauwolfia serpentia* has been reported for zinc and a wide range of other metals, although binding to these metal ions was not demonstrated (Grill *et al.*, 1985, 1987). The structure of these polypeptides suggests that binding to other metal ions is likely, but there appears to be no binding to zinc either *in vivo* or *in vitro* (Reese and Wagner, 1987a), and induced synthesis of these polypeptides has not been detected in response to zinc in other plants (Fujita and Kawanishi, 1987; Kishinami and Widholm, 1987; P. J. Jackson *et al.* unpubl. results). Moreover, inhibition of the synthesis of these polypeptides in cultured plant cells did not lead to zinc sensitivity (Reese and Wagner, 1987b). It therefore appears that the production of poly(γ-glutamylcysteinyl)glycines does not play an important role in zinc tolerance.

Zinc-tolerant plants often accumulate high levels of organic acids. The role these acids play in tolerance is not completely understood; however, the formation of zinc:citrate complexes has been suggested as a contributing factor in zinc detoxification (Godbold *et al.*, 1988). Complexing of zinc by organic acids may lack specificity (Foy *et al.*, 1978), and recent observations on malate synthesis in cell culture of zinc-tolerant or nontolerant ecotypes revealed that the accumulation of malate was unrelated to metal tolerance (Qureshi *et al.*, 1986). Moreover, if these acids play a primary role in zinc tolerance, then tolerant lines must contain qualitative or large quantitative differences in the amounts of the compounds in all parts of the plant or in the parts exposed to the metal ion. In most species, this condition cannot be fulfilled. However, Van Steveninck *et al.* (1987) reported that phytic acid is synthesized in the cytoplasm and the rate of synthesis is enhanced by the formation of zinc:phytate. This process could lead to the detoxification of zinc in the cytoplasm. In some plants, zinc is accumulated in vacuoles. This ion must be complexed immediately after its uptake into cytoplasm. It would then have to be transported to the vacuole by a special transport mechanism. Mathys (1977) proposed a tolerance mechanism involving organic acids as carriers of zinc. Malate is found primarily in the cytoplasm and can bind zinc at this site. This organic acid has been found in much higher concentrations in some tolerant ecotypes than in their zinc-sensitive counterparts. The stability of the zinc:malate complex is sufficient to reduce ionic zinc to a low level in the cytoplasm. It is possible that this complex could be transported through the tonoplast into the vacuole where the zinc could form a complex with oxalate due to the higher stability of the zinc:oxalate complex. The malate could then diffuse back into the cytoplasm where it could again react with zinc. Thus, the concentration of malate within the cytoplasm would not have to increase in the presence of zinc. The availability of a mechanism for transport of the zinc:malate complex into the vacuole would then be an important factor in determining tolerance.

The models outlined above may provide an accurate description of zinc tolerance in higher plants. However, very little information is available to support all aspects of these models. Organic acids appear to play some role in detoxification and zinc complexes are clearly stored in regions of the plant that appear to be metabolically inactive. More research is required before any of these models can be verified. A multidisciplinary approach to this study is required to provide information about tolerance at all levels of plant organization.

H. Other Toxic Trace Metals

Several other metal ions are toxic at high concentrations to all plant and animal life. All metal ions are toxic if the concentration of the free ion exceeds a specific threshold. Toxic trace metals include mercury, lead, uranium, and several others. Some of these ions interfere with root development and photosynthesis (Prasad and Prasad, 1987a,b; Godbold and Hüttermann, 1988), and many other physiological processes are probably susceptible. In the literature there are examples of plant species or ecotypes that are tolerant to many of these toxic trace metal ions (Sarkar and Jana, 1985; Poulter *et al.*, 1985, for examples). In some cases the mechanisms of tolerance appear to be similar to those for the metal ions described above (Poulter *et al.*, 1985). However, just as the physiological basis of toxicity is not confined to a single susceptible site within the plant, tolerance cannot be explained by a single mechanism. Based on current knowledge of the biochemical bases of tolerance to some metal ions, it is possible to propose mechanisms for tolerance to ions that have similar properties. The production of thiol-rich polypeptides appears to be regulated by exposure to several trace metal ions (Grill *et al.*, 1985), and it appears that these compounds have the properties required to bind some of these metals (Grill *et al.*, 1985, 1987; Jackson *et al.*, 1985a, 1987; Steffens *et al.*, 1986). Mercury and other divalent ions might be expected to bind to such thiol-rich compounds. The production of organic acids has also been shown to play a role in tolerance to a number of the different metal ions described above. It is possible that these compounds are also involved in chelation of other ions that have not been tested. They may also be involved in changing the environment of the rhizosphere to minimize the impact of certain ions. It is possible that some species produce large amounts of specific anions that result in the precipitation of the toxic cations as insoluble salts. There are also examples of the reduction of trace metal ions to their metallic form, which results in the precipitation of the metal. This is less toxic to the plant. In some species, the cell wall plays a specific role in metal tolerance, since the toxic ions are chelated to components of the wall. Lead tolerance in *Anthoxanthum odoratum* appears to be associated with the cell wall (Poulter *et al.*, 1985). These walls are a reservoir of a large number of immobilized chemical entities that could play an important role in the chelation of trace metal ions. However, they may lack specificity for the binding of single toxic species. The study of this phenomenon is complex because of the presence of such a large number of chemical species that could potentially play such a role.

III. CONCLUSIONS

A significant amount of information concerning mechanisms of metal tolerance in plants and plant cells has been accumulated. While some important aspects of tolerance have been elucidated, there is no example where all aspects of a specific mechanism of tolerance have been completely characterized. Plants and plant cells synthesize organic acids or polypeptides that chelate metal ions, and this synthesis is a required part of some tolerance mechanisms. However, the synthesis of these compounds alone is not sufficient to explain the difference between sensitive and tolerant ecotypes (Delhaize *et al.*, 1989a, for an example). It is clear that plants accumulate toxic amounts of certain trace metals in chelated forms that are less toxic to the plant. However, the mechanisms by which these complexes are stored are not yet understood. Research has identified several important aspects of metal tolerance, but there is no model species in which the entire mechanism of tolerance has been determined. Efforts to coordinate research to address all aspects of tolerance are required if the biochemical and molecular mechanisms underlying tolerance to toxic concentrations of trace metal ions in plants are to be understood.

It is important to understand the mechanisms of metal tolerance in higher plants, because without this information it is impossible to predict the impact of certain industrial processes on the transport of toxins through the environment. Such an impact ultimately affects human health. Storage of toxic ions in plant tissues available for predation will have a different impact than metal ions excluded from these species.

In addition to providing the necessary data for the production of models that will accurately predict the impact of certain industrial activities on human health and the environment, the study of metal tolerance may provide methods to stabilize already contaminated sites. Certain native species of grasses are being used to stabilize areas contaminated with mine tailings (Gemmel, 1977). Such plants may be valuable for environmental remediation. Alternatively, characterization of the processes plants use to remove and sequester toxic metal ions provide models for the development of new chemical or biological techniques to remove trace metal ions and other toxic chemical compounds from the complex solutions generated by industrial processes.

REFERENCES

Agricola, G. (1556). *In* "De Re Metallica" (Hoover and Hoover, eds.), Min. Mag.) London (Engl. transl., 1912).
Andersson, M. (1988). *Water, Air, Soil Pollut.* 39, 439–462.
Aronow, L., and Kerdel-Vegas, F. (1965). *Nature (London)* 205, 1185–1186.
Asher, C. J., Butler, G. W., and Peterson, P. J. (1977). *J. Exp. Bot.* 28, 279–291.
Awad, A. S., Edwards, D. G., and Milham, P. J. (1976). *Plant Soil* 45, 531.
Bariaud, A., Bury, M., and Mestre, J. C. (1985). *Physiol. Plant.* 63, 382–386.
Bartolf, M., Brennan, E., and Price, C. A. (1980). *Plant Physiol.* 66, 438–441.
Bennetzen, J. L., and Adams, T. L. (1984). *Plant Cell Rep.* 3, 258–261.
Berger, J. M., Jackson, P. J., Robinson, N. J., Lujan, L. D., and Delhaize, E. (1989). *Plant Cell Rep.* 7, 632–635.
Bieleski, R. I. (1970). *Plant Anal. Fert. Probl., Abstr. Int. Colloq., 6th,*
Blamey, F. P. C., Edwards, D. G., and Asher, C. J. (1983). *Soil Sci.* 136, 197–207.
Bradfield, E. G. (1976). *Plant Soil* 44, 495–499.
Bradshaw, A. D. (1984). *In* "Origins and Development of Adaption" (D. Evered and G. M. Collins, eds.), pp. 4–19. Pitman, London.
Brown, P. H., Welch, R. M., and Cary, E. E. (1987). *Plant Physiol.* 85, 801–803.
Brown, T. A., and Shrift, A. (1981). *Plant Physiol.* 67, 1051–1053.

Brown, T. A., and Shrift, A. (1982). *Biol. Rev.* **57**, 59–84.

Burnell, J. N. (1981a). *Plant Physiol.* **67**, 316–324.

Burnell, J. N. (1981b). *Plant Physiol.* **67**, 325–329.

Burnell, J. N., and Shrift, A. (1979). *Plant Physiol.* **63**, 1095–1097.

Bussler, W. (1981). *In* "Copper in Soils and Plants" (J. F. Loneragan, A. D. Robson, and R. D. Graham, eds.), pp. 213–234. Academic Press, New York.

Casterline, J. L., Jr., and Barnett, N. M. (1982). *Plant Physiol.* **69**, 1004–1007.

Cataldo, D. A., Garland, T. R., and Wildung, R. E. (1981). *Plant Physiol.* **68**, 835–839.

Ching, W.-M., Alzner-DeWeerd, B., and Stadtman, T. C. (1985). *Proc. Natl. Acad. Sci. U.S.A.* **82**, 347–350.

Chow, C. M., Nigam, S. N., and McConnell, W. B. (1973). *Can. J. Biochem.* **51**, 489–490.

Colombani, F., Cherest, H., and de Robichon-Szulmajster, H. (1975). *J. Bacteriol.* **122**, 375–384.

Dabin, P., Marafante, E., Mousny, J. M., and Myttenaere, C. (1978). *Plant Soil* **50**, 329–341.

DeBoni, U., Scott, J. W., and Crapper, D. R. (1974). *Histochemistry* **40**, 31–37.

Delhaize, E., Loneragan, J. F., and Webb, J. (1985). *Plant Physiol.* **78**, 4–7.

Delhaize, E., Jackson, P. J., Lujan, L. D., and Robinson, N. J. (1989a). *Plant Physiol.* **89**, 700–706.

Delhaize, E., Robinson, N. J., and Jackson, P. J. (1989b). *Plant Mol. Biol.* **12**, 487–497.

Dilworth, G. L. (1982). *Arch. Biochem. Biophys.* **219**, 30–38.

Dixon, N. E, Gazzola, C., Blakely, R. L., and Zerner, B. (1975). *J. Am. Chem. Soc.* **97** 4131–4133.

Dunnill, P. M., and Fowden, L. (1967). *Phytochemistry* **6**, 1659–1663.

Edelman, L., Czarnecka, E., and Key, J. L. (1988). *Plant Physiol.* **86**, 1048–1056.

Ellis, K. J., Vartsky, D., Zanzi, I., and Cohn, S. H. (1979). *Science* **205**, 323–324.

Ellis, K. J. (1969). *Planta* **88**, 34–42.

Faulstich, H., and Stournasas, C. (1985). *Nature (London)* **317**, 714–715.

Folkeson, L., and Andersson-Bringmark, E. (1988). *Can. J. Bot.* **66**, 417–428.

Folster, H. (1985). *In* "The Chemistry of Weathering" (J. I. Drever, ed.), pp. 197–209. Reidel Publ., New York.

Forström, J. W., Zakowski, J. J., and Tappel, A. L. (1978). *Biochemistry* **17**, 2639–2644.

Fowler, B. A., Hildebrand, C. E., Kojima, T., and Webb, M. (1987). *In* "Metallothionein II" (J. H. R. Kagi and Y. Kagima, eds.), pp. 19–22. Birkhaeuser, Basel.

Fowler, J. L., and Morgan, P. W. (1972). *Plant Physiol.* **49**, 555–559.

Foy, C. D. (1973). *In* "Manganese," pp. 51–76. National Academy of Sciences, National Research Council, Washington, D.C.

Foy, C. D. (1974). *In* "The Plant Root and Its Environment" (E. W. Carson, ed.), pp. 601–642. Univ. Press of Virginia, Charlottesville.

Foy, C. D. (1982). "Acidic Desposition Critical Assessment Document." EPA and North Carolina State University, Raleigh.

Foy, C. D. (1983). *Iowa State J. Res.* **57**, 355.

Foy, C. D., and Campbell, T. A. (1984). *J. Plant Nutr.* **7**, 1365–1388.

Foy, C. D. (1984). "Soil Acidity and Liming," 2nd ed., Agron. Monogr. No. 12.

Foy, C. D., and Fleming, A. L. (1982). *J. Plant Nutr.* **5**, 1313–1333.

Foy, C. D., Chaney, R. L., and White, M. C. (1978). *Annu. Rev. Plant Physiol.* **29**, 511–566.

Foy, C. D., Lee, E. H., and Wilding, S. B. (1987). *J. Plant Nutr.* **10**, 1089–1101.

Fujita, M. (1985). *Plant Cell Physiol.* **26**, 295–300.

Fujita, M., and Kawanishi, T. (1987). *Plant Cell Physiol.* **28**, 379–382.

Garty, J. (1988). *Can. J. Bot.* **66**, 668–671.

Gemmel, R. R. (1977). "Colonization of Industrial Wasteland." Arnold, London.

Gilissen, L. J. W., and Van Staveren, M. J. (1986). *J. Plant Physiol.* **125**, 95–103.

Godbold, D. L., and Hüttermann, A. (1988). *Physiol. Plant.* **74**, 270–275.

Godbold, D. L., Fritz, E., and Hüttermann, A. (1988). *Proc. Natl. Acad. Sci. U.S.A.* **85**, 3888–3892.

Grill, E., Winnacker, E.-L., and Zenk, M. H. (1985). *Science* **230**, 674–676.

Grill, E., Winnacker, E.-L., and Zenk, M. H. (1987). *Proc. Natl. Acad. Sci. U.S.A.* **84**, 439–443.

Grill, E., Winnacker, E.-L., and Zenk, M. H. (1988). *Experientia* **44**, 539–540.

Hecht-Buchholz, C., and Foy, C. D. (1981). *Plant Soil* **63**, 93–95.

Helyar, K. R., and Anderson, A. J. (1971). *Aust. J. Agric. Res.* **22**, 707.

Houtz, R. L., Nable, R. O., and Cheniae, G. M. (1988). *Plant Physiol.* **86**, 1143–1149.

Hsu, B. -D., and Lee, J.-Y. (1988). *Plant Physiol.* **87**, 116–119.

Huang, B., and Goldsbrough, P. B. (1988). *Plant Cell Rep.* **7**, 119–122.

Huang, B., Hatch, E., and Goldsbrough, P. B. (1987). *Plant Sci.* **52**, 211–221.

Huber, R. E., and Criddle, R. S. (1967). *Arch Biochem. Biophys.* **122**, 164–173.

Hunt, C. B. (1972). "Geology of Soils, Classification and Uses." Freeman, San Fransisco, California.

Jackson, P. J., Roth, E. J., McClure, P. R., and Naranjo, C. M. (1984). *Plant Physiol.* **75**, 914–918.

Jackson, P. J., Barton, K., Naranjo, C. M., Sillerud. L. O., Trewhella, J., Watt, K., and Robinson, N. J. (1985a). *Int. Conf. Plant Mol. Biol., 1st,* p. 35.

Jackson P. J., Naranjo, C. M., McClure, P. R., and Roth, E. J. (1985b). *In* "Cellular and Molecular Biology of Plant Stress," (J. L. Key and T. Kosuge, eds.) pp. 145–160. Liss, New York.

Jackson, P. J., Unkefer, C. J., Doolen, J. A., Watt, K., and Robinson, N. J. (1987). *Proc. Natl. Acad. Sci. U.S.A.* **84**, 6619–6623.

Jackson, P. J., Robinson, N. J., and Delhaize, E. (1989). *In* "Metal Ion Homeostasis: Molecular Biology and Chemistry," pp. 337–346. Liss, New York.

Johnson, N. M., Driscoll, Ch. T., Eaton, J. S., Likens, G. E., and McDowell, W. H. (1981). *Geochim. Cosmochim. Acta* **45**, 1421–1437.

Kinraide, T. B. (1988). *Plant Physiol.* **88**, 418–423.

Kishinami, I., and Widholm, J. M. (1987). *Plant Cell Physiol.* **28**, 203–210.

Klan, D. H., Duckett, G., Frankland, B., and Kirkham, J. B. (1984). *J. Plant Physiol.* **115**, 19–28.

Kondo, N., Imai, K., Isobe, M., and Goto, T. (1984). *Tetrahedron Lett.* **25**, 3869–3872.

Lagerwerff, J. V., and Specht, A. W. (1970). *Environ. Sci. Technol.* **4**, 483–486.

Lee, J., Reeves, R. D., Brooks, R. R., and Jaffre, T. (1978). *Phytochemistry* **17**, 1033–1035.

Lindberg, S., and Wingstrand, O. (1985). *Physiol Plant.* **63**, 181–186.

Lipman, C. B., and Mackinney, G. (1931). *Plant Physiol.* **6**, 593–599.

Løbersli, M. E., and Steinnes, E. (1988). *Water, Air, Soil Pollut.* **37**, 25–39.

Lolkema, P. C., and Vooijs, R. (1986). *Planta* **167**, 30–36.

Lolkema, P. C., Doornhof, M., and Ernst, W.H.O. (1986). *Physiol. Plant.* **67**, 654–658.

Magalhães, M. J., Sequeira, E. M., and Lucas, M. D. (1985). *Water, Air, Soil Pollut.* **26**, 1–17.

Mathys, W. (1973). *Flora Jena* **162**, 492–499.

Mathys, W. (1977). *Physiol. Plant.* **40**, 130–136.

Medappa, K. C., and Dana, M. N. (1970). *J. Am. Soc. Hortic. Sci.fl* **95**, 107.

Mengel, K., and Kirkby, E. A. (1982). "Principles of Plant Nutrition," 3rd ed. International Potash Institute, Bern, Switzerland.

Meredith, C. P. (1978). *Plant Sci. Lett.* **12**, 25–34.

Morgan, P. W., Taylor, D. M., and Joman, H. E. (1976). *Physiol. Plant.* **37**, 149–156.

Morimura, S., Takahashi, E., and Matsumoto, H. (1978). *Z. Pflanzenphysiol.* **88**, 395–402.

Mugwira, L. M., Elgawhary, S. M., and Patel, K. I. (1976). *Agron. J.* **68**, 782–787.

Mullins, M., Hardwick, K., and Thurman, D. A. (1985). *In* "Heavy Metals in the Environment" (T. D. Lekkas, ed.), Vol. 2, pp. 43–46. CEP Consultants Ltd., Edinburgh.

Nable, R. A., Houtz, R. L., and Cheniae, G. M. (1988). *Annu. Rev. Plant Physiol.* **35**, 415–442.

Natusch, D. F. S., Wallace, J. R., and Evans, C. A. (1974). *Science* **183**, 202–204.

Ng, B.H., and Anderson, J. W. (1978a). *Phytochemistry* **17**, 879–885.

Ng, B. H., and Anderson, J. W. (1978b). *Phytochemistry* **17**, 2069–2074.

Ng, B. H., and Anderson, J. W. (1979). *Phytochemistry* **18**, 573–580.

Nieboer, E., and Richardson, D. H. (1980). *Environ, Pollut., Ser. B* **1**, 3.

Neitzke, M., and Runge, M. (1985). *Flora (Jena)* **177**, 237–249.

Nigam, S. N., Tu, J., McConnell, W. B. (1969). *Phytochemistry* **8**, 1161–1165.

Nordberg, M., and Kojima, Y. (1979). *In* "Metallothionein" (J. H. R. Kagi and M. Nordberg, eds.), pp. 197–204. Birkhaeuser, Boston, Massachusetts.

Ojima, K., Abe, H., and Ohira, K. (1984). *Plant Cell Physiol.* **25**, 855–858.

Page, A. L., Bingham, F., and Nelson, C. (1972). *J. Environ. Qual.* **1**, 288.

Persson, L. (1969). *Physiol. Plant.* **22**, 959–976.

Peterson, P. J., and Butler, G. W. (1967). *Nature (London)* **213**, 599–600.

Peterson, P. J., and Butler, G. W. (1971). *Aust. J. Biol. Sci.* **24**, 175–177.

Peterson. P. J., and Robinson, P. J. (1972). *Phytochemistry* **11**, 1837–1839.

Petit, C. M., Ringoet, A., and Myttenaere, C. (1978). *Plant Physiol.* **62**, 554–557.

Petolino, J. F., and Collins, G. B. (1985). *J. Plant Physiol.* **118**, 139–144.

Pettersson, S. (1961). *Physiol. Plant.* **14**, 124–132.

Pettersson, S. (1966). *Physiol. Plant.* **19**, 459–492.

Poulter, A., Collin, H. A., Thurman, D. A., and Hardwick, K (1985). *Plant Sci.* **42**, 61–66.

Prasad, D. D. K., and Prasad, A. R. K. (1987a). *J. Plant Physiol.* **127**, 241–249.

Prasad, D. D. K., and Prasad, A. R. K. (1987b). *Phytochemistry* **26**, 881–883.

Price, N. M., and Harrison, P. J. (1988). *Plant Physiol.* **86**, 192–199.

Qureshi, J. A., Hardwick, K., and Collin, A. (1986). *J. Plant Physiol.* **122**, 477–479.

Rauser, W. E. (1984). *J. Plant Physiol.* **116**, 253–260.

Rauser, W. E., and Ackerley, C. A. (1987). *Can. J. Bot.* **65**, 643–646.

Rauser, W. E., and Curvetto, N. R. (1980). *Nature (London)* **287**, 563.

Rauser, W. E., and Winterhalder, E. K. (1985). *Can. J. Bot.* **63**, 58–63.

Rauser, W. E., Hartman, H.-J., and Weser, U. (1983). *FEBS Lett.* **164**, 102–104,

Reese, R. N., and Roberts, L. W. (1985). *J. Plant Physiol.* **120**, 123–130.

Reese, R. N., and Wagner, G. J. (1987a). *Plant Physiol.* **84**, 574–577.

Reese, R. N., and Wagner, G. J. (1987b). *Biochem. J.* **241**, 641–647.

Reese, R. N., Mehra, R. K., Tarbet, E. B., and Winge, D. R. (1988). *J. Biol. Chem.* **263**, 4186–4192.

Robb, J., Busch, L., and Rauser, W. E. (1980). *Ann. Bot. (London)* [N.S.] **46**, 43–50.

Robinson, N. J., and Jackson, P. J. (1986). *Physiol. Plant.* **67**, 499–506.

Robinson, N. J., and Thurman, D. A. (1986a). *Proc. R. Soc. London., Ser. B* **227**, 493–501.

Robinson, N. J., and Thurman, D. A. (1986b). *Planta* **169**, 192–197.

Robinson, N. J., Ratliff, R. L., Anderson, P. J., Delhaize, E., Berger, J. M., and Jackson, P. J. (1988). *Plant Sci.* **56**, 197–204.

Rufty, T. W., Miner, G. S., and Raper, C. D., Jr. (1979). *Agron. J.* **71**, 638–644.

Sarkar, A., and Jana, S. (1985). *Water, Air, Soil Pollut.* **27**, 15–18.

Scheller, H. V., Huang, B., Hatch, E., and Goldsbrough, P. B. (1987). *Plant Physiol.* **85**, 1031–1035.

Schultz, C. L., and Hutchinson, T. C. (1985). *In* "Heavy Metals in the Environment" (T. D. Lekkas, ed.), Vol. 2, pp. 51–54. CEP Consultants Ltd., Edinburgh.

Sela, M., Tel-Or, E., Fritz, E., and Hüttermann, A. (1988). *Plant Physiol.* **88**, 30–36.

Shaw, W. H., and Anderson, J. W. (1974). *Biochem. J.* **139**, 37–42.

Shortle, W. C., and Smith, K. T. (1988). *Science* **240**, 1017–1018.

Shrift, A., and Ulrich, T. (1969). *Plant Physiol.* **44**, 893–896.

Sirkar, S., and Amin, J. V. (1974). *Plant Physiol.* **54**, 539 543.

Sirkar, S., and Amin, J. V. (1979). *Indian J. Exp. Biol.* **17**, 618–619.

Sliwkowski, M. X., and Stadtman, T. C. (1987). *Proc. Natl. Acad. Sci. U.S.A.* **85**, 368–371.

Stadtman, T. C. (1980). *Annu. Rev. Biochem.* **49**, 93–110.

Stadtman, T. C. (1987). *FASEB J.* **1**, 375–379.

Steffens, J. C., Hunt, D. F., and Williams, B. G. (1986). *J. Biol. Chem.* **261**, 13879–13882.

Stobart, A. K., Griffiths, W. T., Ameen-Bukhari, I., and Sherwood, R. P. (1985). *Physiol. Plant.* **63**, 293–298.

Street, J. J., Sabey, B. R., and Lindsay, W. L. (1978). *J. Environ. Qual.* **7**, 286–290.

Suhayda, C. G., and Haug, A. (1986). *Physiol. Plant.* **68**, 189–195.

Takijima, Y., Katsumi, F., and Koizumi, S. (1973a). *Soil Sci. Plant Nutr.* **19**, 183–193.

Takijima, Y., and Katsumi, F. (1973b). *Soil Sci. Plant Nutr.* **19**, 235–244.

Taylor, G. J., and Foy, C. D. (1985). *Can. J. Bot.* **63**, 2181–2186.

Thornton, F. C., Schaedle, M., Raynal, D. J., and Zipperer, C. (1986). *J. Exp. Bot.* **37**, 775–785.

Trelease, S. F., DiSomma, A. A., and Jacobs, A. L. (1960). *Science* **132**, 618.

Tukendorf, A., Lyszcz, S., and Bazynski, T. (1984). *J. Plant Physiol.* **115**, 351–360.

Turner, R. G. (1969). *In* "Ecological Aspects of the Mineral Nutrition of Plants" (I. H. Rorison, ed.), p. 399. Blackwell, Oxford.

Ulrich, B., Mayer, R., and Khanna, P. K. (1979). *Schriftenr. Forstl. Fak. Univ. Goettingen* **58**, 1–279.

Ulrich, T., and Shrift, A. (1968). *Plant Physiol.* **43**, 14–20.

Unsworth, M. H., and Harrison, R. M. (1985). *Nature (London)* **317**, 674.

Vallee, B. L., and Wacker, W. E. C. (1970). *In* "The Proteins" (B. L. Vallee and W. E. C. Walker, eds.), Vol. 5, pp. 25–60. Academic Press, New York.

Van Assche, F., and Clijsters, H. (1986a). *J. Plant Physiol.* **125**, 355–360.

Van Assche, F., and Clijsters, H. (1986b). *Physiol. Plant.* **66**, 717–721.

Van Bruwaene, R., Kirchmann, R., and Impens, R. (1984). *Experientia* **40**, 43–52.

Van Steveninck, R. F. M., Van Steveninck, M. E., Fernando, D. R., Horst, W. J., and Marschner, H. (1987). *J. Plant Physiol.* **131**, 247–257.

Vlamis, J., Williams, D. E., Corey, J. E., Page, A. L., and Ganje, T. J. (1985). *Soil Sci.* **139**, 81–87.

Wagatsuma, T., and Ezoe, Y. (1985). *Soil Sci. Plant Nutr.* **31**, 547–561.

Wagner, G. J., and Trotter, M. M. (1982). *Plant Physiol.* **69**, 804–809.

Wagner, G. J., Nulty, E., and LeFevre, M. (1984). *J. Toxic. Environ. Health* **13**, 979–989.

Watabe, T., Uchida, S., and Kamada, H. (1984). *J. Radiat. Res.* **25**, 274–282.

Weigel, H. J. (1985). *J. Plant Physiol.* **119**, 179–189.

Weigel, H. J. and Jäger, H. J. (1980). *Plant Physiol.* **65**, 480–482.

Weinstein, L. H., Kaur-Sawhney, R., Rajam, M. V., Wettlaufer, S. C., and Galston, A. W. (1986). *Plant Physiol.* **82**, 641–645.

Woolhouse, H. W. (1983). *Encycl. Plant Physiol., New Ser.* **12B**, 245–300.

Woolhouse, H. W., and Walker, S. (1981). *In* "Copper in Soils and Plants" (J. F. Loneragan, A. D. Robson, and R. D. Graham, eds.), pp. 236–267. Academic Press, Sydney, Australia.

Yasumura, S., Vartsky, D., Ellis, K. J., and Coh, S. H. (1980). *In* "Cadmium in the Environment" (J. O. Nriagu ed.), pp. 12–34. Wiley, New York.

Yokota, A., Shigeoka, S., Onishi, T., and Kitaoka, S. (1988). *Plant Physiol.* **86**, 649–651.
Yost, K. J., Miles, L. J., and Parsons, T. W. (1980). *Environ. Int.* **3**, 473–484.
Zhao, X.-J., Sucoff, E., and Stadelmann, E. J. (1987). *Plant Physiol.* **83**, 159–162.
Zinoni, R., Birkmann, A., Leinfelder, W., and Bock, A. (1987). *Proc. Natl. Acad. Sci. U.S.A.* **84**, 3156–3160.

CHAPTER 11

Spectrophotometric Detection of Plant Leaf Stress

Harold W. Gausman
Jerry E. Quisenberry

U.S. Department of Agriculture
Agricultural Research Service,
Cropping Systems Research Laboratory,
Lubbock, Texas

Spectrophotometrically made measurements of reflectance, transmittance, and absorptance of single leaves can often be utilized to detect plant stress or damage caused by insects, diseases, drought, ozone, mineral toxicity, nutrient deficiency, adverse temperature, growth regulators, and soil salinity. Generally, three wavebands are useful: (1) a portion of the visible light region (0.50–0.75 μm) affected by chlorophyll absorptance of red light; (2) the near-infrared light-reflectance region (0.75–1.35 μm) affected by internal leaf structure and dehydration; and (3) the near-infrared light water-absorptance region (1.35–2.50 μm) affected by leaf water content. (Usually, stressed leaves exhibit higher light reflectance than nonstressed leaves.) The ultraviolet light region (0.28- to 0.32-μm waveband) and the blue light region (0.45-μm wavelength) also show promise to detect plant stress.

Moreover, techniques such as fluorescence detection of water stress are promising.

I. INTRODUCTION

When a plant's leaf intercepts incoming radiation (photons) at a critical angle, a portion of the light is absorbed. (The term *light* is used here to denote the 0.35- to 2.50-μm waveband, which includes the longer wavelength ultraviolet, visible, and near-infrared electromagnetic radiation.) The amount of absorption depends on the energy (wavelength) of the photons involved; photons with the shortest wavelength (high-energy ultraviolet and visible light) are involved in photosynthesis and photochemical reactions such as photomorphogenesis, phototropism, and chlorophyll synthesis, whereas photons with the longest wavelength (low-energy infrared light) influence heating processes, evaporation, and transpiration. Consequently, alterations in the concentration of the leaf chloroplasts' pigments, internal leaf structure, and leaf tissue water content are largely responsible for inducing differences among various plant tissues in light absorptance.

A portion of the incoming photons are also reflected when they impinge on a plant leaf at a critical angle. Specular (mirror) reflectance occurs at the leaf cuticle, whereas diffuse reflectance originates from light scattering (multiple reflection) mainly within the leaf mesophyll. Light that is neither absorbed nor reflected is transmitted through the leaf mesophyll and out of the epidermis.

Spectrophotometric measurements of leaf reflectance, particularly in the visible and near-infrared light regions, are useful to previsually investigate or detect effects of various stresses on plants. These could include low and high temperature tolerance, water deficiency, nutrient deficiency and toxicity, saline conditions, various plant maladies, and atmospheric pollution. Moreover, the effects of exogenously applied chemicals to plant leaves should be detectable either in terms of stress or possibly in stimulated growth. Spectrophotometric plant leaf measurements have the potential to facilitate the screening of chemicals for biological applications.

II. BACKGROUND

Willstätter and Stoll (1918) explained leaf light reflectance and transmittance on the basis of critical reflection of visible light at the cell wall–air interface of spongy mesophyll tissue. Sinclair (1968) hypothesized that leaf reflectance derives from the diffuse characteristics of plant cell walls. Gausman *et al.* (1970) quantitatively related near-infrared light reflectance to the number of intercellular air spaces within the leaf mesophyll.

Reflectance of light from leaves occurs internally because it can be reduced by infiltrating leaves with water (Pearman, 1966; Moss, 1951) or with oil, and it was predicted that internal discontinuities other than air-cell interfaces were responsible for a part of the near-infrared light reflectance by a leaf (Woolley, 1971). This premise was confirmed because refractive index discontinuities among cell membranes, crystals, cell walls, and surrounding protoplasm did increase the reflectance of near-infrared light at the 0.85-μm wavelength (Gausman, 1977).

An experiment was conducted to determine the refractive index of plant leaves by vacuum infiltration (Gausman *et al.*, 1976). Leaves of black-eyed pea *(Pisum sativum)*, cucumber *(Cucumis sativus)*, string bean *(Phaseolus vulgaris)*, and tomato *(Lycopersicon esculentum)* plants were vacuum-infiltrated with distilled water and various mixtures of oil and hexane to provide several refractive indices. Infiltrated leaves reflected less light than non-infiltrated leaves over the 0.5- to 2.5-μm waveband because cell wall–air interfaces were partially eliminated. Minimal reflectance should occur when the average refractive index of plant cell walls is matched by the infiltration fluid. An average value of 1.425 approximated the refractive index of plant cell walls for the four plant genera.

Diffuse reflectance and transmittance of a compact leaf such as corn *(Zea mays)*, a leaf impregnated with water, and an immature cotton leaf *(Gossypium hirsutum)* (Gausman *et al.*, 1969) was predicted from a plate theory (Allen *et al.*, 1969). Cotton leaf maturation is characterized by cell division and expansion that increased light reflectance and reduced light transmittance of the leaf. Generalization of the plate theory (flat-plate model) to include the effect of intercellular air spaces led to the void area index (VAI) concept of a leaf. Consequently, a typical plant leaf can be specified by four optical parameters: (1) the equivalent water thickness (EWT), a number that specifies the amount of water in a leaf; (2) the void area index (VAI), a measure of the intercellular air space in a leaf; (3) the effective index of refraction n; and (4) the effective absorption coefficient k (Gausman *et al.*, 1970). Over the 1.35- to 2.50-μm waveband, the absorption spectra of leaves are like those of pure liquid water. Leaf reflectance differences among plant leaves over the 0.50- to 1.35-μm waveband are caused principally by Fresnel reflections at external and internal leaf surfaces and by plant pigment absorption. Reflectance over the 1.35- to 2.50-μm waveband is influenced largely from Fresnel reflections and water absorption. The leaf's effective index of refraction can be approximated by a cubic equation.

A leaf with an EWT specified by D can be subdivided conceptually into N compact layers of individual thickness D/N, which are subsequently piled back to the original thickness D. The thickness of the air plates that separate these compact layers is a matter of indifference but will be taken as infinitesimal to facilitate calculations.

A leaf of EWT specified by D will be regarded as a pile of N compact layers of individual thickness D/N. Transmittance t and reflectance r of a single leaf can be expressed at a given wavelength by the relations:

$$\frac{t}{a - a^{-1}} = \frac{r}{b - b^{-1}} = \frac{1}{ab - a^{-1}b^{-1}} \qquad (11.1)$$

where a and b are parameters to be determined by experiment. If the single leaf described by Eq. (11.1) is considered to be a pile of N layers, Eq. (11.1) becomes the following:

$$\frac{t}{a_o - a_o^{-1}} = \frac{r}{b_o^{N} - b_o^{-N}} = \frac{1}{a_o b_o^{N} - a_o^{-1} b_o^{-N}} \qquad (11.2)$$

where a_o and b_o are new parameters to be determined. As indicated by Allen and Richardson (1968), Eq. (11.2) was derived by Stokes in 1862 on the assumption of an integral number of piled transparent plates, and Ingle in 1942 showed that N need not be confined to integral values. The transformation between a, b, and a_o, b_o of Eqs. (11.1) and (11.2) can be written from inspection by the relations:

$$a = a_o \qquad (11.3)$$
$$b = b_0^N$$

The transmittance t_o and reflectance r_o of a single compact layer can be obtained by setting $N = 1$ in Eq. (11.2). Thus:

$$\frac{t_o}{a_o - a_o^{-1}} = \frac{r_o}{b_o - b_o^{-1}} = \frac{1}{a_o b_o - a_o^{-1} b_o^{-1}} \qquad (11.4)$$

Eliminate a_o and b_o from Eq. (11.4) by means of Eq. (11.3) to obtain the relations:

$$\frac{t_o}{a - a^{-1}} = \frac{r_o}{b^{1/N} - b^{-1/N}} = \frac{1}{ab^{1/N} - a^{-1}b^{-1/N}} \qquad (11.5)$$

Generalization of the plate theory to a leaf with intercellular spaces is not complete provided that the quantity N can be determined. Unfortunately, the effective index of refraction n and the quantity N are confounded; that is, unless the effective index of refraction n is assumed, the number N cannot be determined and vice versa.

The scattering coefficients of a single leaf and an individual compact layer can be specified by the respective relations:

$$S = [2a/(a^2 - 1)] \log b,$$
$$S_o = [2_{ao}/(a_o^2 - 1)] \log b_o \tag{11.6}$$

Substitute Eq. (11.3) into Eq. (11.6) and combine them to obtain the relation:

$$s/s_o = N \tag{11.7}$$

The value N is obtained in practice by the quotient $\bar{S}/\bar{S}_o$, where $\bar{s}$ of a given leaf is the average value of s measured over the 0.75- to 1.35-μm waveband spectral range and $\bar{s}_o = 0.61823$ is the average value of five infiltrated citrus leaves measured over the same spectral range. The VAI calculated by the value $N - 1$ obtained from Eq. (11.7) is arbitrary. The procedure is justified only by consistency of the results obtained.

Optical parameters can be subjected to statistical pattern recognition techniques to discriminate among kinds of leaves. Moreover, the electromagnetic signatures predicted from measured optical constants can be tested against quantitative reflectance measurements from ground or airborne sensors. Also, single-leaf spectral data can be used as an input in mathematical models to predict multiple-leaf reflectance with reasonable accuracy (Lillesaeter, 1982).

III. BASIC CONTRIBUTIONS OF LABORATORY SPECTROPHOTOMETRIC MEASUREMENTS

Single-leaf reflectance measurements have contributed greatly to an understanding of the interaction of electromagnetic radiation with plant leaves and other plant components such as leaf petioles, leaf sheaths, heads of grass, and inflorescences (Gausman, 1985). This had included spectrophotometric measurements of plant leaf reflectance, transmittance, and absorptance, generally over the 0.50- to 2.5-μm (sometimes 0.40- to 2.6-μm) waveband.

Laboratory measurements have been used to characterize visible light spectra (0.45- to 0.75-μm waveband), near-infrared light spectra (0.75- to 1.35-μm waveband), and infrared water-absorption light spectra (1.35- to 2.50-μm waveband) for biological factors such as leaf pigment concentration, pubescence, and senescence. Probably the best indicator of altered visible light optical properties is a change in reflectance at approximately the 0.55-μm wavelength caused by variations in leaf pigment concentration, primarily chlorophyll: High light reflectance is usually associated with low total chlorophyll concentration, and conversely, low light reflectance is correlated with high total chlorophyll concentration.

Near-infrared light is scattered or reflected from plant leaves by internal refractive discontinuities within the leaf mesophyll. The most important discontinuity is the cell wall–air space interface. If near-infrared light travels at a critical angle from a hydrated cell wall with a refractive index of 1.425 to an air space with a refractive index of 1.0, the near-infrared light is scattered or reflected. This is easily demonstrated by using vacuum infiltration to replace the air within leaves with a liquid. After liquid infiltration of the leaf, reflectance becomes minimal, presumably when the refractive index of the liquid matches the refractive index of the cell wall.

Attention to leaf maturity during sample collection is mandatory if ensuing spectrophotometric measurements are to be meaningful in comparing, for example, spectral reflectance differences of stressed and nonstressed leaves. In this respect, young leaves have a more compact internal structure within their mesophyll than do fully expanded nonstressed leaves. Hence, it is not always possible to identify the cause of a light spectrum that seems to represent a typical compact leaf mesophyll structure, because stress reflectance results may be confounded with leaf maturity.

Generally, leaves of stressed plants have a lower near-infrared light reflectance than leaves of nonstressed plants. Sometimes, however, spectrophotometric measurements indicate that diseased leaves have a higher near-infrared light reflectance than "normal" leaves. This effect is probably caused by dehydration of the diseased leaf, which causes damaged cells to collapse in such a manner that the number of air voids within the leaf mesophyll increases and, consequently, light scattering increases (Sinclair, 1968).

Internal discoloration of leaves and a black surface coating, or both, (see Hart and Myers, 1968) will cause a decrease in near-infrared light reflectance. The discoloration within the leaf mesophyll can be caused by chlorophyll saponification and by oxidation and polymerization of polyphenol oxidase to form a brown pigmentation. This resulting alteration in light reflectance is not caused by a change in the number of air spaces within the mesophyll; near-infrared light absorptance is increased by chemical light absorption either within the leaf mesophyll or within the coating on the leaf surface, or possibly both. Practically, internal leaf discoloration may be important for ozone or smog pollution detection.

Subcellular particles within leaves interact with electromagnetic radiation and increase near-infrared light reflectance (Gausman, 1977). This effect is now thought by Rock *et al.* (1986) to be important in facilitating the detection of forest tree damage from space (Gausman, 1987). However, the percentage of contribution of subcellar particles to the total near-infrared light reflectance from a plant leaf is not yet known.

From 90 to 95% of the ultraviolet radiation that impinges on a plant is absorbed by chemical constituents within the leaves.

IV. STRESS DETECTION

It is important to monitor plant leaf maturity to facilitate the use of spectro-photometric measurements to detect plant stress. Stressed leaves usually have lower reflectance than nonstressed leaves, if the leaves are the same chronological age. Physiological leaf age is difficult to determine.

A. Water and Temperature

Drought or temperature stress influences the water content of plant leaves. Frozen sour orange leaves (*Citrus aurantium*) had lower reflectance over the 0.50- to 2.5-μm waveband than nonfrozen leaves because freezing apparently destroyed the semipermeability of their cell membranes and, subsequently, intercellular sap migrated to intercellular spaces. Filling of air spaces with sap decreased the number of light-scattering (Peynado *et al.*, 1979), hydrated cell wall–air interfaces; consequently, reflectance was decreased. Reflectance measurements showed significant differences between noninjured and freeze-injured leaves, regardless of their water-soaked or nonwater-soaked appearance. Therefore, reflectance measurements should be useful in detecting cell membrane leakage, supposedly injury, within citrus leaves.

Effects of drought or temperature stress, or both, can also be evaluated within the 1.35- to 2.50-μm waveband water-absorption region. Wavebands around the 1.65- or 2.20-μm wavelengths can be used to evaluate leaf water content (Walter *et al.*, 1982), and detect general leaf freeze damage (Escobar *et al.*, 1983) and cell membrane freeze damage (Peynado *et al.*, 1979).

Early research showed that drying (dehydration) of soybean and corn leaves caused marked increases in their reflectance throughout the 0.5- to 2.6-μm waveband, particularly in the 1.3- to 2.6-μm wave band (Hoffer and Johannsen, 1969). Moreover, a comparison of reflectance spectra of corn leaves from plants grown with sandy and clay soil of varying moisture contents showed that there were important differences to consider in differentiating plant leaf spectra from soil spectra.

Leaf dehydration greatly increases the spectrophotometrically measured light reflectance over the 0.5- to 2.5-μm waveband. Reflectance increased as relative turgidity decreased below values of 80% at selected 0.54-, 0.85-,

1.45-, and 1.64-μm wavelengths (Thomas *et al.*, 1966). Relative turgidity is used to measure plant water stress; it is the actual leaf water content expressed as a percentage of the turgid or saturation water content. Regression equations were calculated (Thomas *et al.*, 1971) to express the incident light reflectance from the upper (adaxial) single-leaf surfaces as a function of relative turgidity and water content. Reflectances at the 1.45- and 1.95-μm wavelengths were related to the leaf relative turgidity of water content. Because of variations in internal leaf structure associated apparently with water availability during leaf development, however, the ability to predict leaf water status from reflectance measurements was poor. With cotton, the greatest reflectance change occurred when the relative turgidity was below 70%, and the leaves were visibly wilted. Within the relative turgidity range from 70 to 80%, reflectance changes were small, and they were not always definable for predictive purposes because of variation among leaves of field-grown cotton plants, caused by age (maturation) differences and osmotic stresses. Corn (*Zea mays*), grain sorghum (*Sorghum bicolor*), and soybean (*Glycine max*) leaf reflectances were highly linearly correlated with relative water content at two strong water-absorbing wavelengths, 1.45 and 1.95 μm, and two wavelengths of lower absorptivity, 1.10 and 2.20 μm (Carlson *et al.*, 1971).

B. Salinity

Morphological studies indicated that plants from saline environments had thicker leaves, more developed palisade parenchyma, smaller intercellular spaces, and fewer stomata per unit (Lesage, 1890; Harter, 1908; Hayward and Bernstein, 1958) than plants from nonsaline areas. Salinization also depressed cell division in leaves of cotton (*G. hirsutum*) (Strogonov, 1964) and reduced the rate of cell enlargement and of protein and RNA synthesis in bean leaves (*P. vulgaris*) (Nieman, 1965). Spectrophotometric studies showed that leaves of cotton plants affected by soil salinity had reduced light reflectance and increased light transmittance as compared with unaffected leaves of the same chronological age. The salinity-stressed leaves were stunted with a more compact cell arrangement than nonstressed leaves.

C. Nutrient Deficiency or Toxicity

A spectrophotometer was used to measure diffuse reflectance from upper (adaxial) leaf surfaces of sweet pepper (*Capsicum annuum,* cabbage *(Brassica oleracea),* and spinach *(Spinacia oleracea)* leaves to quickly estimate their nitrogen status (Thomas and Oerther, 1972). Leaf light absorptance in the visible region is primarily dependent on chlorophylls *a* and *b* and carotenoid (carotene and xanthophyll) concentrations in components (grana) of the chloroplasts (Thomas and Gausman, 1977). Green leaves absorb 75–90% of the blue light (about the 0.45-μm wavelength) and red light (about the 0.68-μm wavelength) of the visible spectrum. Absorptance is smallest in the waveband around the 0.55-μm wavelength, where a green light reflectance peak of usually less than 20% occurs. Reflectance was inversely correlated with the leaf nitrogen content of three plant species. Regression equations were developed to express reflectance as a function of the leaf nitrogen content of greenhouse-grown plants and were used to estimate the nitrogen content of field crops. With field-grown sweet peppers, for example, the difference between Kjeldahl-determined and reflectance-estimated nitrogen content was less than 0.7%

Spectrophotometric measurements can reveal visible light reflectance differences between leaves of chlorotic (iron deficient with high visible light reflectance) and green ("normal" with much lower visible light reflectance) grain sorghum plants *(S. bicolor)* (Gausman *et al.,* 1975a). However, it has been difficult to distinguish among single-leaf spectra for several nutrient deficiencies (Gausman *et al.,* 1973) or among single leaf spectra for nutrient deficiencies and a nutrient deficiency spectrum (Escobar and Gausman, 1978).

There has been considerable speculation about the influence of the *red-edge effect* in interpreting results. The red-edge effect consists of a sharp rise in the reflectance curve between the 0.675-μm wavelength within the visible portion of the electromagnetic radiation spectrum and the beginning of the near-infrared plateau (0.75-μm wavelength) (Horler *et al.,* 1983). Shifts in the position and slope of the red edge for example, can occur with nutrient toxicity stress on plants (Rock *et al.,* 1986). If the shift is toward the blue end of the electromagnetic radiation spectrum, the phenomenon shift is only a 0.007 to 0.040-μm change; this slight change is undetectable with infrared photography. High-spectral resolution instrumentation would be necessary to detect the shift.

Changes in leaf chlorophyll content are useful in detecting stress spectrophotometrically. For example, a deficiency of plant nutrients affected chloroplast ultrastructure in sunflowers *(Helianthus annus),* including a change in the ratio between chlorophyll *a* and *b* during potassium deficiency (Milivojevic *et al.,* 1982). Therefore, spectral measurements might be useful in detecting specific nutrient deficiencies and evaluating the photosynthetic potential of plants.

D. Plant Maladies

Spectrophotometric measurements can be used to detect plant maladies, such as diseases, nematode damage, and insect damage. We consider only a few examples here.

Colwell (1956) was one of the first investigators to suggest that infrared film could record any disease that interfered with the internal reflection of light within leaves. Keegan *et al.,* (1956) studied the effects of stem rust *(Puccinia graminis tritici)* and leaf rust (*P. triticina* or *P. rubigover tritici)* of wheat *(Triticum aestivum)* on light reflectance. The data showed that severe infestation (as compared with low levels) caused a rounding of the near-infrared reflectance plateau's shoulder over the 1.00- to 0.75-μm waveband.

Cellular discoloration within leaves may be useful in detecting nonvisual symptoms of plant maladies. Often, this technique is referred to as *previsual detection.* For example, potato *(Solanum tuberosum)* late blight *(Phytophthora infestans)* could be detected by aerial photography from 3 to 5 days before visual symptoms became apparent (Manzer and Cooper, 1967); tobacco *(Nicotiana tabacum)* ringspot virus *(N.* spp. virus) could be detected about 1 day before visual symptoms were evident (Burns *et al.,* 1969); and ozone-damage leaf areas of cantaloupe plants (*Cucumis sativus)* were detected photographically 16 hr before the damage was visible (Gausman *et al.,* 1978). Conversely, beetle damage could not be predicted (Heller, 1968), and too much biological variability interfered with previsual detection of tree disease (Meyer, 1967).

To study the effects of internal damage on light energy spectra, cotton leaves *(G. hirsutum)* were infiltrated with anhydrous ammonia (Cardenas *et al.,* 1968–1970). Spectrophotometric laboratory measurements on anhydrous ammonia–treated leaves showed more absorptance and less reflectance and transmittance than untreated leaves over the 0.75- to 1.35-μm waveband and the 0.50- to 0.75-μm waveband. Apparently, the brownish discoloration caused by the anhydrous ammonia increased leaf opaqueness. Moreover, other conditions that caused decreased reflectances (plateau rounding) were rust infection on "Westar" wheat leaves, *(T. aestivum)* benzene vapor on cotton leaves, natural freezing of seagrape leaves *(Coccoloba unifera),* ammonia treatment of cotton leaves, hair removal by shaving of velvet plant leaves *(Gynura aurantiaca),* and leaf freeze injury (Gausman and Cardenas, 1969).

The occurrence of rizomania disease was investigated in sugar beet *(Beta vulgaris)* fields in West Germany (Sahwald, 1981). An increase in reflectance of up to 100% was recorded for diseased plants, as compared with that for healthy plants, at the 0.55-, 0.59-, and 0.68-μm visible light wavelengths. The large increase in visible light reflectance was caused by the difference in canopy geometry between the diseased and healthy sugar beet plants. The leaves drooped on the diseased plants, causing their whitish-appearing petioles to be exposed more to the remote sensor than the petioles of healthy plant leaves. Therefore, the visible light reflectance of the petioles of diseased plant leaves was compared with the visible light reflectance of healthy plant leaves. This difference in reflectance was very vivid between the respective images on infrared color photographs.

E.   Atmospheric Pollution

Ozone-damaged leaves develop necrotic areas that dehydrate rapidly. Thus, the detection of ozone damage with remote sensing techniques would probably evaluate changes in leaf water content.

Effects of ozone damage on the reflectance and photographic responses of cantaloupe plant *(Cucumis melo)* leaves and canopies were studied to determine the best wavelengths to detect ozone damage and to determine if ozone damage could be previsually detected (Gausman *et al.*, 1978a). Mean spectrophotometrically measured light reflectances at the 0.55- and 0.65-μm wavelengths in the visible region (0.50- to 0.75-μm waveband) among the control and the light and severely ozone-damaged leaves was significantly greater ($P = 0.01$) than for the other treatments.

The reflectances for the leaves of all treatments were different statistically ($P = 0.01$) for the 1.45-, 1.65-, 1.95-, and 2.20-μm wavelengths in the near-infrared water-absorption region (1.35- to 2.50-μm waveband). As ozone damage increased, reflectance increased because of leaf dehydration (Gausman *et al.*, 1979). Therefore, it might be possible to detect ozone-damaged plants within the water-absorption region of the infrared spectrum (1.35- to 2.50-μm waveband) because leaf dehydration greatly increases reflectance within this region.

F.   Plant Growth Regulators

Mepiquat chloride (1,1-dimethylpiperidinum chloride) caused treated cotton leaves *(G. hirsutum)* to become thicker than untreated leaves, which increased near-infrared light reflectance (Richardson and Gausman, 1982). Also, mepiquat chloride-treated leaves had a higher chlorophyll concentration than untreated leaves. Subsequently, visible light reflectance was decreased. Near-infrared light reflectance measurements are useful too. For example, a bioregulator that reduces leaf expansion might be detectable because a compact leaf usually has less reflectance than a fully expanded leaf.

Moreover, mepiquat chloride enhanced the tolerance of undesirable low and high temperatures (Richardson and Gausman, 1982; Gausman *et al.*, 1981). This was easily detectable by reflectance measurements.

V.   BIOCHEMICAL AND BIOLOGICAL FACTORS AFFECTING LEAF OPTICAL PROPERTIES

A.   Ultraviolet Light Interactions

Ultraviolet (UV) light or radiation is reflected, transmitted to a limited extent, and absorbed by plant leaves. Partial destruction of the stratosphere's ozone layer by pollutants like nitrogen oxides and chlorofluoromethanes would increase the amount of solar UV radiation that reaches the earth's surface (Vu *et al.*, 1982). Increased solar UV-B radiation (0.28- to 0.32-μm waveband) could cause detrimental effects in different plant species, including a decrease in growth, photosynthesis, and forage or grain production (Biggs, 1973).

Crop plant leaves usually absorb more than 90% of UV radiation reaching their surface, with essentially no transmittance and less than 10% reflectance (Caldwell, 1968; Allen *et al.*, 1975). Plants have apparently developed mechanisms to compensate for potentially harmful sunlight UV radiation. These mechanisms include formation of pigment, photoreactivation, movement of cell organelles to other positions, and shifting of UV-sensitive metabolic processes from day to night. However, UV radiation is apparently essential for flavonoid synthesis, including a sequence of many anabolic activities (Wellman, 1976). Thick epidermises attenuated more UV radiation than did thinner epidermises of six plant species (Gausman and Escobar, 1982). Apparently, thick epidermises have more compounds such as the flavonoids to enhance UV-radiation absorptance.

Solar UV-radiation distribution in terrestrial plant communities has been reviewed (Allen *et al.*, 1975), and model predictions have been developed that describe the range of UV-B radiation regimes expected in plant communities under the current stratospheric ozone content and under 25%. Therefore, a technique is needed to remotely detect UV radiation-damaged plants. Consequently, cotton plants *(G. hirsutum)* were exposed to UV radiation for 7 days and compared with controls that received only natural light. Light reflectance was spectrophotometrically measured in the laboratory over the 0.5- to 2.5-μm waveband on nonexcised leaves immediately after treatment and again 7 days later. Leaves of cotton plants exposed to UV radiation had less near-infrared reflectance at the 0.85-μm wavelength than did the control leaves. The UV radiation-treated leaves were thinner, smaller in area, and had a different mesophyll structure than the control leaves. Thus, reflectance measurements made at the 0.85-μm wavelength in the near-infrared band might be useful to detect UV radiation-damaged cotton plants.

B.   Sooty Mold Detection

Hart and Myers (1968) measured total diffuse reflectance of citrus leaves *(Citrus* spp.) that were lightly, moderately, or heavily coated or were free of sooty-mold fungus, *Capnodium citri*, over the 0.5- to 2.5-μm waveband. Citrus leaves that were free of sooty mold reflected 58 and 53% of the light at the 0.77- and 1.3μm wavelengths, respectively, whereas leaves that were heavily coated with sooty mold reflected 9 and 23% at the respective wavelengths. A light coating of mold reduced reflectance values 14% at the 0.77-

μm wavelength and 2% at the 1.30-μm wavelength; a medium coating reduced reflectance values 29 and 10%, respectively.

The reduction in reflectance induced by sooty-mold deposits on citrus leaves provides for the detection of insect pests such as brown soft scale, *Coccus hesperidum*, with infrared aerial color photography. The brown soft scale and a number of other homopterous insects produce honeydew that serves as a host medium for the sooty-mold fungus that blackens citrus foliage.

Attenuated total reflectance of citrus leaves heavily coated with or free of sooty-mold fungus was also measured over the 2.5- to 40.0-μm waveband with a spectrophotometer equipped with a multiple reflection unit (Gausman and Hart, 1974). Citrus leaves free of sooty-mold fungus deposit reflected less radiation than heavily coated at the 3.0- and 3.5μm wavelengths and at every wavelength within the 6.0- to 40.0-μm waveband. Apparently, sooty-mold deposits are lower in amides, amines, and methylated compounds than "clean" leaves, which may have caused the reflectance differences between them at the 3.0- and 3.5-μm wavelengths. Moreover, a very strong absorption band was found within the 6.0- to 40.0-μm waveband; this absorption was probably caused by the CH_2-0 bond. Although absolute reflectance of infrared radiation from leaves is low (0.05–0.10%) from a range of 3 to 25 μm (Gates ant Tantraporn, 1952), discrimination among citrus leaves with and without sooty-mold fungus deposits may be possible by making infrared radiation measurements at the 3.0- and 3.5-μm wavelengths and at every wavelength interval from 6.0 to 40.0 μm.

C. Leaf Pubescence

Shull (1929) concurred with Coblentz (1913) that hairs on *Verbascum thapsus* (common mullein) did not increase light reflectance to any marked extent in comparison with leaves of nonhairy plants. Shull also found that this was true with *Abutilon theophrasti* (velvet leaf). Tomentose hairiness, however, on the undersurface of *Populus alba* (silver-leaf poplar) and *Magnolia acuminata* (cucumber tree), increased reflectance. Moss (1951) found that the hairy lower surface of *P. alba* leaves, as compared with their upper surface, reflected about 15% more incident light in the 0.4- to 0.7-μm waveband.

Billings and Morris (1951) studied the leaf reflectance of *Eurotia lanata* (white sage), whose leaves are densely covered with white stellate hairs. They concluded that leaf hairs are of great importance in the reflectance of visible light, but they are not necessarily of as much value in reflecting near-infrared light.

Gates and Tantraporn (1952) thought that if a glabrous leaf surface has a high reflectivity, the presence of leaf hairs would decrease reflectivity because of the scattering and entrapment of radiation within the leaf's hairy envelope. This mechanism was thought to be more effective in the near-infrared than in the visible light region. They used three plant species: *V. thapsus* (common mullein), *Asclepias syriaca* (milkweed), and *Elaeagnus angustifolia* (Russian olive).

Pearman (1966) measured light reflectance from the upper surface of leaves of *Arctotheca nivea* (sand dune species) before and after leaf hair removal. Hair removal reduced the average leaf reflectance from 31.7 to 15.0%, when measured at 0.02-μm-wide intervals over the 0.34- to 0.62-μm waveband. He concluded that hairs present many interfaces to incoming radiation that scatter light and decrease the amount of light entering the leaf, thereby decreasing absorption.

Gausman and Cardenas (1969) removed hairs from the upper surfaces of the leaves of the velvet plant *(Gynura aurantiaca)* with an electric razor. The leaves are covered with purple velvet-like hairs that contain anthocyanin. The hairs are multicellular and unbranched. Hairiness increased total and diffuse reflectance over the 0.75- to 1.00-μm waveband but decreased total and diffuse reflectance over the 1.0- to 2.5-μm waveband. Hair removal did not influence total or diffuse reflectance within the visible light waveband ranging from 0.5 to 0.7 μm. This did not agree with the results of Billings and Morris (1951) or Pearman (1966) whose studies indicated that leaf pubescence enhanced the reflectance of visible light. However, the hairs of *Gynura aurantiaca* (velvet plant) as compared, for example, with the white stellate hairs of *Eurotia lanata* (Billings and Morris, 1951), are purple in color because they contain anthocyanin. The anthocyanin absorption band, although somewhat pH dependent, occurs within the 0.50- to 0.55-μm waveband (Gilliam *et al.*, 1962). Thus, light absorption by anthocyanin may modify the spectrophotometrically measured reflectance of light, which normally reaches its maximum peak in the visible portion of the spectrum at the 0.55-μm wavelength.

Hair removal increased absorptance of near-infrared light over the 0.75- to 1.00-μm waveband. The maximum increases were 4.4 and 4.2% at the 0.75- and 0.80-μm wavelengths, respectively. Theoretically, unshaven leaves were highly transparent to near-infrared light in the 0.75- to 1.00-μm waveband. After hair removal, leaves apparently became more opaque and light absorptance was increased. The increased opacity was probably caused by exudate discoloration on the cut stamps after hair removal. Apparently, phenol oxidase caused the discoloration or browning of the exudate (Bonner and Galston, 1952). This effect also causes rounding of the near-infrared reflectance plateau.

Pubescence can be used in remote sensing to detect plants with pubescent leaves such as the silverleaf sunflower plant (*H. argophyllus*) (Gansman *et al.*, 1977b), a weed in the sandy soils of south and southeast Texas. The appendages of a young silverleaf sunflower plant are densely white-tomentose. This pubescence increased reflectance greatly over the 0.5- to 2.5-μm waveband as compared with that of the sparsely hairy leaves of common sunflower (*H annuus*).

Szwarcbaum (1982) showed that leaf pubescence has an important influence on spectral properties of the leaves of three Mediterranean shrubs—*Salvia tribola*, *Cistus salviifolius*, and *C. incanus*. Since transpiration cannot always overcome the plant's heat load, the three shrubs have apparently developed leaf characteristics that help to cool them. These include a decrease in leaf surface area as a result of producing smaller leaves and folding the leaf margins, a difference in leaf properties caused by different hair densities, and the ability to expose the lower leaf surface under conditions of water stress. The more pubescent leaves of *C. incanus* absorbed more energy in the 0.4 to 0.5-μm waveband than the less pubescent leaves of *C. salviifolius*, even though their reflectance was similar.

D. Plant Leaf Senescence

Senescence is the deterioration in plant leaves, flowers, fruits, stems, and roots that ends their functional life (Thimann, 1980; Salisbury and Ross, 1969). In many perennial plants, the aboveground vegetation dies yearly but the crown and roots remain viable. Most herbaceous annual plants have a progressive senescence of their leaves from the oldest to the youngest, followed by the death of the stems and roots.

During leaf senescence, starch, chlorophyll, and protein and nucleic acid (RNA) components are degraded, and the catabolic products may be translocated to anabolically active areas of plants. Fall coloration is caused partly by the unmasking of yellow and orange carotene and red anthocyanin pigments when the green chlorophyll pigments are lost. Leaf senescence is hastened by high temperature, drought, a short photoperiod, nutrient deficiency, and other stresses.

During leaf senescence, their light reflectance usually increases markedly in the green part of the visible light spectral region, peaking at the 0.55-μm wavelength because of chlorophyll degradation (Knipling, 1967). If an abundance of anthocyanin or carotene pigments are present after the loss of chlorophyll, there may be relatively high reflectances in the red and near-blue region of the visible spectrum.

Leaf senescence decreases near-infrared light reflectance (Knipling, 1967) over the 0.75- to 1.35-μm waveband. With the leaves of some plant species, particularly forest trees and cereal crops, the near-infrared plateau, at about the 0.75-μm wavelength, is reduced and rounded off considerably, also characteristic of leaves with damaged cells (Cardenas *et al.*, 1968–1970).

VI. INNOVATIONS AND NEW TECHNOLOGY

A. Infrared Water Absorption Region

Young leaves contain less water than mature leaves because their immature cells are primarily photoplasmic with little vacuolate water storage (Landegaårdh, 1966). During cell growth, cell water-filled vacuoles develop that may later coalesce to form a central sap cavity; the protoplasm covers only the inside of the cell wall in a thin layer.

Water infiltration of leaves decreased absorption throughout most of the visible region and apparently sharpened the band encompassing the 0.68-μm wavelength (Moss, 1951). Pearman (1966) reported a decrease in reflectance from intracellular water infiltration. Normal drying of leaves increased reflectance for the visible wavelength. Rabinowitch (1951) reported that light scattering within leaves was decreased by water injection into their air spaces.

Leaf dehydration increases the spectrophotometrically measured light reflectance over the 0.5- to 2.5 μm waveband. Reflectance was found to increase as relative turgidity decreased below values of 80% at selected 0.54-, 0.85-, 1.65-, and 1.45-μm wavelengths (Thomas *et al.*, 1966b; 1971). Relative turgidity is used to measured plant water stress; it is the actual leaf water content expressed as a percentage of the turgid or saturation water content. Regression equations were calculated (Thomas *et al.*, 1971) to express the incident light reflectance from the upper (adaxial) single-leaf surfaces as a function of relative turgidity and water content. Reflectance at the 1.45- and 1.95-μm wavelengths were related to the leaf relative turgidity or water content. With cotton, the greatest reflectance change occurred when the relative turgidity was below 70%, and the leaves were visibly wilted. Within the relative turgidity range from 70 to 80%, reflectance changes were small, and they were not always definable for predictive purposes because of variation among leaves of field-grown cotton plants caused by age (maturation) differences and osmotic stresses. Corn (*Z. mays*), grain sorghum (*S. bicolor*), and soybean (*G. max*) leaf reflectances were highly linearly correlated with relative water content at two strong water-absorbing wavelengths, 1.45 and 1.95 μm, and at two wavelengths of lower absorptivity, 1.10 and 2.20 μm (Carlson *et al.*, 1971).

The linear correlation of leaf water content with reflectance is strongest in the near-infrared region. Sinclair (1968) believed that tissue of dehydrated leaves collapse in such a manner that the number of air voids increases in the leaf mesophyll and, therefore, near-infrared reflectance increases. Dadykin and Bedenko (1961), however, related oak (*Quercus*) leaf reflectance to different moisture regimes in the 0.4- to 0.8-μm waveband.

Johannsen (1969) measured the reflectance of corn and soybean leaves from plants grown at decreasing soil moisture contents. The water absorption bands centered at the 1.45- and 1.95-μm wavelengths were inversely related to leaf moisture. The green color and chlorophyll absorption responses (0.53 and 0.64 μm, respectively) showed a high negative linear correlation with leaf moisture. This indicated that changes in leaf moisture content quickly affected the leaf spectral responses to pigment concentrations. Cotton leaf air volume also affected visible light reflectance (Gausman *et al.*, 1975b).

Hoffer and Johannsen (1969) showed that a healthy, turgid green leaf exhibited a close relation between water absorption and reflectance. In wavelengths where water absorption was high, leaf reflectance was low. This was most apparent in the primary water-absorption bands centered at the 1.45- and 1.95-μm wavelengths.

Light reflectance peaks at the 1.65- and 2.20-μm wavelengths followed by the 1.45- and 1.95-μm wavelength water-absorption bands, respectively, have decreased leaf reflectance with an increased leaf water content (Gausman *et al.*, 1977a). These peaks are useful for estimating leaf water content, plant species identification, and separation of soil from vegetation.

Succulent plants have water-storage tissue in their mesophyll (Fahn, 1967). Therefore, they have a higher water content and absorb more radiation in the near-infrared water-absorption region (1.35- to 2.50-μm wave band) than nonsucculent plants. Peperomia *(Peperomia obtusifolia)* is an example of a succulent plant that has water-storage tissue called the hypodermis. This water-storage layer absorbed energy and caused the absence of the typical near-infrared light reflectance peak at about the 2.20-μm wavelength from the upper (adaxial) leaf surface. This phenomenon may be useful in (Gausman *et al.*, 1978b) distinguishing succulent plants from crop and woody plants by using a sensor band to encompass either the 1.65- or 2.20-μm wavelengths. These wavelengths were predicted to be useful in the future for plant species discrimination by remote sensing (Gausman *et al.*, 1978b). Moreover, either the 1.65- or 2.20-μm wavelength is promising for the evaluation of freeze damage to cell membranes (Peynado *et al.*, 1979) or the assessment of leaf water content (Walter *et al.*, 1982).

B. Videography

A video system has been developed that displays results of interactions of near-infrared radiation with plant leaves within the 0.78- to 1.10-μm waveband (Escobar *et al.*, 1982; Gausman *et al.*, 1983; Everitt *et al.*, 1986). The system has a video camera, video monitor, cassette player, and cassette recorder. Three examples from the literature that have been shown to affect light reflectance were demonstrated successfully with the video system: leaf water infiltration, leaf maturation, and leaf stacking. The video system has potential use in facilitating teaching, research, and remote sensing applications.

C. Fluorescence Detection

Fluorescence measurement is a promising tool to detect plant stress, particularly water stress (McFarlane *et al.*, 1980). Burke and Quisenberry (1983) conducted an intensive investigation to determine if chlorophyll a fluorescence transients would provide a rapid, nondestructive technique to analyze the adaption of photosynthetic light quanta conversion during severe water stress. The cotton *(G. hirsutum)* genotype Lubbock Dwarf was planted within a rainout shelter that was used to keep water off the plots during the growing season. Selected rows were irrigated throughout the experiment with a drip-irrigation system.

The irrigated Lubbock Dwarf plants had a consistently lower fluorescence P/O ratio (P, maximum fluorescence level; O, initial fluorescence level) than the water-stressed plants. A typical P/O ratio for irrigated plants was about 1.3, while stressed plants routinely had a ratio of 1.6. A high P/O ratio shows that there is a better photosynthetic light quanta conversion; plants that exhibit a P/O of 1.0 are not photosynthetically active. The $P - T/T$ ratio (P, maximum fluorescence level; T, steady-state fluorescence level) was also utilized in analyses of the light quanta conversion, and the results supported the data provided by the P/O ratios. Common $P - T/T$ ratios were 0.9 and 2.0 for irrigated and water-stressed plants, respectively.

The relationship between the fluorescence P/O ratio and the photosynthetic water-use efficiency was also analyzed. A positive correlation between the P/O ratio and the water-use efficiency of Lubbock Dwarf leaves was observed.

After 3 months of growth under stressed conditions, plants in the shelter were irrigated with 2 in. water, and the P/O ratios were determined 48 hr after irrigation. Both irrigated and previously water-stressed plants had identical fluorescence and P/O ratios similar to those obtained for irrigated plants throughout the study.

Chlorophyll a fluorescence transients can provide a rapid indication of the stress-induced alteration of the photosynthetic light quanta conversion capabilities. The cotton genotype Lubbock Dwarf does undergo a modification of its photosynthetic system, which increases the efficiency of light quanta utilization under water-stressed conditions. And finally, fluorescence P/O ratios can be used in a Lubbock Dwarf planting to select plants that have increased water-use efficiency.

D.   Carbon Dioxide Concentration

Aircraft-mounted sensors were used to measure carbon dioxide exchange above a corn field, forest, and lake during midday conditions (Brach *et al.*, 1982). Mean carbon dioxide absorption values were consistent with ground-based observations. Therefore, such information could be used to quantitatively evaluate carbon dioxide source and sink distributions in the biosphere, to correlate satellite with near-surface measurement data, and to monitor crop performance such as phytomass production. The detection of a difference in oxygen concentration might be a very early indicator of the onset of a plant stress condition. This would have important implications in management of agricultural resources and in selection of plants with inherent resistance to stress conditions.

E.   Detection of Herbicidal Effects

The detection of herbicidal effects on plants is a relatively new and promising use of single-leaf reflectance measurements. Some of the pioneering research was conducted by Walter and Koch (1981), who reported the effects of atrazine (1.44 L/ha rate) and 2,4-D amine (0.72 L/ha rate) on beans *(P. vulgaris)* and maize *(Z. mays)*. Beans were very sensitive to both herbicides; maize was not. Two days after treatment the effects of both herbicides on leaf light reflectance were detectable spectrophotometrically, even though there were no noticeable visual differences between treated and nontreated leaves.

F.   Estimating Constituents of Agricultural Products

The use of near-infrared reflectance is a promising tool to measure various constituents in plant materials and agricultural products. Three selected examples are estimating nutrient content of dehydrated vegetables (Park *et al.*, 1982), analyzing nitrogen and oil content of plant materials (Starr *et al.*, 1981); and analyzing forage crop quality (Norris *et al.*, 1976).

Near infrared analysis is based on the utilization of energy absorbed in the near-infrared region of the electromagnetic spectrum by chemical groups that are characteristic of major constituents of the plant material or agricultural product. After proper calibration of the instrument using samples with known concentrations of the desired component to be measured, interference by other constituents is kept minimal by using multiple wavelength readings, which are incorporated or integrated into multiple regression equations for prediction purposes.

VII. CONCLUSION

Single-leaf spectral measurements made in the laboratory have contributed greatly to an understanding of the mechanism of the interaction of electromagnetic radiation with plant leaves and other plant components such as leaf petioles, leaf sheaths, heads of grasses, inflorescences, and so on. This has included spectrophotometric measurements of leaf reflectance, transmittance, and absorptance, generally over the 0.5- to 2.5-μm (sometimes 0.4- to 2.6-μm) waveband.

The laboratory spectrophotometric measurements have been used to characterize visible light spectra (0.45- to 0.75-μm waveband), near-infrared light spectra (0.75- to 1.35-μm waveband), and infrared, water-absorption light spectra (1.35- to 2.5-μm waveband) for biological factors such as leaf pigment concentration, pubescence, and senescence. Probably the best indicator of changes in leaf optical properties was a difference in reflectance at the 0.55-μm wavelength caused by variations in leaf pigment concentration: High reflectance was associated with low total chlorophyll concentration; and conversely, low reflectance was correlated with high chlorophyll concentration.

Near-infrared light is scattered or reflected from leaves by refractive index discontinuities. The most important discontinuity is the cell wall–air space interface. If near-infrared light travels at a critical angle from a hydrated cell wall with a refractive index of about 1.425 to an open air space with a refractive index of 1.0, the near-infrared light is scattered or reflected. This can be easily demonstrated by replacing air in leaves with a liquid by vacuum infiltration. Reflectance becomes minimal if the refractive index of the liquid presumably matches the refractive index of the cell wall.

It is extremely important to consider leaf maturity if spectrophometric measurements are to be made to compare the spectral properties of stressed and nonstressed leaves. Young leaves have a more compact structure than nonstressed leaves; stressed leaves are not fully expanded and have a more compact structure than fully expanded nonstressed leaves.

Generally, stressed plant leaves have a lower near-infrared light reflectance than nonstressed leaves. However, spectrophotometric measurements often show that diseased leaves have higher near-infrared light reflectance than "normal" leaves, possibly caused by leaf dehydration.

Internal discoloration of leaves or a black coating on their surfaces, or both, will cause a decrease in near-infrared light reflectance. The discoloration can be caused by saponification of chlorophyll and the oxidation and polymerization of polyphenol oxidase to a brown pigmentation. The absorptance of near-infrared light is increased by internal chemical changes or by chemicals in the black coating on leaf surfaces.

Subcellular particles of leaves interact with light by increasing near-infrared reflectance. However, their contribution to the total amount of near-infrared light reflectance is probably minor.

From 90 to 95% of ultraviolet radiation is absorbed by leaf epidermises.

Single-leaf measurements have been used successfully to a very limited extent in the detection of stresses such as low and high temperature, soil salinity, nutrient deficiency or toxicity, plant maladies, and atmospheric pollution. However, the single-leaf measurements have been used rather extensively as basic information to develop irrigation scheduling, study plant canopy interactions, estimate phytomass, estimate constituents of agricultural products, and facilitate desirable plant selection in plant breeding programs.

There are new and promising stress-detection techniques, such as videography, fluorescence detection, and measurement of carbon dioxide concentration.

More research is needed on the effects of leaf water content, pigmentation, senescence, and pubescence on leaf spectral properties. Also, the use of other regions besides the 0.5- to 2.5-μm waveband should be investigated. With the availability of more sophisticated instrumentation, more subtle differences at various wavelengths of light reflectance should be detectable among plant leaves.

ACKNOWLEDGMENTS

We thank Ms. Lynne Szymanowski and Ms. Linda Graves for their efforts in the typing and editing of the manuscript.

REFERENCES

Allen, L. H., Jr., Gausman, H. W., and Allen, W. A. (1975). *J. Environ. Qual.* **4**, 285–294.
Allen, W. A., and Richardson, A. J. (1968). *J. Opt. Soc. Am.* **58**, 1023–1028.
Allen, W. A., Gausman, H. W., Richardson, A. J., and Thomas, J. R. (1969). *J. Opt. Soc. Am.* **59**, 1376–1379.
Biggs, R. H. (1973). *Proc. Conf. Clim. Impact Assess. Program, 2nd, of Transportation 1972*, Rep. No. DOT-TSC-OST-73-4, Part 1, pp. 394–398.
Billings, W. D., and Morris, R. J. (1951). *Am. J. Bot.* **38**, 327–331.
Bonner, J., and Galston, A. W. (1952). "Principles of Plant Physiology." Freeman, San Francisco, California.

Brach, E. J., Eesjardins, R. L., Alvo, P., and Schuepp, P. H. (1982). *Science* **216**, 733–735.

Burke, J. J., and Quisenberry, J. E. (1983). *Proc. Beltwide Cotton Prod. Res. Conf.*, p. 60.

Burns, F. E., Starzyk, M. J., and Lynch, D. L. (1969). *Trans. Ill. State Acad. Sci.* **62**, 102–105.

Caldwell, M. M. (1968). *Ecol. Monogr.* **38**, 243–268.

Cardenas, R., Gausman, H. W., Allen, W. A., and Schupp, M. (1968–1970). *Remote Sens. Environ.* **1**, 199–202.

Carlson, R. E., Yarger, D. N., and Shaw, R. H. (1971). *Agron. J.* **63**, 486–489.

Coblentz, W. W. (1913). Natl. *Bur. Stand. Bull.* **9**, 283–325.

Colwell, R. N. (1956). *Hilgardia* **26**, 223–286.

Dadykin, V. R., and Bedenko, V. P. (1961). *Dokl. Acad. Sci. USSR, Bot. Sec. (Engl. Transl.)* **134**, 212–214.

Escobar, D. E., and Gausman, H. W. (1978). *J. Rio Grande Valley Hortic. Soc.* **32**, 81–88.

Escobar, D. E., Bowen, R. L., Gausman, H. W., and Cooper, G. R. (1982). *AgRISTARS, EW-U2-04351; JSC-18563.*

Escobar, D. E., Bowen, R. L., Gausman, H. W., and Cooper, G. R. (1983). *J. Rio Grande Valley Hortic Soc.* **36**, 61–66.

Everitt, J. H., Escobar, D. E., Blasquez, C. H., Hussey, M. A., and Nixon, P. R. (1986). *Photogram. Eng. Remote Sens.* **52**, 1655–1660.

Fahn, A. (1967). "Plant Anatomy." Pergamon, New York.

Gates, D. M., and Tantraporn, W. (1952). *Science* **115**, 613–616.

Gausman, H. W. (1977). *Remote Sens. Environ.* **6**, 1–9.

Gausman, H. W. (1985). *Grad. Stud.—Tex. Tech Univ.* **29**, 1–79.

Gausman, H. W. (1987). In "Color Aerial Photography and Videography in the Plant Sciences and Related Fields" (J. H. Everitt, ed.), pp. 5–20. Am. Soc. Photogram. Remote Sens., Falls Church, Virginia.

Gausman, H. W., and Cardenas, R. (1969). *Bot. Gaz. (Chicago)* **130**, 158–162.

Gausman, H. W., and Escobar, D. E. (1982). *Remote Sens. Environ.* **12**, 485–490.

Gausman, H. W., and Hart, W. G. (1974). *J. Econ. Entomol.* **67**, 479–480.

Gausman, H. W., Allen, W. A., and Cardenas, R. (1969). *Remote Sens. Environ.* **1**, 19–22.

Gausman, H. W., Allen, W. A., Cardenas, R., and Richardson, A. J. (1970). *Appl. Opt.* **9**, 545–552.

Gausman, H. W., Escobar, D. E., and Rodriguez, R. R. (1973), *Proc. Bienn. Workshop Color Aerial Photogr. 4th*, pp. 13–27.

Gausman, H. W., Gerbermann, A. H., and Wiegand, C. L. (1975a). *Photogram. Eng. Remote Sens.* **41**, 177–181.

Gausman, H. W., Thomas, J. R., Escobar, D. E., and Berumen, A. (1975b). *J. Rio Grande Valley Hortic. Soc.* **29**, 109–114.

Gausman, H. W., Allen, W. A., Escobar, D. E. (1976). *Appl. Opt.* **13**, 109–111.

Gausman, H. W., Knipling, E. D., and Escobar, D. E. (1977a). *Photogram. Eng. Remote Sens.* **43**, 1183–1185.

Gausman, H. W., Menges, R. M., Escobar, D. E., Everitt, J. H., and Bowen, R. L. (1977b). *Weed Sci.* **25**, 437–440.

Gausman, H. W., Escobar, D. E., Rodriguez, R. R., Thomas, C. L., and Bowen, R. L. (1978a). *Photogram. Eng. Remote Sens.* **44**, 481–485.

Gausman, H. W., Escobar, D. E., Everitt, J. H., Richardson, A. J., and Rodriguez, R. R. (1978b) *Photogram. Eng. Remote Sens.* **44**, 487–491.

Gausman, H. W., Namken, L. N., Stein, E., Leamer, R. W., Walter, H., Rodriguez, R. R., and Escobar, D. E. (1979). *J. Rio Grande Valley Hortic. Sci.* **33**, 105–112.

Gausman, H. W., Escobar, D. E., and Rodriguez, R. R., (1981). *Plant Growth Regul. Soc. Am. Bull.* **9**, 6–8.

Gausman, H. W., Escobar, D. E., and Bowen, R. L. (1983). *Remote Sens. Environ.* **13**, 363–366.

Gilliam, A. E., Stern, E. S., and Jones, E. R. H. (1962). An Introduction to Electronic Absorption Spectroscopy in Organic Chemistry. Arnold, London.

Hart, W. G., and Myers, V. I. (1968). *J. Econ. Entomol.* **61**, 617–624.

Harter, L. L. (1908). *U.S. Dep. Agric. Bur. Plant Ind. Bull.* **134**.

Hayward, H. W., and Bernstein, L. (1958). *Bot. Rev.* **24**, 584–635.

Heller, R. C. (1968). *Proc. Int. Symp. Remote Sens. Environ. 5th, 1967*, pp. 387–434.

Hoffer, R. M., and Johannsen, C. J. (1969). *In* "Remote Sensing in Ecology" (P. O. Johnson, ed.), pp. 1–16. Univ. of Georgia Press, Anthens.

Horler, D. N., Dockray, M., and Barber, J. (1983). *Int. J. Remote Sens.* **4**, 273–288.

Johannsen, C. J. (1969) Ph. D. Thesis, Purdue Univ. Library, Lafayette, Indiana.

Keegan, H. J., Schleter, J. C., Hall, W. A., Jr., and Hass, G. M. (1956). *Nat'l. Bur. St. and (V.S.), Rep.* **4591**, 1–161.

Knipling, E. B. (1967). *U.S. C. F. S. T. T., AD Rep.* **AD-652–679**, 1–24.

Lesage, P. (1890). *Rev. Gen. Bot.*, pp. 168–170.

Lillesaeter, O. (1982). *Remote Sens. Environ.* **12**, 247–254.

Lundegårdh, O. (1966). "Plant Physiology." Am. Elsevier, New York.

Manzer, F. E., and Cooper, G. R. (1967). *Maine Agric. Exp. St., Bull.* **646**, 1–14.

McFarlane, J. C., Watson, R. D., Theisen, A. F., Jackson, R. D., Ehler, W. L., Pinter, P. J., Jr., Idso, S. B., and Reginato, R. J. (1980) *Appl. Opt.* **19**, 3287–3289.

Meyer, M. P. (1967). *Proc. Workshop Infrared Color Photogr. Plant Sci.*, Part V, pp. 5–7.

Milivojevic, D., Borivoje, K., and Miloje, S. (1982). *Period. Biol.* **84**, 160–162.

Moss, R. A. (1951). Ph. D. Thesis, Iowa State Univ. Library, Ames.

Nieman, R. H. (1965). *Plant Physiol.* **40**, 156–161.

Norris, K. H., Barnes, R. F., Moore, J. E., and Shenk, J. S. (1976). *J. Anim. Sci.* **43**, 889–897.

Park, Y. W., Anderson, M. J., and Mahoney, A. W. (1982). *J. Food Sci.* **47**, 1558–1561.

Pearman, G. I. (1966). Aust. *J. Biol. Sci.* **19**, 97–103.

Peynado, A., Gausman, H. W., Escobar, D. E., Rodriguez, R. R., and Garza, M. V. (1979). *Cryobiology* **16**, 63–68.

Rabinowitch, E. I. (1951). "Photosynthesis and Related Processes," 603–1173, 1211–2022. Wiley, (Interscience), New York. New York.

Richardson, A. J., and Gausman, H. W. (1982), *Remote Sens. Environ.* **12**, 501–507.

Rock, B. N., Vogelmann, J. E., Williams, D. L., Vogelmann, A. F., and Hoshizaki, T. (1986). *BioScience* **36**, 439–445.

Salisbury, F. B., and Ross, C. (1969) "Plant Physiology." Wadsworth, Belmont, California.

Sanwald, E. F. (1981). *Spectral Signatures Objects Remote Sens., Int. Colloq.*, Avignon, France, pp. 201–208.

Shull, C. A. (1929), *Bot. Gaz.* **87**, (Chicago) 583–607.

Sinclair, T. R. (1968). M. S. Thesis, Purdue University, Lafayette, Indiana.

Starr, C., Morgan, A. G., and Smith, D. B. (1981). *J. Agric. Sci.* **97**, 107–118.

Strogonov, B. P. (1964). *Physiological Basis of Salt Tolerance of Plants."* Daniel Davey & Co., New York.

Szwarcbaum, I. (1982). *Plant Sci. Lett.* **26**, 47–56.

Thimann, K. V. (1980). "Senescence in Plants." CRC Press, Boca Raton, Florida.

Thomas, J. R., and Gausman, H. W. (1977) *Agron. J.* **69**. 799–802.

Thomas, J. R., and Oerther, G. F. (1972). *Agron. J.* **64**, 11–13.

Thomas, J. R., Myers, V. I., Heilman, M. D., and Wiegand, C. L. *Proc. Symp. Remote Sens. Environ.*, *4th*, pp. 305–312.

Thomas, J. R., Namken, L. N., Heilman, M. D., and Wiegand, C. L. (1966). *Proc. Symp. Remote Sens. Environ., 4th.*

Thomas, J. R., Namken, L. N., Oerther, G. F., and Brown, R. G. (1971). *Agron. J.* **63**, 845–847.

Vu, C. V., Allen, L. H., Jr., and Garrard, L. A. (1982). *Environ. Exp. Bot.* **22**, 465–473.

Walter, H., and Koch, W. (1981). *Spectral Signatures Objects Remote Sens., Int. Colloq.,* Avignon, France, pp. 225–234.

Walter, H., Gausman, H. W., Escobar, D. E., Rodriguez, R. R., and Rittig, F. R. (1982). *J. Rio Grande Hortic. Soc.* **35**, 27–33.

Wellman, E. (1976). *Photochem. Photobiol.* **24**, 659–660.

Willstätter, R., and Stoll, A. (1918) "Unterschunger über die Assimilation der Kohlensäure," pp. 122–127. Springer-Verlag, Berlin.

Woolley, J. T. (1971), *Plant Physiol.* **47**, 656–662.

Index